Progress in Metallic Alloys
http://dx.doi.org/10.5772/61725
Edited by Vadim Glebovsky

Contributors
Yong W. Kim, Ryan Cress, Vadim Glebovsky, Galina Abrosimova, Valeriy P. Korzhov, Wojciech Polkowski, Arnoldo Bedolla Jacuinde, Shumin Wang, José David Villegas Cárdenas, Victor Manuel López Hirata, Carlos Alberto Camacho Olguin, Maribel Saucedo-Muñoz, Antonio De Ita De La Torre, Takuo Sakon, Gurrappa Injeti

Published by InTech
Janeza Trdine 9, 51000 Rijeka, Croatia

Publishing Process Manager Andrea Koric
Technical Editor SPi Global
Cover InTech Design team

First published October, 2016
Printed in Croatia

Additional hard copies can be obtained from orders@intechopen.com

Progress in Metallic Alloys, Edited by Vadim Glebovsky
p. cm.
Print ISBN 978-953-51-2696-6
Online ISBN 978-953-51-2697-3

For EU product safety concerns:
IN TECH d.o.o., Prolaz Marije Krucifikse Kozulić 3, 51000 Rijeka, Croatia,
info@intechopen.com or visit our website at intechopen.com.

Meet the editor

Prof. Vadim Glebovsky was born in 1936 in Sverdlovsk/ Urals, USSR. He graduated from the Urals Technical University in physical chemistry and metallurgy of steels in 1961. He received his PhD from the Baikov Institute of Metallurgy, Moscow with Acad. Alex Samarin on the study of the high-temperature reactions and distribution of sulphur-35 between liquid iron-carbon alloys and slags during a HF levitation in vacuum/CO (1964). He then went to the Research Center of Heavy Machinery for four years. During this time, he was a postdoctoral fellow, working on high-quality steels and alloys. In 1969, he joined the staff of the Materials Science Department in Chernogolovka, where he is currently a Senior Scientist/Professor at the Institute of Solid State Physics, RAS. During forty five years he was a leader of the laboratory of metallurgical processes. Prof. Glebovsky's scientific interests included theoretical and experimental studies of growing single crystals of high-purity refractory metals and alloys as well of the deep purification processes via a liquid state. Prof. Glebovsky has more than 300 scientific publications, among them one book on levitation melting of alloys, four book chapters and two edited books on various aspects of new materials processing.

Contents

Preface IX

Section 1 **Advanced Topics in Metallic Alloys 1**

Chapter 1 **Introductory Chapter: Preferential Sputtering and Oxidation of Nb-Ta Single Crystals Studied by LEIS 3**
Vadim Glebovsky

Chapter 2 **Statistical Physics Modeling of Disordered Metallic Alloys 17**
Ryan P. Cress and Yong W. Kim

Chapter 3 **Amorphous and Nanocrystalline Metallic Alloys 45**
Galina Abrosimova and Alexandr Aronin

Chapter 4 **Assessment of Hardness Based on Phase Diagrams 85**
Jose David Villegas Cárdenas, Victor Manuel López Hirata, Carlos Camacho Olguin, Maribel L. Saucedo Muñoz and Antonio de Ita de la Torre

Chapter 5 **Differential Speed Rolling: A New Method for a Fabrication of Metallic Sheets with Enhanced Mechanical Properties 111**
Wojciech Polkowski

Section 2 **New Alloys for Special Application 127**

Chapter 6 **The Superconducting Tape of Nb3Al Compound 129**
V.P. Korzhov

Chapter 7 **Niobium in Cast Irons 187**
A. Bedolla-Jacuinde

Chapter 8 **Indium Phosphide Bismide 221**
Liyao Zhang, Wenwu Pan, Xiaoyan Wu, Li Yue and Shumin Wang

Chapter 9 **Selecting Appropriate Metallic Alloy for Marine Gas Turbine Engine Compressor Components 247**
Injeti Gurrappa, I.V.S. Yashwanth and A.K. Gogia

Chapter 10 **Magnetocaloric and Magnetic Properties of Meta-Magnetic Heusler Alloy Ni41Co9Mn31.5Ga18.5 265**
Takuo Sakon, Takuya Kitaoka, Kazuki Tanaka, Keisuke Nakagawa, Hiroyuki Nojiri, Yoshiya Adachi and Takeshi Kanomata

Preface

Metal alloys are macroscopic homogeneous metallic materials, consisting of a mixture of two or more chemical elements with a predominance of metal components. The alloys are one of the major structural materials. The technique used for more than five or six thousand alloys. The solid state alloys can be homogeneous or heterogeneous. The alloys may be present as interstitial solid solutions, substitution solid solutions, chemical compounds and simple substances as crystallites. The properties of alloys are completely determined by their crystal structure or phase microstructure. The alloys exhibit metallic properties, such as electrical conductivity, thermal conductivity, metallic luster and ductility. Such a detailed list of seemingly simple things would be surprising if in every word it have not been hidden centuries of research, mistakes, achievements and discoveries. If desired, anybody could write an exciting-romantic-adventure novel, describing the history of the particular alloys and their role in human life.

Until recently, the term "metal" was associated with the term "crystal", whose atoms were arranged in space strictly orderly fashion. In the middle of the last century, scientists have discovered the metal alloys having no crystalline structure, i.e. amorphous metal alloys with a disordered arrangement of atoms in space. Metals and alloys with disordered arrangement of atoms became known as amorphous metal glasses, paying tribute to the analogy that exists between the disordered structure of a metal alloy and inorganic glass. Opening of amorphous metals made a great contribution to the science of metals, significantly changing our ideas about them. It was found that amorphous metals are very different in their properties from the metal crystals, which are characterized by an ordered arrangement of atoms. Formation of an amorphous structure of metals and alloys lead to fundamental changes in the magnetic, electrical, mechanical, and other superconducting properties. Some of them were very interesting both for science and for practice. The emergence of amorphous alloys - it is not only the result of scientific research being conducted in materials science and physics of metals. Virtually every group of metal alloys, such as iron-based or titanium-based alloys, have a long and interesting history.

In general, metallic alloys are the interdisciplinary subject or even an area that cover physics, chemistry, material science, metallurgy, crystallography, etc. This book, which You, dear readers, hold in your hands or watch on your PC monitor, is devoted to this old/new subject – the metallic alloys. The primary goal of this book was to provide coverage of advanced topics and trends of R&D of metallic alloys. The chapters of this book were contributed by the respected and well-known researchers in this area. They have presented the up-to-date developments of the metallic alloys technologies. This book consists of two blocks filled with 10 chapters which provide the researches in many aspects of the metallic alloys including the studies of amorphous and nanoalloys, modeling of disordered metallic alloys, superconducting alloys, differential speed rolling, meta-magnetic Heusler alloys, etc. We hope that You, dear readers, find this book interesting and helpful for your work and studies. If so, this could be the best pleasure and reward for us.

I would like to thank all of the authors for their chapters contributed to this book. It is my great pleasure to acknowledge the friendly assistance of Ms Andrea Koric, who continuously showed high professionalism and readiness to support the writing-up this book from its very beginning to the final format. I also would like to acknowledge my lovely children Nastya and Kirill, for their patience and love, throughout all my years of studying science.

Professor Vadim Glebovsky
Institute of Solid State Physics RAS
Chernogolovka, Russia

Section 1

Advanced Topics in Metallic Alloys

Chapter 1

Introductory Chapter: Preferential Sputtering and Oxidation of Nb-Ta Single Crystals Studied by LEIS

Vadim Glebovsky

Additional information is available at the end of the chapter

http://dx.doi.org/10.5772/65016

Metal alloys—macroscopically homogeneous metallic materials consist of a mixture of two or more chemical elements with a predominance of metal components. The alloys are one of the major structural materials. The technique uses more than five or six thousand alloys. The solid-state alloys can be homogeneous or heterogeneous. The alloys may be presented as interstitial solid solutions, substitution solid solutions, chemical compounds, and simple substances as crystallites. The properties of alloys are completely determined by their crystal structure or phase microstructure. The alloys exhibit metallic properties, such as electrical conductivity, thermal conductivity, metallic luster, and ductility. Such a detailed list of seemingly simple things would be surprising if in every word it has not been hidden in the centuries of research, mistakes, achievements, and discoveries. If desired, anybody could write an exciting-romantic-adventure novel, describing the history of the particular alloys and their role in human life.

Until now, the term "metal" was more or less associated with the term "crystal," whose atoms are arranged in space in a strictly orderly fashion. In the middle of the last century, scientists discovered the metal alloys having no crystalline structures, that is, amorphous metal alloys with a disordered arrangement of atoms in space. Metals and alloys with disordered arrangement of atoms became known as amorphous metal glasses, paying tribute to the analogy that exists between the disordered structure of a metal alloy and an inorganic glass. Discovering amorphous metals made a great contribution to the science of metals, significantly changing our ideas about them. It was found that amorphous metals are very different in their properties from the metal crystals, which are characterized by an ordered arrangement of atoms. Formation of an amorphous structure of metals and alloys lead to fundamental changes in the magnetic, electrical, mechanical, and even superconducting properties. Some of them were very interesting both for science and for application. The emergence of amorphous alloys—it is not the single result of scientific research being conducted in materials science and physics of metals. Virtually every group of metal alloys, such as iron-based or titanium-based alloys, have a long and interesting history.

In general, metallic alloys are the interdisciplinary subject or even an area that cover physics, chemistry, material science, metallurgy, crystallography, *etc.* This book, which you, dear readers, are holding in your hands or watching on your PC monitor, is devoted to this old/new subject—the metallic alloys. The primary goal of this book is to provide coverage of advanced topics and trends of R&D of metallic alloys. The chapters of this book are contributed by the respected and well-known researchers in this area. They have presented the up-to-date developments of the metallic alloys technologies. The book consists of 10 chapters divided into two sections of the metallic alloys including the studies of amorphous and nanoalloys, modeling of disordered metallic alloys, superconducting alloys, differential speed rolling, meta-magnetic Heusler alloys, *etc.* We hope that you, dear readers, will find this book interesting and helpful for your work and studies. If so, this could be the best pleasure and reward for us.

As scientific editor of this book, I had to read all chapters and more than once, especially if the chapter does not meet the standards adopted in the publishing house. In particular, this could be due to a deviation from the scope of the manuscript or its translation, scientific content or quality of the so-called similarity (plagiarism) of a manuscript. I was a bit lucky —the authors of submitted manuscripts were, as a rule, consistent with accepted standards, although there were also some deviations. So, part of the manuscripts had an increased volume (text, figures) that was solved through negotiations between publishers and authors. A CrossCheck program, through which the manuscripts are analyzed, records all matches with publications in all editions, and within a reasonable time. In our case, there are no borrowing from the "other people's" publications (which is a real plagiarism), but only self-citation, when the manuscript contains pieces from own articles. Sometimes the index of self-citation is very rude, and the authors have been asked to correct the situation. Once the authors did not agree to fix the text and took their manuscript back, which we met with a great regret, because the manuscript contained a very interesting scientific content, and could be, if corrected, one of the best chapters of the book.

I would like to thank all of the authors of this book for their contributed chapters. It is my great pleasure to acknowledge the friendly assistance of Ms Andrea Koric, who continuously showed high professionalism and readiness to support the writing of the book from its very beginning to the final format. I also would like to express my deep thanks to my lovely daughter Nastya and my son Kirill, for their patience and love, throughout all my years in science.

At this point, I would like to finish the formal part of my "Introductory chapter: Preferential sputtering and oxidation of Nb-Ta single crystals studied by LEIS." and switch to my research contributions to metal alloys. In different periods of my scientific life, I had to deal with a variety of metals and alloys: the iron-carbon alloys and different steels, many alloys for thin-film metallization based on high-purity single-crystalline refractory metals Mo, W, Nb, Ta, Ti, the systems Pt/Pd and W/Ti for microelectronics, different alloys for bio-implants, single-crystalline alloys of Nb-Ta. About the preferential sputtering and oxidation of these alloys studied by low-energy ion scattering, I would like to tell in the Introductory chapter.

Currently, Nb-Ta alloys are used in many fields of science and technology: in the electric appliances and electronics, in the chemical industry for the manufacture of chemical apparatuses, in the rocket technology for the manufacture of the nozzle heads, and others. Nb superconducting alloys are used in heavy duty atomic accelerators for manufacturing windings magnets for hot plasma reflectors, lasers, and other nuclear power plants. It is also known the use of the alloys in aviation technology for manufacturing uncooled turbine blades in jet engines, and others. Nb-/Ta-based alloys currently provide performance products at temperatures up to 1300°C and based on Ta up to 1700°C. Despite the higher melting temperature of alloys based on Ta, they are less common than Nb-based alloys. There are several reasons for that; the main is the scarcity and high price of Ta. Therefore, in recent years, began to attract the attention to new ideas: In the manufacture of these elemental metals, they are not separate; indeed, why separate them, if they are always related to each other in nature and, therefore, supplement each other in alloys excellently.

Ta has a unique feature—it is the only metal that has a biological compatibility with a living tissue. Metal, named after the mythological martyr, has an interesting mission to the mankind —it came to the aid of man, his living tissues. In reconstructive surgery and neurosurgery, Ta began to be used during World War II: The replacement of the damaged parts of the skull, bound broken bones, replacing the small bones with the wire and the metal strips. Ta yarn and mesh used for the replacement of muscle tissues, and as a basis for the growth of new tissues. A metal mesh is used to reinforce the walls of the abdominal cavity, with the help of a thin Ta wire to stitch tendons and damaged nerves. A similar property—a biocompatibility—is a characteristic of Ta-Nb alloys. A lower density compared with Ta alloys that makes them promising. Ta may come to replace stainless steel, gold, and other conventional alloys because, unlike traditional metallic materials for implants, the human body perceives Ta and Nb-Ta alloys, not as a foreign body, but as your own bone. Perhaps, said about Nb-Ta alloys looks more like a hymn to the glory of these metals and their alloys. So let it be—I really admire their unusual physical properties, capabilities, and believe in the enormous potential in the nearest future.

As well known, Nb and Ta, having similar lattice parameters, crystallize in a similar body-centered cubic lattice. Both metals have similar chemical and physical properties. Thus, the Nb-Ta system should have a continuous range of the substitutional solid solutions (alloys) [1]. Moreover, the pure Nb and Ta and their alloys can be produced as single crystals with a known crystallography. The physical and chemical interaction of oxygen with Nb and Ta can be studied by the methods of a surface analysis such as low-energy electron diffraction (LEED) or Auger electron spectroscopy (AES), having a larger sensitivity depth than LEIS [2–13]. The interaction of oxygen with metal surfaces is important in catalysis, corrosion, and growth. A series of single crystals (110) of Nb-Ta alloys has been studied by LEIS for obtaining quantitative information about the single crystals of Nb-Ta alloys during their interaction with oxygen. In this chapter, the results of the LEIS experiments on the single crystals of several Nb-Ta alloys and the elemental Nb and Ta are presented. The contents of Nb and Ta and alloys of the surface oxygen in the upper layer of the surface may be quantified by LEIS, that is a surface analysis technique with extremely high sensitivity and selective atomic layer to the outer surface [14].

When the matrix effects are absent, the composition can be quantified by calibration of surface [15]. As an editor of the book, I would like to present the study of these alloys, which could be a main part of the introductory chapter. The study covers several more or less traditional techniques (levitation melting, EB floating zone growing single crystals of refractory metals, X-ray Laue characterization of single crystals, recrystallization for growing massive single crystals of alloys, elemental characterization by ICP MS, and others) and UHV techniques for studying upper layers of single crystals (LEIS, LEED, SIMS). By techniques used as well by the aims and results, this study is also traditional. A part of experiments is done in ISSP RAS, Chernogolovka, another part of the study is done together with Prof. Hidde Brongersma in TUE, Eindhoven.

1. Experimental

The alloys of Nb and Ta are obtained by mixing the pure elemental powders in a desired ratio by using a high-frequency levitation melting. This method is crucible less—the metal sample melts in an electromagnetic field formed by a conic inductor. The radio-frequency electromagnetic field provides a uniform mixing of both metals in the liquid state [16]. To form cylindrical cast rods, the melt is cast into a cylindrical water-cooled copper mold. Single crystals of these alloys are grown by electron beam floating zone melting which provides refining material together with a uniform distribution of both elements in the volume [17]. Single crystals of both metals and three Nb-Ta alloys are grown with a growth rate of 3 mm min^{-1} using the specially prepared single-crystalline seeds of three main crystallographic orientations—(111), (110), and (100). For this part of the study, the discs of the (110) plane index are cut off by electro-erosion and then mechanically and chemically polished. X-ray Laue back reflection is used to a crystallographic check of the as grown crystals and final discs. It was discovered that $Nb_{0.75}Ta_{0.25}$, $Nb_{0.5}Ta_{0.5}$, and $Nb_{0.25}$ $Ta_{0.75\,25}$ alloys could not be grown directly from the melt as crystals. In order to grow crystals of these alloys, recrystallization is used which consists of a strain deformation followed by a high-temperature annealing (up to the melting temperature of alloys). For this study, the following groups of single crystals are grown in different volume composition: Nb, $Nb_{0.75}Ta_{0.25}$, $Nb_{0.5}Ta_{0.5}$, $Nb_{0.25}Ta_{0.75}$, and Ta. Rutherford backscattering spectroscopy (RBS) is used to check the composition of the volume. Contents of both metals in alloys are analyzed also by ICP MS.

2. SIMS and SNMS experiments

Before LEIS experiments, the alloys are studied by SIMS and SNMS. These measurements are made by a 200 quadruple mass spectrometer Leybold SSM. The base pressure in the system is in a low range of 10^{-10} mbar. The Ar^+ primary beam with energy of 5 keV is used. Static SIMS spectra of the surface are recorded with a current density of 50 cm^{-2}, and with a typical acquisition time of 200 s. This leads to a total dose of 5×10^{13} ions cm^{-2}, which is a static limit. Bulk analyses by SIMS and SNMS are performed with a higher current density of 5 mA cm^{-2}, and an acquisition time of 1 h to improve a signal-to-noise ratio. The SNMS is emitted by ionized forms of a post-60 eV electron beam. The samples are of the (100) surface orientation, in order to eliminate the effect of differences in density between various lattice planes [18]. Firstly, it is shown that prolonged sputtering is important for obtaining meaningful SIMS spectra with dimers and trimers of Nb. Next, the SIMS and SNMS spectra of Nb, Ta, and NbTa alloys are compared. Positive SIMS spectra of the as grown Nb single crystal in the mass range from 75 to 315 atomic mass units (amu) are measured. Since the spectrum is measured in a static mode, it shows the composition of the surface. The Nb^+ peak is at 93 amu, and it dominates the spectrum. Contaminants can be seen in the form of cluster ions such as NbC^+ (105 amu), NbN^+ (107 amu), NbO^+ (109 amu), NbF^+ (112 amu), and NbO_2^+ (125 amu). The presence of hydrogen, which is easily dissolved by these metals, is represented by NbH^+ peak at 94 amu. In the higher mass range, a small surface contamination by Ta is visible in the peaks of Ta^+ (181), TaO^+ (197 amu), and TaO_2^+ (213 amu). Of some interest is a dimer Nb_2^+. However, this peak has a low intensity, because it is very sensitive to the surface cleanliness. In a spectrum of the same Nb crystal after 30 min of sputter with 5 keV Ar^+ ions at a density of 5 μA cm^{-2}, there are high intensities of Nb_2^+ and Nb_3^+ clusters, while clusters which are typical of impurities have a much lower intensity than in the first spectrum (without long sputtering). During etching, various secondary ion signals are recorded. Several characteristic intensity ratios are registered. Carbon and oxygen are removed within the first 5 min of the etching process, which corresponds to a removal of several tens of atomic layers. Simultaneously with the removal of impurities, dimers increase the intensity of Nb on the order of magnitude, and thus, Nb trimer peak appears. Small peaks of NbC^+ and NbO^+ remain but correspond to carbon and oxygen concentrations below the limit of detection of Auger electron spectroscopy. The conclusion is made that the long-term etching to achieve sputter balance is essential to obtain stable spectra, which are really representative of the bulk composition.

3. LEIS experiments

The unique properties of LEIS in combination with new instrumental developments allow conducting research in emerging areas of science and technology. **Figure 1** shows some of the characteristics of LEIS in comparison with such widely used analytical techniques such as Auger electrons spectroscopy (AES), X-ray photoelectron spectroscopy (XPS), secondary ion mass spectroscopy (SIMS). It is clear that none of these techniques has any such high depth sensitivity as LEIS to the topmost atomic layer. The treatment of information obtained by means of analytical methods for studying the surface is quite complicated. While SIMS method has the highest sensitivity to alkali metals, LEIS is far more sensitive to the noble metals and especially to metals with high atomic masses [19].

A target surface in LEIS is irradiated with a monoenergetic beam of inert gas ions with energy in the range of 1–5 keV. Upon reaching the surface of the target, an ion undergoes one or more collisions with the target atoms. The ion scattering spectroscopy investigates the energy distribution of the primary ions, backscattered in a vacuum. The de Broglie wavelength for ions with energy of 1 keV is very small compared with the interatomic distances on the surface.

Thus, in contrast to the scattering of electrons and phonons, the majority of ion scattering phenomena can be quite accurately described by the methods of classical mechanics. The ions are scattered by the Coulomb interaction between the (shielded) nucleus of the ion and atom. Under normal experimental conditions, this interaction is important only for distances of less than 0.05 nm. This is a good approximation for the assumption that at any given moment an ion interacts with a single atom. Since the time of interaction ($\sim 10^{-15}$ s) is very small compared with the characteristic time for the phonons ($\sim 10^{-12}$ to 10^{-13} s), the target atom can be considered as a free atom. In the process of scattering, an ion loses some of its kinetic energy. Energy losses can also be accurately calculated in the approximation of elastic scattering. LEIS experiments are conducted using the scattering apparatus ions (**Figure 2**). Primary ions are formed in the ion source and directed perpendicular to the target surface. Ions are dispersed to 1440 target atoms and energy is analyzed by a cylindrical mirror analyzer. Using very pure ion beams is essential to obtain a low level of background spectra. The nominal base pressure in the vessel is in a low range of 10^{-10} mbar and can be controlled by a quadruple mass spectrometer. The device is equipped with a source of ions for sputtering at a grazing angle of 15°.

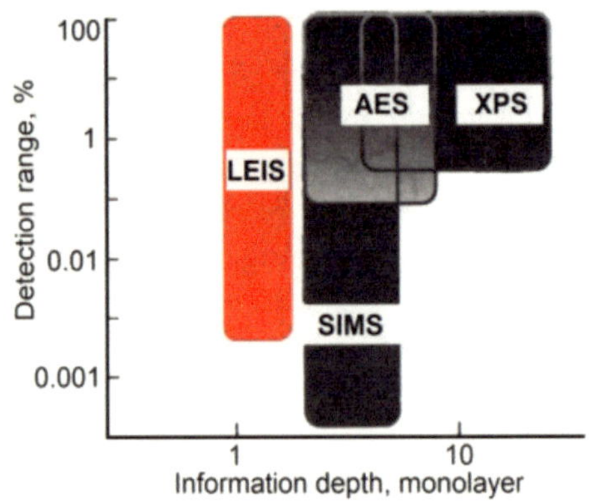

Figure 1. Comparison of LEIS with SIMS, XPS, AES.

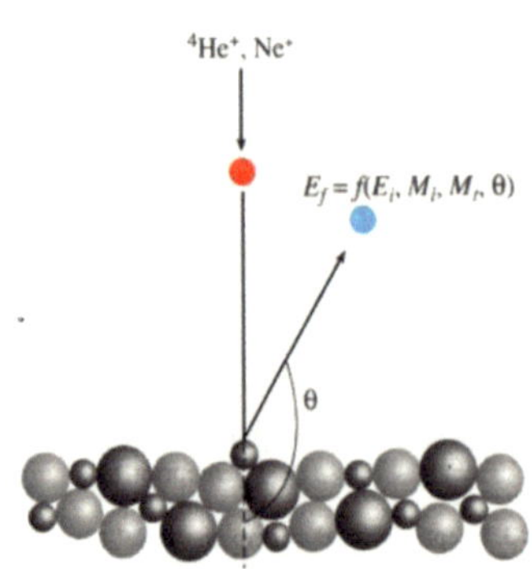

Figure 2. LEIS experiment.

4. Adsorption of oxygen and sputtering

Surfaces of Nb-Ta alloys are purified by the Ar ion-sputtering cycles at room temperature and annealed at 800 K. This temperature is too low to remove all the defects. In addition, it is impossible to remove all the oxygen in this way, but it is effective to remove surface contaminants (carbon, nitrogen, hydrogen). For achieving atomically clean surfaces, there are necessary to have the annealing temperatures above 2000 K; thus, it seems our annealing is not yet available. Oxygen (99.995%) is supplied to a vessel with a dose of 30–50 L (Langmuir), which is high enough to saturate the surface. **Figure 3** shows the typical LEIS spectra of clean and oxygen-covered Nb-Ta alloy. By coating the surface of Nb-Ta with oxygen, Nb and Ta peaks are screened, and consequently, the Nb and Ta intensities decrease. The following procedure allowed us to study this effect in more detail. The sample is first saturated with oxygen. Then oxygen is evacuated, and LEIS spectra measured with 1.5 keV $^4He^+$ primary ion beam which also provides a slower removal of oxygen from the surface by sputtering. The procedure is repeated for a reproducibility check. Thus, LEIS spectra of alloys and metals with different oxygen coverage can be obtained. The intensities for Nb and Ta depend linearly on the intensity of oxygen. Some changes in the primary ion beam time can be corrected by calibration against a clean Cu surface. Thus, the final composition effects on secondary electron emission and the effective current target can be avoided.

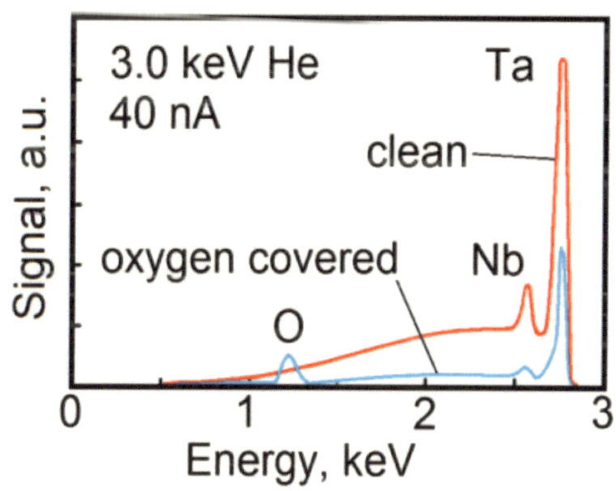

Figure 3. Typical LEIS spectra of the oxygen-coated and clean surfaces of (110)$Nb_{0.75}Ta_{0.25}$. The ion $^4He^+$ with energy 3.0Â keV, output current of 40 nA. To reduce measurement time, the oxygen-coated sample is only measured in the range of interest.

5. Quantification of Nb and Ta at the surface

In **Figure 4**, the linear dependence shows that there are no matrix effects for these ion-atom combinations. Removing oxygen by sputtering, apparently does not affect the scattering process and ion fractions of the adjacent atoms of Nb and Ta. Only more atoms of Nb and Ta are exposed to the primary ion beam, which corresponds to an increase in the Nb and Ta signals. Such behavior differs from that of the secondary ion mass spectrometry (SIMS), where part of the ion-sputtered particles changes drastically by the presence of oxygen. To obtain Nb and Ta signals for Nb_xTa_{1-x} alloys without oxygen, the lines in **Figure 4** are extrapolated to zero oxygen coverage. It is interesting that a linear relationship has been obtained when plotting the extrapolated Ta signals as a function of the corresponding extrapolated Nb signals, if no matrix effects present in these LEIS experiments [19, 20]. Results in **Figure 5** show that this prediction is performed within an experimental error. Deviations of approximately 15% in the linear relationship between Nb and Ta signals can be result of several reasons. Positioning and focusing system should be made individually for every sample. The signals are calibrated with a standard Cu specimen. Both dimensions have errors of a few percent. The bulk material is of very high purity; however, adsorption and segregation can change the situation and increase the content of impurities on the surface.

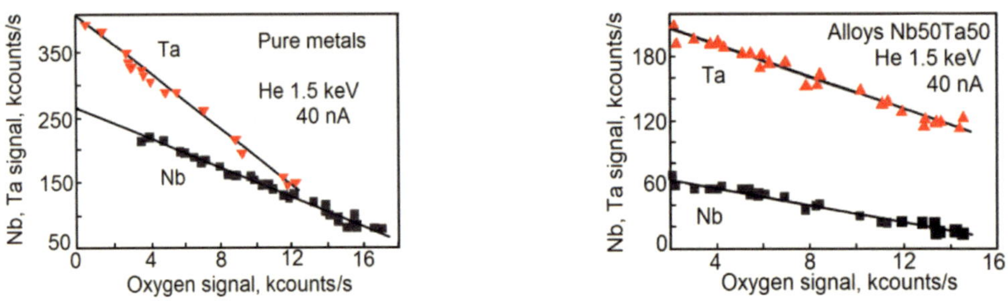

Figure 4. Nb and Ta signals vs. oxygen signals for LEIS on pure metals and $Nb_{0.5}Ta_{0.5}$.

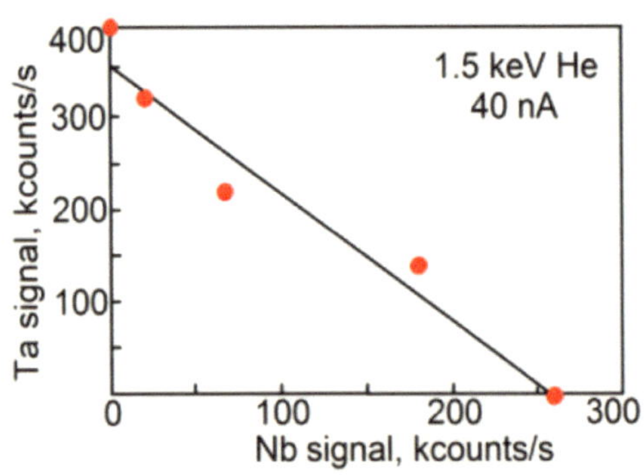

Figure 5. The peak intensity Ta vs. Nb alloys for Nb_xTa_{1-x} system without oxygen.

As for carbon, with a low sensitivity in LEIS, it is difficult, if possible, to detect. Different patterns could have different contents of impurity atoms on the surface. Because of the low temperatures of sputtering and annealing, the surface of different samples could not be a perfect (110) plane. With changing the structure of the sample surface, Nb and Ta densities become lower than that of a higher density packaged (110) plane. Determination of the peak intensity of Nb in a LEIS spectrum is not a simple task for such alloys because Nb peak overlaps Ta peak of a low-energy tail. A special oven has been used for high-temperature (2000°C) annealing the samples in the preparation chamber. A linear curve in **Figure 5** can be used to calculate the surface composition for clean Nb_xTa_{1-x} alloys, since the signals for both metals are proportional to their content on the surface. The experimental data are plotted and make a straight line through the experimental points to the beginning of the graph coordinates, taking the intersection with a linear curve. The Nb and Ta surface contents are found by dividing the transferred Nb and Ta signals by the signals of pure Nb and Ta, respectively. The surface contents of the alloys calculated in the described way are shown in **Figure 6**. The surface of samples is clearly enriched in Ta. These alloys have very high melting temperatures (2690–3270 K); thus, thermally activated surface segregation can be neglected at room temperature. Nb-Ta alloys are the ideal systems for an experimental determination of the role of the mass difference on the preferential sputtering of both metals from the matrix. Sigmund's theory [20] gives the ratio R of the sputter yield Y_{Nb} of Nb to that of Ta Y_{Ta}: where N_{Nb} and N_{Ta} are the atomic concentrations (number of atoms per unit volume), M_{Nb} and M_{Ta}, the atomic masses, and U_{Nb} and U_{Ta}, the surface binding energies of Nb and Ta, respectively. The exponent m, which is currently expected to about 1/6, is a parameter characteristic of the interaction potential.

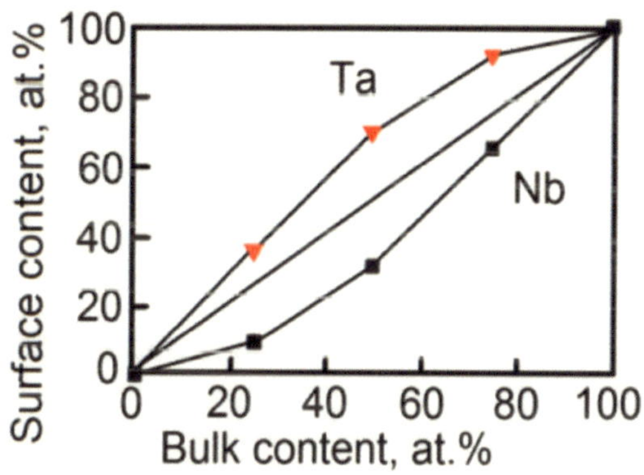

Figure 6. The concentrations of Nb(Ta) on the surface against Nb(Ta) in the volume (in at.%). Sigmund's model (central line) for preferential sputtering is shown for comparison.

$$R = Y_{Nb} / Y_{Ta} = N_{Nb} / N_{Ta}(M_{Ta} / M_{Nb})^{2m}(U_{Ta} / U_{Nb})^{1-2m} \tag{1}$$

The ratio of the surface binding energies of Ta and Nb is estimated to be equal to the ratio enthalpies of evaporation of these elements (1,09) [21]. Since sputtering is mainly limited to atoms by the outermost layer, the preferential sputtering of Nb should lead to enrichment in Ta in the upper layer with a factor of 1.3. However, the observed enrichment is even higher than predictions based on the preferential sputtering (**Figure 6**). Since our setup does not allow for removal of oxygen when heated, it is likely that oxygen-induced segregation in combination with a primary sputter can be a reason for the observed effect.

6. Quantification of oxygen

The quantification of the oxygen signal may be done using a calibration with respect to the surface of Ni(100)–O*c*(2 × 2). As known, this is a very stable structure that is obtained when the surface is saturated with oxygen, has oxygen coverage of half of a monolayer, and corresponds to 8×10^{14} atoms of oxygen per 1 cm^{-2}. For calibrations, oxygen adsorption on Ni(100) is studied by LEIS under the same conditions as in the experiments with Nb-Ta alloys (**Figure 7**), again using signal of pure Cu for normalization.

In **Figure 7**, the maximum coverage of oxygen on Ni(100) corresponds to the density of oxygen atoms 8 × 1014 atoms cm^{-2} and gives the oxygen signal of 7.3 × 103 counts s^{-12}. A linear decrease of the Ni signal with increasing the oxygen signal demonstrates the lack of matrix effects. The quantification of the maximum oxygen concentration in Nb-Ta samples using this calibration is possible. Dividing the oxygen density by the metal density, that is 13.0×10^{14} and 12.9×10^{14} atoms cm^{-2} for Nb(110) and Ta(110), respectively, provides values of an oxygen coverage. The results are shown in **Table 1**.

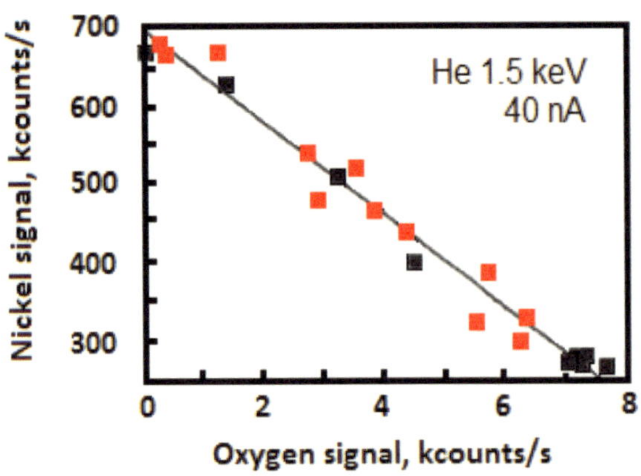

Figure 7. Ni peak intensity as compared to the O peak intensity to the surface of the Ni(100). Various oxygen coverages obtained by sputter (red) and by monitoring the oxygen exposure (black).

Sample	Surface composition		Maximum oxygen density	Maximum oxygen/metal
	%Nb	%Ta	(10^{15} at cm^{-2})	ratio
Nb	100	–	1.82	1.41
$Nb_{0.75}Ta_{0.25}$	65	35	1.72	1.33
$Nb_{0.5}Ta_{0.5}$	31	69	1.51	1.17
$Nb_{0.25}Ta_{0.75}$	9	91	1.43	1.10
Ta	–	100	1.30	1.00

Table 1. Quantitative composition of the (110)Nb-Ta alloys by LEIS.

An oxide growth on Ta (110) and Nb (110) is described by the formation TaO(111) [3] and NbO (111) [8], respectively. The oxygen/metal ratio of 1.0 that we get with LEIS for Ta(110) is very close to these studies. For Nb(110), however, it is found the oxygen/metal ratio of 1.4, which is higher than the value of Ta. The oxygen coverage on Nb is higher because of better shielding of Nb as compared with Ta. Covering and shielding the oxygen atoms on the Nb-Ta alloys increases with increasing the Nb content. For Nb, it is expected that the surface contains more oxygen than Ta surface. Hu et al. [12] reported the existence of two Nb oxides (NbO and Nb_2O_5, detected via XPS) on Nb(110) surface bared to 3000 L of oxygen. Haas et al. [3] observed that the solubility of oxygen in Nb greater than in Ta (4.5% and 3%, respectively). Also, the structure difference between Ta and Nb oxides on the surface can produce differences in the exposed oxygen density on surfaces of Nb or Ta.

Preferential sputtering and oxidation of three single-crystalline (110)Nb_xTa_{1-x} alloy (x = 0.25, 0.5, 0.75), together with single crystals of pure Nb and Ta, are studied by LEIS. After sputter cleaning, LEIS showed Ta enrichment on the surface of all NbTa alloys, indicating Nb preferential sputtering. This is in a reasonable agreement with theory. After contact with oxygen, linear relationships between O and Nb and Ta signals indicate that the matrix does not affect the LEIS signals for these systems. LEIS is very useful for collecting quantitative information about the composition of the outer layer of the surface of the alloys. Nb-Ta alloys differ from those in the bulk. The oxygen coverage on NbTa alloys after exposure to oxygen has been determined with an accuracy of about 15% after calibration using a maximum coverage of oxygen in the known Ni system (100)–Oc (2 × 2). The maximum surface oxygen concentration is defined as 13×10^{15} atoms cm^{-2} for Ta(110) and 18×10^{15} atoms cm^{-2} for Nb(110), which corresponds to the oxygen coverage of 1.0 and 1.4, respectively. The maximum oxygen coverage of the alloys increases with the Nb content.

Author details

Vadim Glebovsky

Address all correspondence to: glebovs@issp.ac.ru

Institute of Solid State Physics RAS, Chernogolovka, Russia

References

[1] L H Bennett, T W Massalski, B C Giessen. Alloy phase diagrams. North-Holland. Amsterdam (1983).

[2] J E Boggio, H E Farnsworth. Low-energy electron diffraction and photoelectric study of (110) tantalum as a function of ion bombardment and heat treatment. Surf Sci 1964; 1: 399–406.

[3] T W Haas, A G Jackson, M P Hooker. Adsorption on niobium (110), tantalum (110), and vanadium (110) surfaces. J Chem Phys 1967; 46: 3025–3033.

[4] B Sewell, D F Mitchell, M Cohen. A kinetic study of the initial oxidation of a Ta(110) surface using oxygen k_α X-ray emission. Surf Sci 1972; 29: 173–188.

[5] H H Farrell, H S Isaacs, M Strongin. The interaction of oxygen and nitrogen with the niobium (100) surface: I. Morphology. Surf Sci 1973; 38: 31–52.

[6] H H Farrell, M Strongin. The interaction of oxygen and nitrogen with the niobium (100) surface: II. Reaction kinetics. Surf Sci 1973; 38: 18–30.

[7] R Panlei, M Bujor, J Bardolie. Continuous measurement of surface potential variations during oxygen adsorption on the (100), (110) and (111) faces of niobium using mirror electron microscope. Surf Sci 1977; 62: 589–609.

[8] G Ertl, M Plancher. A statistical model for oxygen adsorption on heterogeneous metal surfaces. Appl Surf Sci 1980; 6: 453–463.

[9] M Grundner, J Halbntter. On the natural Nb_2O_5 growth on Nb at room temperature. Surf Sci 1984; 136: 144–154.

[10] A V Titov, H Jagodzinski. Structure of O layers on the Ta(100) surface. Surf Sci 1985; 152/153: 409–418.

[11] N Shamir, U Atzmony, J A Schultz, M H Mintz. Hydrogen-oxygen interrelations on a niobium surface. J Vac Sci Technol A 1987; 5: 1024.

[12] Z P Hu, Y P Li, M R Ji, J X Wu, The interaction of oxygen with niobium studied by XPS and UPS—ADS. Solid State Commun 1989; 71: 849.

[13] C Surgers, H V Lohneysen. Effect of oxygen segregation on the surface structure of single-crystalline niobium films on sapphire. Appl Phys A 1992; 54: 350–354.

[14] H H Brongersma, P Groenen, J-P Jacobs. Application of low energy ion scattering to oxidic surfaces. In: Science of Ceramic Interfaces, J Novotny (Ed.). Material Science Monographs 8, Elsevier Science B.V., Amsterdam (1994) pp 113–182.

[15] P A J Ackermans, G C R Krutzen, H H Brongersma. The use of a calibration in low-energy ion scattering: Preferential sputtering and S segregation in CuPd alloys. Nucl Instr Meth B 1990; 45: 384–389.

[16] V G Glebovsky, V T Burtsev. Levitation Melting of Metals and Alloys. Metallurgia Publ House. Moscow (1974) 174 p.

[17] V N Semenov, B B Straumal, V G Glebovsky, W Gust. Preparation of Fe-Si single-crystals and bicrystals for diffusion experiments by the electron-beam floating-zone technique. J Cryst Growth 1995; 180: 151.

[18] H J Borg, J W Niemantsverdrit, H H Brongersma, V G Glebovsky. A SIMS/SNMS study of high-purity Nb_xTa_{1-x} alloys. J Surf Phys 1992; 13: 32–36.

[19] L C A van den Oetalaar, J P Jacobs, M J Mietus, H H Brongersma, V N Semenov, V G Glebovsky Quantitative surface analysis of NbTa alloys by low–energy ion scattering. Appl Surf Sci 1993; 70/71: 79–84.

[20] P Sigmund, M W Sckerl. Momentum asymmetry and the isotope puzzle in sputtering by ion bombardment. Nucl Instr Meth B 1993; 82: 242–254.

[21] F R de Boer. R Boom, W C M Mattens, A R Miedema, A K Niessen. Cohesion and structure. In: Cohesion in metals: transition metal alloys, F R de Boer, D G Pettifor (Eds.). North Holland. Amsterdam 36 (1988) pp 385 and 539.

Chapter 2

Statistical Physics Modeling of Disordered Metallic Alloys

Ryan P. Cress and Yong W. Kim

Additional information is available at the end of the chapter

http://dx.doi.org/10.5772/64837

Abstract

The great majority of metallic alloys in use are disordered. The material property of a disordered alloy changes on exposure to thermal, chemical, or mechanical forcing; the changes are often irreversible. We present a new first principle method for modeling disordered metallic alloys suitable for predicting how the morphology, strength, and transport property would evolve under arbitrary forcing conditions. Such a predictive capability is critically important in designing new alloys for applications, such as in new-generation fission and fusion reactors, where unrelenting harsh thermal loading conditions exist. The protocol is developed for constructing a coarse-grained model that can be specialized for the evolution of thermophysical properties of an arbitrary disordered alloy under thermal, stress, nuclear, or chemical forcing scenarios. We model a disordered binary alloy as a randomly close-packed (RCP) assembly of constituent atoms at given composition. As such, a disordered alloy specimen is an admixture of nanocrystallites and glassy matter. For the present purpose, we first assert that interatomic interactions are by repulsion only, but the contributions from the attractive part of the interaction are restored by treating the nanocrystallites as nanoscale pieces of a single crystalline solid composed of the same constituent atoms. Implementation of the protocol is discussed for heating of disordered metals, and results are compared to the known melting point data.

Keywords: nanocrystallite size distribution, glassy state atoms, simulated alloy specimen, thermal forcing, melting point

1. Introduction

Under thermal, mechanical, or chemical forcing, disordered metallic alloy specimens may change in their thermophysical properties, such as thermal diffusivity, electrical resistivity, spectral emissivity, and many other properties. The degree to which such modifications are materialized depends on both the intensity and duration of the forcing. In the case of a thermal forcing, the temperature serves as the control parameter of forcing in reference to the specimen's intrinsic threshold properties, such as the melting point. The modification has serious consequences in utilization of metallic alloys in high-temperature and high-stress processes. Examples are found in nuclear reactors, chemical reactors, pyrometallurgical processes, and others. Thermophysical properties of alloys drift away from the design values, compromising the performance metrics as well as even leading to material failures.

The questions are why and how such a forcing modifies the material's basic thermophysical properties. Two characteristic features highlight alloy modifications due to thermal forcing: one, enrichment of the more mobile atoms near the alloy surface, which has been observed in direct Measurement; and two, the morphological transformation as quantified in terms of the nanocrystallite size distribution [1, 2]. Both of the features influence the transport of mass, momentum, and energy because the exact details of the pathways for transport of flux quanta across a surface are determined by them. The latter feature is a precursor to alloy melting, and we show that the associated morphological transformation can be theoretically treated. This theoretical treatment will lead to a better understanding of the changing factors that influence the thermophysical properties of the alloy.

We focus on identifying the basic physical mechanisms that affect thermophysical properties of simple metallic alloys and incorporating their coarse-grained formulations, or their simplest representations, into the alloy model. The goal is to render the construction of a realistic model of any arbitrary disordered alloy easy and simple. We hypothesize that the changes in the alloy's thermophysical properties are mediated by the changes in the size distribution function of nanocrystallites due to re-equilibration of nanocrystallites in size distribution at elevated temperatures. Transport of excitations through a thermally forced disordered alloy specimen would involve two different material media, crystalline versus glassy, whose physical sizes have been modified due to thermal forcing, and transmission of excitations across the interfaces between them has also been modified. The rates of excitation transport through the specimen would thus be changed as a result of the modifications of the distribution function of the nanocrystallites. It has been shown for a number of different alloys that the thermal forcing results in changes of the specimen's elemental composition profile as a function of depth from the surface, distinctly different from the bulk composition [3–5].

The theoretical insight into the state of the atomistic structure of a disordered binary alloy can help quantify the contributions from the structural properties of the alloy specimen to the transport of thermal excitations through the alloy. After setting up the theoretical model of how this structure would change as a function of temperature, we can proceed with predicting how the thermal transport properties would be affected by the morphological changes and move on to mapping out the changing thermophysical properties. The theoretical prediction of how such modifications would materialize will go a long way toward developing new materials and forecasting the modes of structural failures in existing materials.

Available experimental data on the thermal conductivity of solids vary widely. This is in part due to difficulties in making accurate measurement of the thermal conductivities of solids and in part due to problems in physical and chemical characterization of the test specimens of such solids. In the case of disordered binary alloys, these complications lead to serious gaps in experimental data in conflict with the thermodynamic property data that are available. Experimental uncertainties in measurement can arise from many different sources, including poor sensitivity of measuring instruments and sensors, specimen contamination, stray heat flows that are unaccounted for, and incorrect form factors of the test specimens. Perhaps, the most serious problem is in the indeterminate nature of the alloy composition itself, as experimentalists are often unsure of the elemental composition of their specimens [6]. More recent studies on the properties of binary alloys have focused on binary alloys as polycrystalline materials. In the context of binary alloys, the polycrystalline model suggests crystallite grains separated by grain boundaries. In such a material, the physics of the interfaces between nanocrystallites tend to dominate the transport properties of the alloy [7]. Because of the high density of grain boundaries, recent efforts have focused on controlling of the formation of grain boundaries in order to produce more stable binary alloys for high temperature processes [8].

2. Modeling of disordered alloys as a RCP assembly of constituent atoms

Since the introduction of complex metallic alloys as a material, the question has been on how the atoms in such metals arranged themselves. In this context, Bernal first imagined the alloy as a random assembly of hard spheres [9]. The radii of the spheres would correspond to the atomic radii of the constituent atoms within the alloy. Studying of the packing of spheres has a long history because of its ability to serve as a simple but useful model for a variety of physical systems [10]. The molecular nature of fluids and glasses has also enjoyed the physical visualization by hard sphere packing [11, 12]. Dense packing of hard spheres is generally separated into three subclasses: ordered close packing, random close packing, and random loose packing. Ordered close packing in three dimensions reveals periodic symmetry arising from a unit cell structure and accounts for the highest density of hard spheres [13]. Random close packing has historically been studied experimentally by filling a container with hard spheres at random and shaking the container to achieve a maximum random packing [14]. Random loose packing is the result of not shaking the container creating a less dense version of random close packing [15].

Berryman formalized the concept of random close packing (RCP). In order to have a randomly close-packed structure, it was required that all spheres be arranged at random and that the structure filled a volume at maximum density where all spheres are in contact with more than one other sphere [16]. Berryman also reported a packing fraction of 0.64 on average for RCP in three dimensions. He found that random loose packing is in some sense less fundamental than the concept of random close packing, as the definition of random loose packing requires a minimum density below which the configurations of the structure are unstable and therefore not "packed." As packing fraction increases, the phase of the matter being simulated changes. The lowest branch corresponds to the liquid phase, a packing fraction of around 0.49 corresponds to the freezing point. Packing fractions greater than 0.49 correspond to a solid phase. We are interested in the metastable branch equivalent to a mixed phase of disordered glassy arrangements and ordered nanocrystallites. This branch will provide a foundation for the structure of disordered alloys. The metastable branch is an extension of the liquid phase and extends to the random close-packed state [17].

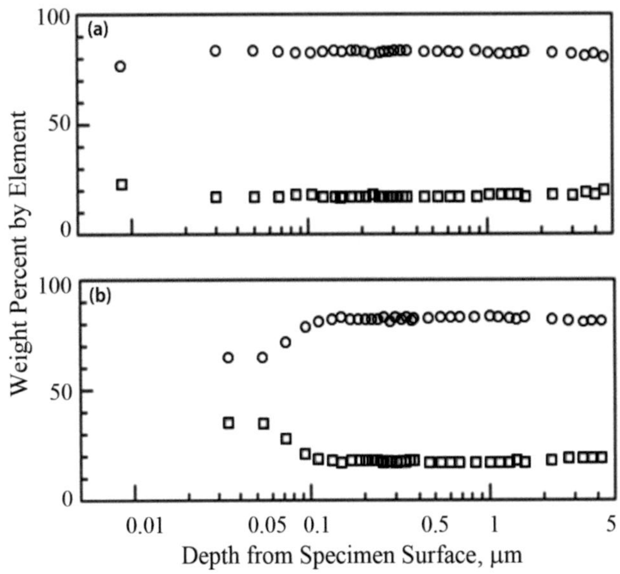

Figure 1. Measured alloy composition of a 80W%Ni-20W%Cr Nichrome ribbon specimen by element (nickel in circles and chromium in squares) as a function of depth from the surface by means of the method of laser-produced plasma spectroscopy: (a) fresh specimen at room temperature before thermal forcing (top); and (b) fresh specimen cooled to room temperature after a 15-hour heating in vacuum at 1100 K (bottom) [1–5].

The pivotal experiment in developing the structural model of disordered binary alloys was the study of alloy composition as a function of depth from the surface. **Figure 1** shows the measured elemental composition of a Nichrome specimen by the method of laser-produced plasma (LPP) plume spectroscopy [5]. The laser pulse heats the surface, launching a thermal diffusion front heading into the bulk. As the power density is increased, ablation of surface atoms takes place, initiating a delayed ablation front that is also directed into the bulk. When the power density exceeds about 1×10^{10} W/cm^2, these two fronts travel at the same velocity in lock step and the atomic plume from the target surface becomes a very high density plasma plume that is in local thermodynamic equilibrium. As such, the elemental composition of the plasma plume becomes representative of the alloy specimen in elemental composition. In its afterglow regime of the plasma, it is possible to make quantitative measurement of the elemental populations as a function of depth. **Figure 1** shows the two sets of measurement for the two specimens taken out of a same batch of Nichrome alloy, one fresh specimen at room temperature and another fresh specimen at room temperature but after 15 hours of heating in vacuum at 1100 K. We see that the thermal forcing has modified the near-surface composition profile dramatically. The chromium enrichment at the specimen surface means two aspects in their transport: one, the availability of excess mobile atoms, by virtue of thermal forcing at elevated temperature; and two, preferential drift of chromium atoms over nickel atoms toward

http://dx.doi.org/10.5772/64837

the surface due to their mass difference, in the presence of the Coulomb force between an atom and its image charges at the specimen surface.

The key question here is where the sources of excess mobile atoms are. We postulate that the excess mobile particles arise from dissociation of nanocrystallites within the disordered alloy specimen under thermal forcing at elevated temperatures. The dissociation adds to the populations of glassy state atoms, and such atoms are less tightly bound to glassy state clusters than to nanocrystallites. Both chromium and nickel atoms drift toward the surface but by virtue of its slightly smaller mass, enrichment of chromium atoms results at the surface. The thickness of the chromium-enriched region near the surface grows in proportion to the length of the time of thermal forcing and the forcing temperature [1].

3. Nanocrystallite size distribution

The order within an RCP model can be quantified by means of the degree of crystallinity in the structure. The degree of crystallinity is the probability that an atom in the structure is part of nanocrystallites rather than being part of glassy state clusters. The basic building block of a nanocrystallite in three dimensions is a tetrahedron composed of four spheres [18]. To define the structure of a disordered binary alloy specimen, we proceed first to determine the distribution of nanocrystallites by size in two dimensions. The distribution function in two dimensions is then transformed into three dimensions. The normalization constant of the distribution function is found by requiring that the integral of the distribution over size can be equated to the degree of crystallinity multiplied by the total number of atoms in the specimen. The nanocrystallite size distribution in three dimensions is found by transforming the distribution in two dimensions into one in three dimensions.

As it turns out, the degree of crystallinity of a structure has a strong dependence on alloy composition fraction. The structure we use as the basis for the arrangement of atoms in a binary alloy is first measured in two dimensions using a simulated two-dimensional (2D) assembly of spheres. Alloys of different compositions are constructed by mixing the spheres of two different diameters. Here, we assert that the primary physics that controls the size of crystalline assembly is the repulsive part of the inter-particle interaction potential, ignoring the attractive part at this stage. The attractive interaction plays significant roles in capturing the symmetry property of the alloy's nanocrystallites. This is fully realized when we compose the theoretical model of the disordered alloy in atomic dimensions; we make use of the known crystalline structure of the alloy according to the established solid-state database of the particular crystalline assembly of atoms as a disordered solid specimen. More details will be given when we present the specific example of $AuCu_3$.

The simulated assembly of the hard spheres in two dimensions is randomly reinitialized by random close packing and analyzed for nanocrystallites by means of digital photography. This sequence of measurement for characterization of disorder in an alloy specimen is repeated many times to find a statistically stable nanocrystallite size distribution. Two different size spheres were mixed into a single layer within a rectangular 2D container with a transparent base, according to given alloy composition. The mixture is randomized each time by tilting the baseplate of the assembly about the horizontal plane back and forth for five times. A high-resolution digital image is taken of the assembly under diffuse illumination from below after each randomization routine. The images are analyzed for determination of particle positions and identify the nanocrystallites. We identify a nanocrystallite as an assembly of spheres, where all constituent spheres of the nanocrystallite are in contact with at least two other spheres. This basic rule is applied throughout to identify nanocrystallites of different sizes at each alloy composition.

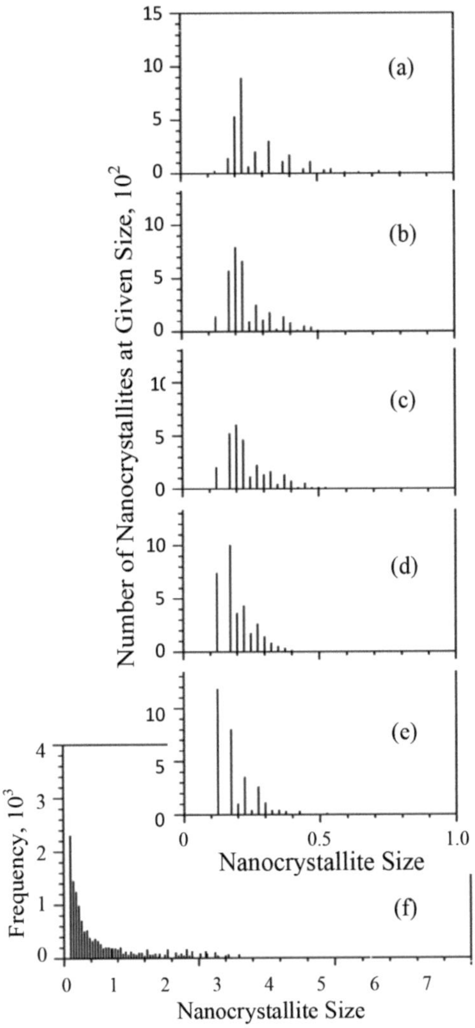

Figure 2. The frequency of occurrence of nanocrystallites as a function of nanocrystallite size for six different compositions: (a) 20W%Small:80%Large; (b) 40%S:60%L; (c)50%S:50%L; (d) 75%S:25%L; (e) 90%S:10%L; and (f) 100%S:0%L. The nanocrystallite size is given in terms of the radius of the smallest circular area into which all particles of the nanocrystallite can be fit in.

http://dx.doi.org/10.5772/64837

A nanocrystallite of a certain number of spheres can be made up of many different combinations of small and large spheres. For simplicity, the nanocrystallite size was identified by an effective radius, which is determined by weighted mean value of radii of spheres making up a nanocrystallite at given alloy composition. Experimental results are shown in **Figure 2**. They show a strong composition dependence on both the degree of crystallinity and the nanocrystallite size distribution.

4. Monte-Carlo simulation of disordered RCP alloys

Direct measurement of the nanocrystallite size distribution is an extremely time-consuming exercise. In order to help ease the process of applying the alloy modeling process, we have developed a Monte-Carlo code technique for simulating a disordered RCP binary alloy specimen for any given alloy composition. An ensemble of these numerically simulated alloy specimens is analyzed to obtain the nanocrystallite size distribution function for the alloy that is comparable to the measurement of the type shown in **Figure 2**. The Monte-Carlo code is structured to reproduce the suite of the measured nanocrystallite size distribution functions at six different alloy compositions.

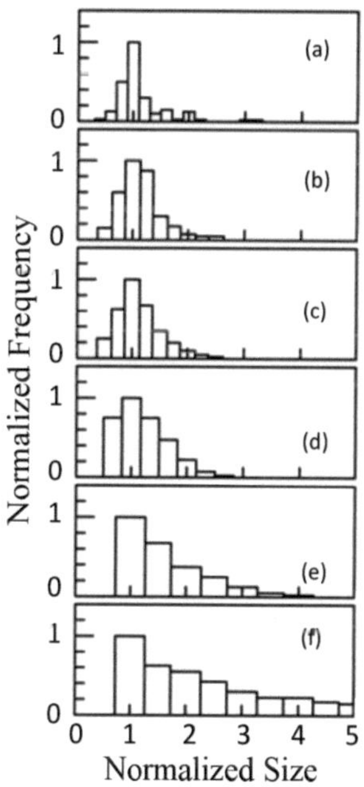

Figure 3. The normalized nanocrystallite size distribution function is shown for six different alloy compositions. The normalized frequency of occurrence is relative to the maximal value, and the normalized nanocrystallite size is normalized with respect to the maximal radius, i.e., the most populous size in two dimensions. The measured degree of crystallinity γ, which is defined as the probability that an atom is part of nanocrystallites in the specimen, is also tabulated: (a) 20W%Small:80%Large, γ = 0.456; (b) 40%S:60%L, γ = 0.479; (c) 50%S:50%L, γ = 0.307; (d) 75%S:25%L, γ = 0.316; (e) 90%S:10%L, γ = 0.486; and (f) 100%S:0%L, γ = 0.640.

We have found that the Monte-Carlo code must be subjected to two basic rules as follows. The alloy building process begins with three-particle crystallite as a seed in two dimensions or four-particle crystallite in three dimensions. The next particle is selected randomly according to given alloy composition. In two dimensions, the selected particle is placed next to the seed nanocrystallite. When the particle is placed in one of the three crystal points the seed crystallite grows in size by one. The crystal points are located between two particles that are in contact with particles of the nanocrystallite. When it is placed at any other point, the glassy state medium grows larger by one. After many runs of the numerical simulation, we have compiled the probability for placing the new particle into a crystal point in such a way that the measured nanocrystallite size distribution functions are replicated. We have found that the probability that replicates the nanocrystallite size distribution functions of **Figure 3** is influenced by the degree of crystallinity of the specimen. But its dependence on alloy composition is found to be very weak and thus ignored.

Rule one is that each new particle introduced into the alloy specimen being constructed be selected according to the probability of being placed into a crystal point, as shown in **Figure 4**. The probabilities that are most successfully replicating the measurements of **Figure 3** are shown as a function of the degree of crystallinity.

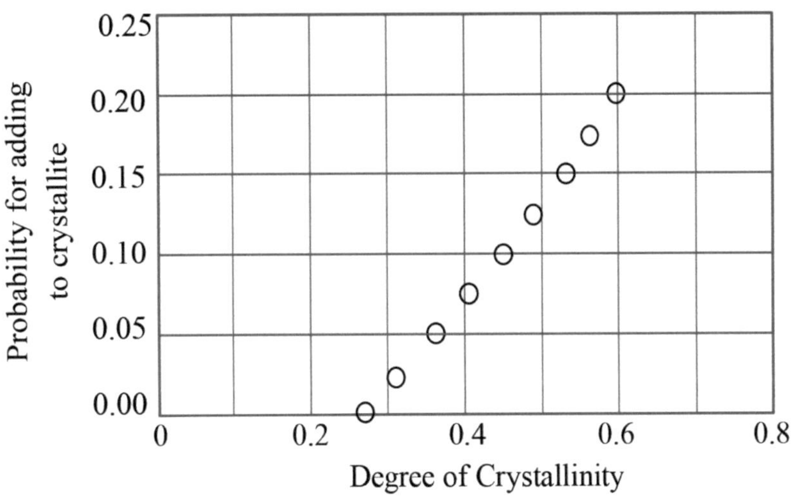

Figure 4. The probability that a new particle introduced into the numerically simulated alloy medium is placed at one of the crystal points of the nanocrystallite under numerical construction.

As one proceeds with construction of an alloy specimen of given composition by numerical simulation, the outer edges of the specimen invariably develop fingered growth fronts. These patterns appear entirely stochastically, and if left unattended, the simulated specimen becomes filled with numerous large-scale defects of open voids. Rule two is to choke off the growth of such large-scale defects by inserting a particle as soon as the gap between two nearest neighbor particles becomes equal to, or larger than, the diameter of the smallest particles in the pool of particles. An example of a numerically simulated monodisperse alloy is shown in **Figure 5**.

http://dx.doi.org/10.5772/64837

Figure 5. A sample specimen of a disordered alloy as generated by Monte Carlo code simulation.

The key data such as the nanocrystallite size distribution function can be extracted from the simulated specimens of the type shown in **Figure 5**. The simulation is repeated for a large number of times to conduct "the experiment." It is also conceivable to generalize the procedure to acquire "the experimental data" in three dimensions by simulating the alloy specimens in three dimensions according to the same rules of alloy construction that have been invoked for the construction in two dimensions.

5. Dissociation equilibrium of nanocrystallites under thermal forcing

In the present approach, a disordered alloy specimen is modeled as a random mixture of nanocrystallites and glassy atoms at room temperature and constrained by the degree of crystallinity at given alloy composition. When the specimen is forced at an elevated temperature, the constituent nanocrystallites and glassy state atoms undergo excitations in the form of phonons, positional displacements, and structural transformations within the bounds of Maxwell-Boltzmann statistics. At some point in the forcing at a fixed temperature, small movements of the constituent glassy state atoms and nanocrystallites can result in fluctuations in the mass density of the specimen. As the temperature is raised, the rates of these excitations increase to the extent that the size distribution of nanocrystallites is bound to undergo significant changes. This means that thermal dissociation of atoms from the nanocrystallites within the medium takes place into the populations of glassy state atoms. Such inelastic processes require energies to overcome the activation energy each host structure demands of constituent atoms within the nanocrystallites.

This general coarse-grained physical picture has been examined for simulated disordered specimens in two dimensions by means of a laboratory experimental set-up. An alloy specimen is simulated in two dimensions within a 2D specimen cell, consisting of a transparent conducting baseplate bounded by a rectangular aluminum frame resting on it. The specimen cell is filled with steel spheres and is driven by two mutually orthogonal linear drive motors under computer control. Two independent stepping motors are used for this purpose. Each motor is driven chaotically according to a sequence of random numbers, which are delta-autocorrelated, i.e., the successive displacements of the sequence remain uncorrelated. The operation of the set-up has been tuned such that the distribution of particle velocities in the cell obeys the Maxwell-Boltzmann statistics exactly. When the amplitude of the stepper displacements is increased by a constant factor, the Gaussian velocity profile has been found to broaden proportionately. We have thus succeeded in producing a "mechanical oven" in which thermal forcing can be effected at different "temperatures" on the simulated alloy specimens.

The response of the specimens with 2D RCP structures to thermal forcing in the mechanical oven experiment has been investigated, and this was described in previous work [5]. Observations from a series of experiments with real alloy specimens have been viewed in the light of the simulated thermal forcing experiment. The conclusions have formed the solid basis of the microscopic physical processes that take place within each alloy specimen. What was seen in the simulated thermal forcing experiment was that the degree of crystallinity of the RCP structure decreased as a function of effective temperature due to dissociation of nanocrystallites within the specimen at elevated temperature.

Thus, we begin first principle modeling of thermal dissociation of nanocrystallites by means of the law of mass action [19]. At room temperature, the structure of a disordered alloy specimen consists a population of nanocrystallites having a certain size distribution function suspended in the sea of glassy-state atoms. The exact functional form of the size distribution is controlled by alloy composition, and so is the degree of crystallinity. The structure of an individual alloy specimen results from random close packing of glassy-state atoms with nanocrystallites, as randomly selected from the ensemble of nanocrystallites having the designated size distribution for the particular given alloy composition. We note that the random assembly of the specimen is carried out in three dimensions. This means that a suitable procedure for transformation of the nanocrystallite size distribution in two dimensions into three dimensions must be established at some later stages of the theoretical development.

In order to model the thermal dissociation of a nanocrystallite in the alloy, we consider a nanocrystallite containing i-particles. We are interested in calculating the percentage of i-size crystallites that will thermally dissociate j times, losing j atoms from its surface. For each step of thermal dissociation, a nanocrystallite loses one atom from its surface, which then becomes part of the glassy matter. To lose j-atoms from the surface of a nanocrystallite would require j-reaction equations of thermal dissociation:

http://dx.doi.org/10.5772/64837

$$c_i \to c_{i-1} + A_g$$
$$c_{i-1} \to c_{i-2} + A_g$$
$$\vdots$$
$$c_{i-j+1} \to c_{i-j} + A_g$$

Here, c_i represents i-atom nanocrystallite and A_g the glassy-state atom as reactants. Each of these reactions may be expressed in the following form:

$$\sum_{i=reactant} v_i A_i = 0 \tag{1}$$

where A_i denotes i-th reactant and v_i is the number of i-th reactants that are needed to complete the reaction.

The assertion that only the atoms on the surface of the nanocrystallite undergo thermal dissociation is based on the fact that removal of a single atom from the surface would incur the lowest energy cost in the low temperature heating environment. A nanocrystallite breaking off chunks of atoms at a time would require much higher energy and therefore much less likely. With this system of reaction equations for each i-atom nanocrystallite, we have established the dissociation pathway along which each i-atom nanocrystallite becomes i atoms in the glassy matter medium.

The equation of state for the system of nanocrystallites and glassy state atoms can be expressed in the form of total volume occupied by all nanocrystallites and glassy state atoms. For instance, the thermal expansion of the specimen can be written out in terms of the alloy's thermal expansion coefficient of nanocrystallites ξ_c and of the glassy-state medium ξ_g:

$$V(N_0, T) = V_{C0}[1 + \xi_C(T - T_{rm})]^3 + V_{g0}\left[1 + \xi_g(T - T_{rm})\right]^3 \tag{2}$$

The equation of state depends on the system's state at room temperature. The alloy specimen will be constrained by the total number of atoms N_0, which is conserved:

$$N_0 = N_g + \sum_{i=2}^{i_{max}} D(R_i, T) N_i i \tag{3}$$

Here, N_g is the number of glassy-state atoms, $D(R_i, T)$ is the nanocrystallite size distribution, and N_i is the number of i-atom nanocrystallites in the specimen. The initial volume occupied by all nanocrystallites and the initial volume occupied by glassy-state atoms are, respectively,

$$V_{c0} = \sum_{i=1}^{i_{max}} D(R_i, T)\frac{4\pi R_i^3}{3} \tag{4}$$

$$V_{g0} = N_g \frac{4\pi R_g^3}{3} \tag{5}$$

where R_i is the nanocrystallite radius and R_g is the radius of the average volume needed for a single glassy-state atom. The morphology of the alloy specimen is represented by the degree of crystallinity, and the crystallite size distribution changes the individual parameters in each of these equations dictating the evolution of the alloy specimen as a function of temperature.

The chemical potential, i.e., the energy needed to increase the number of the *i*-th reactant species in the system by one, is found from the Gibbs free energy $F(T, p, N_1, N_2, ...)$ of the system:

$$\mu_i = \frac{\partial F}{\partial N_i} = -k_B T \ln\left[q_i / N_i\right] \tag{6}$$

T is the specimen temperature, p the pressure and N_i the total number of *i*-th species within the specimen, either a nanocrystallite or a glassy state atom. k_B is the Boltzmann constant. q_i is the canonical partition function for the *i*-atom nanocrystallite. The reaction equilibrium satisfies the law of mass action, which may be written in the following form:

$$\prod_{i=reactant} \left[q_i / N_i\right]^{\nu_i} = 1 \tag{7}$$

There are (i-1) dissociation steps for an *i*-atom nanocrystallite. For each dissociation step, there is a reaction equation of the form of Eq. (1), and the corresponding law of mass action may be written in the following form:

$$\frac{q_i}{q_{i-1} q_A} = \frac{N_i}{N_{i-1} N_A} \tag{8}$$

We will later attend to the canonical partition function for each reactant in the reaction equation for thermal dissociation. In order to address the temperature dependence of the nanocrystallite populations by size we must solve the system of the law of mass action equations simultaneously. The dissociation of an atom from an *i*-atom nanocrystallite increases the population of (*i*-1)-atom nanocrystallites by one, which in turn affects the reaction equilibrium of the (*i*-1)-atom nanocrystallites with similar consequences on the populations of the smaller nanocrystallites. To help manage the simultaneous nature of the rather large number of reaction equilibria involved, we introduce the dimensionless degree of dissociation as follows:

$$\alpha_{i,j} = \frac{N_{i,i-1} + N_{i,i-2} + \cdots + N_{i,i-(j-1)}}{N_{i,i-1} + N_{i,i-2} + \cdots + N_{i,i-j}} \tag{9}$$

Here, $\alpha_{i,j}$ is defined as the ratio of the number of all initially *i*-atom nanocrystallites that have been dissociated once through *j* times to the number of all initially *i*-atom nanocrystallites that have been dissociated once through (*j*–1) times. Since the distribution of nanocrystallites by size will be determined by the nanocrystallite size distribution at given atomic composition, each crystallite size will be populated at a certain number N_i. In order to track the evolution of the number of *i*-atom nanocrystallites as a function of temperature, we introduce the degree of dissociation. The degree of dissociation is the percentage of *i*-atom nanocrystallites that will dissociate *j* times. This expresses the remaining number of *i*-atom nanocrystallites after an increase in temperature, while also providing information on the size of the resulting nanocrystallites after the dissociation. In order to write the law of mass action equation in terms of the degree of dissociation, the number of *i*-atom nanocrystallites is replaced with the set of degrees of dissociation for the dissociation steps the nanocrystallites must undergo:

$$\frac{\left(1-\alpha_{i,j}\right)}{\alpha_{i,j}\left(1-\alpha_{i,j-1}\right)} = \frac{N_A q_{i-j}}{q_{i-j-1} q_A} \tag{10}$$

Let us consider, for example, dissociation of a 10-particle nanocrystallite. This will present nine stages of dissociation to consider and a set of nine coupled equations to calculate its degrees of dissociation, as shown below:

$$\frac{\left(1-\alpha_{10,1}\right)}{\alpha_{10,1}\left(1-\alpha_{10,2}\right)} = \frac{N_A q_9}{q_8 q_A}$$

$$\frac{\left(1-\alpha_{10,2}\right)}{\alpha_{10,2}\left(1-\alpha_{10,3}\right)} = \frac{N_A q_8}{q_7 q_A}$$

$$\vdots$$

$$\frac{\left(1-\alpha_{10,9}\right)}{\alpha_{10,9}} = \frac{N_A q_1}{q_2 q_A}$$

The structure of these coupled equations provides the method for computing the degrees of dissociation. It requires that the partition function for each nanocrystallite as well as the glassy state atoms in the alloy be known. Not only that these equations are coupled but also that each equation contains two undetermined degrees of dissociation. Each pair of successive equations shares a common degree of dissociation. An exception is that the last equation in the sequence of dissociation steps contains only one unknown degree of dissociation to be solved. It would seem possible to calculate the degree of dissociation in the last equation, if the right-hand side of the equation is fully prescribed. The solution may be carried into the next equation to solve for the remaining unknown degree of dissociation if its right-hand side is prescribed. This procedure can be continued for the full set of dissociation equations, provided that the total number of glassy-state atoms in the specimen at the given temperature is known.

The total number of glassy-state atoms is not known, however. At room temperature, the population of glassy state atoms in the alloy specimen is given by the degree of crystallinity. When significant thermal dissociation of nanocrystallites commences, N_A, the number of glassy state particles in the specimen, increases with temperature. The way the dissociation equations are presented above, N_A appears on the right-hand side of each equation so that the entire system of dissociation equations for the alloy specimen can be solved by the trial and error method. That is, first guess a value for $N_A^{(0)}$, solve for all $\alpha_{i,j}$'s and calculate $N_A^{(1)}$; continue until $N_A^{(l)}$ agrees with $N_A^{(l+1)}$ within a preset error. Here, l is an integer that tracks the number of iterations.

As nanocrystallites dissociate, the asymptotic value of $N_A^{(l)}$ will change as a function of temperature. This number of glassy-state atoms can then be recalculated using the set of degrees of dissociation and compared with the initial value to assess the self-consistency of the computation. If not in agreement within the preset error criterion, the process is reinitialized and computation is repeated until a satisfactory agreement is attained.

To calculate the coupled law of mass action equations, it is necessary to write the canonical partition functions for each of the reactants involved in the thermal dissociation reaction equation. The partition function of a reactant includes eigenstates according to all the degrees of freedom the reactant has, be it a nanocrystallite or a glassy state atom. However, we make a note of the fact that the canonical partition functions appear in each of the degrees of dissociation equations contains the partition functions in the form of the ratio of two partition functions—that is, the partition function of a nanocrystallite before, and after, a single-step dissociation. The partition function for a single i-particle nanocrystallite may be written out as follows:

$$q_i = \left(q_{trans} q_{rot} q_{vib}\right)_i$$

or,

http://dx.doi.org/10.5772/64837

$$\frac{q_{i-j}}{q_{i-j-1}}=\left(\frac{q_{i-j}}{q_{i-j-1}}\right)_{trans}\left(\frac{q_{i-j}}{q_{i-j-1}}\right)_{rot}\left(\frac{q_{i-j}}{q_{i-j-1}}\right)_{vib}$$

The subscripts *trans, rot,* and *vib* denote, respectively, translational, rotational, and vibrational degrees of freedom. The rotational and vibrational motions of the nanocrystallites are dominated by the crystalline structure, and we approximate the ratio to be unity to find:

$$\frac{q_{i-j}}{q_{i-j-1}}\approx\left(\frac{q_{i-j}}{q_{i-j-1}}\right)_{trans}$$

The translational degrees of freedom give rise to the partition function of i-particle nanocrystallite at temperature T as follows:

$$\left(q_{trans}\right)_i=V\left(\frac{2\pi iMk_BT}{h^2}\right)^{3/2} \tag{11}$$

V is the volume of the alloy specimen, m is the mass of the atom in the glassy matter, T is the temperature of the specimen, k_B is Boltzmann's constant, and h is Planck's constant. Thus, we write Eq. (11) as follows:

$$\frac{\left(1-\alpha_{i,j}\right)}{\alpha_{i,j}\left(1-\alpha_{i,j-1}\right)}=\frac{N_A}{V}\left(\frac{i-j+1}{i-j}\right)^{3/2}\left(\frac{2\pi Mk_BT}{h^2}\right)^{-3/2}\exp\left(\frac{D_{i-j+1}}{k_BT}\right) \tag{12}$$

where M denotes atomic mass. In the event that the specimen is a binary alloy of given composition, M would be the average atomic mass, weighted at the alloy composition. D_{i-j+1} is the dissociation potential for a particle on the periphery of the (i–j) particle nanocrystallite. It measures the zero of the energy scale of the nanocrystallite after the dissociation of a single particle relative to that of the nanocrystallite before dissociation. In other words, it is the energy needed to remove an atom from the surface of the (i–j+1) atom nanocrystallite by thermal dissociation.

6. Calculation of the dissociation potential

The most important part of the law of mass action computation is the energy of thermal dissociation or the dissociation potential. A particle on the surface of a nanocrystallite is bound to the surface an attractive potential, and it must overcome this potential energy to be dissociated from the surface to become a glassy state atom.

The final key parameter needed for calculation of the series of law of mass action equations is the potential energy experienced by an atom on the surface of an i-atom nanocrystallite or the dissociation potential $D_{0,\,i}$. For each nanocrystallite of the alloy, we calculate this potential according to the unit cell information for the alloy specimen as a single crystal and the Lennard-Jones parameters for the interacting atom pairs contained in the nanocrystallite of the alloy. Our first approximation is to treat each i-atom nanocrystallite to be spherical in shape. Using the structural information for the alloy as a single crystal, we begin with a cubic sample of the crystalline alloy and sculpt a spherical nanocrystallite of radius r by chiseling away atoms farther than a radius r from the center of the sample. In this manner, the full set of spherical nanocrystallites, as specified by the nanocrystallite size distribution function for the alloy, is generated.

The surface atoms in each of these spherical nanocrystallites are considered in the determination of the dissociation potential. To find the dissociation potential, we must take sum of all of the interatomic interaction potential contributions from every other atom in the nanocrystallite. Using the Lennard-Jones potential, we write the interaction potential for i-th atom due to N-atom nanocrystallite as follows:

$$V_i = \sum_{j=1}^{N} 4\varepsilon_j \left[\left(\frac{\sigma_j}{r_{ij}} \right)^{12} - \left(\frac{\sigma_j}{r_{ij}} \right)^{6} \right] \tag{13}$$

For most of the metallic elements, Lennard-Jones potential constants σ_j and ε_j have already been computed through quantum chemistry calculations [20]. In a binary nanocrystallite, these quantum chemistry values cover only the like atom interactions. To account for potential interactions between unlike atoms, as indicated by i and j that are unequal, we make use of the Kong combination rules for Lennard-Jones potential parameters [21]:

$$\epsilon_{ij}\sigma_{ij}^{\;6} = \left(\epsilon_{ii}\sigma_{ii}^{\;6} \epsilon_{jj}\sigma_{jj}^{\;6} \right)^{\frac{1}{2}} \tag{14}$$

$$\epsilon_{ij}\sigma_{ij}^{\;12} = \frac{\epsilon_{ii}\sigma_{ii}^{\;12}}{2^{13}} \left[1 + \left(\frac{\epsilon_{jj}\sigma_{jj}^{\;12}}{\epsilon_{ii}\sigma_{ii}^{\;12}} \right)^{\frac{1}{13}} \right]^{13} \tag{15}$$

As illustrated in **Figure 6** to calculate the total interaction potential, we move the surface atom radially outward from the center of the nanocrystallite and calculate the total interaction potential at each radial position. We repeat this calculation for all of the surface atoms and take an average over all of the surface atoms. For a binary nanocrystallite, there are different dissociation potentials for different atom pairs. The dissociation potential is calculated for all of the three possible pairs of atoms, but the final value of the dissociation potential is assigned with the weighted average of the two atom types according to atomic composition of the binary alloy specimen. Nanocrystallites of different sizes are built and analyzed in the similar manner.

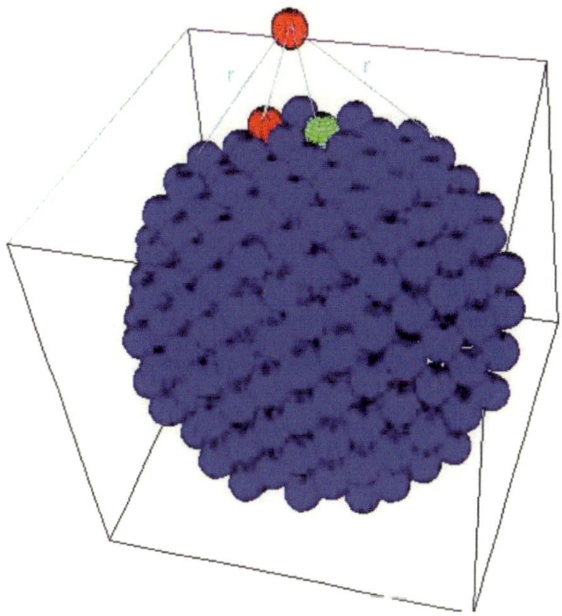

Figure 6. The dissociation potential for a single particle on the surface of a nanocrystallite is the energy needed to free a surface atom against attraction by all atoms in the nanocrystallite. The work done to move the atom in red to infinity is computed. The computation is repeated for all possible surface positions and pairings of surface atoms with the rest of the atoms within the nanocrystallite

For the alloy of $AuCu_3$ computation, we have used $\sigma_j = 2.6367\text{x}10^{-10}$ m and $\varepsilon = 5152.9$ K for gold and $\sigma_j = 2.3374\text{x}10^{-10}$ m and $\varepsilon = 4733.5$ K for copper [20]. The computed dissociation potential grows with increasing size of the nanocrystallite as shown in **Figure 7**, and this can be fitted by an analytical function. The resulting dissociation potential function can then be incorporated into the system of simultaneous algebraic equations for the large number of degrees of dissociation needed in the law of mass action reaction equilibrium computation. The dissociation potential grows to an asymptotic value for large nanocrystallites. To fit the dissociation potential as a function of nanocrystallite size, it is necessary to look at how the dissociation potential differs from the asymptotic value. When the values of the dissociation potential are subtracted from the asymptotic value and plotted on a log-log scale, a linear relation between the dissociation potential and nanocrystallite size is seen. To get the best fit possible for this functional dependence, a constant is added to optimize the fit. This fit has the functional form of

$$D(i) = D(Large\,i\,limit) - A * i^B + Constant \tag{16}$$

An example of the fitting procedure for the alloy $AuCu_3$ can be seen in **Figures 7** and **8**. **Figure 8** shows a straight-line fit of the computed dissociation energy as a function of nanocrystallite size in number of particles that are contained in each crystallite.

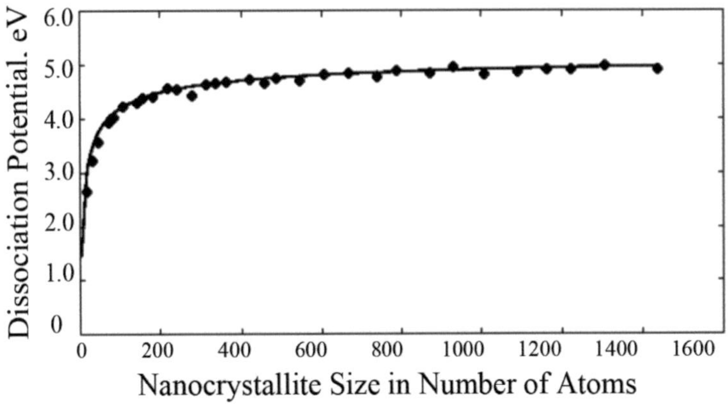

Figure 7. Energy required for removal of a single surface atom from a nanocrystallite by thermal dissociation as a function of nanocrystallite size for the $AuCu_3$ alloy.

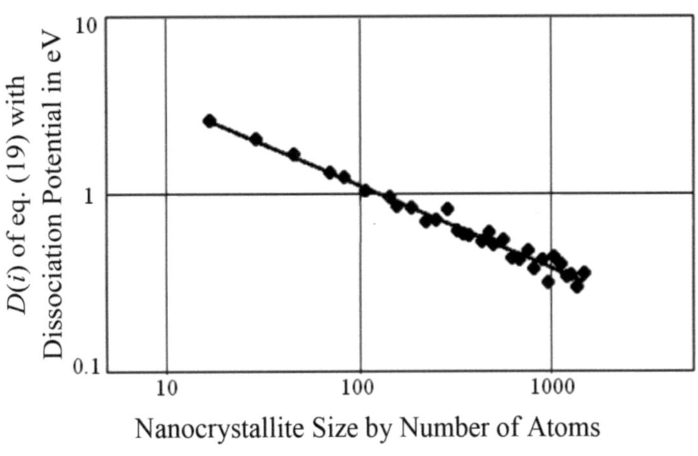

Figure 8. The computed dissociation potential of **Figure 7** as a function of nanocrystallite size is transformed into *D(i)* of Eq. (9), and *D(i)* is plotted as a straight line against nanocrystallite size for the $AuCu_3$ alloy.

7. Theoretical prediction of disorder in $AuCu_3$ at elevated temperatures

We consider the case of the binary alloy, $AuCu_3$, which has an alloy composition of 75% copper (small) to 25% gold (large) with the respective atomic radii of gold and copper. The structure of the alloy is known based on the 3D nanocrystallite size distribution. In order to determine the number of atoms contained in each nanocrystallite size bin, we need to know the total number of atoms in the alloy specimen to be modeled. Using $AuCu_3$ as an example again, we have the atomic mass of each of the two different atoms in the alloy: $m_{Cu} = 6.655 \times 10^{-23} g$ and $m_{Au} = 3.27 \times 10^{-22} g$. From the mass density, we know the relationship between the number of particles in the alloy specimen and the volume of the specimen, where the mass density of the alloy $\rho_{AuCu_3} = 10.9946$ g/cm³ ($AuCu_3$ parameters are obtained from reference [22]).

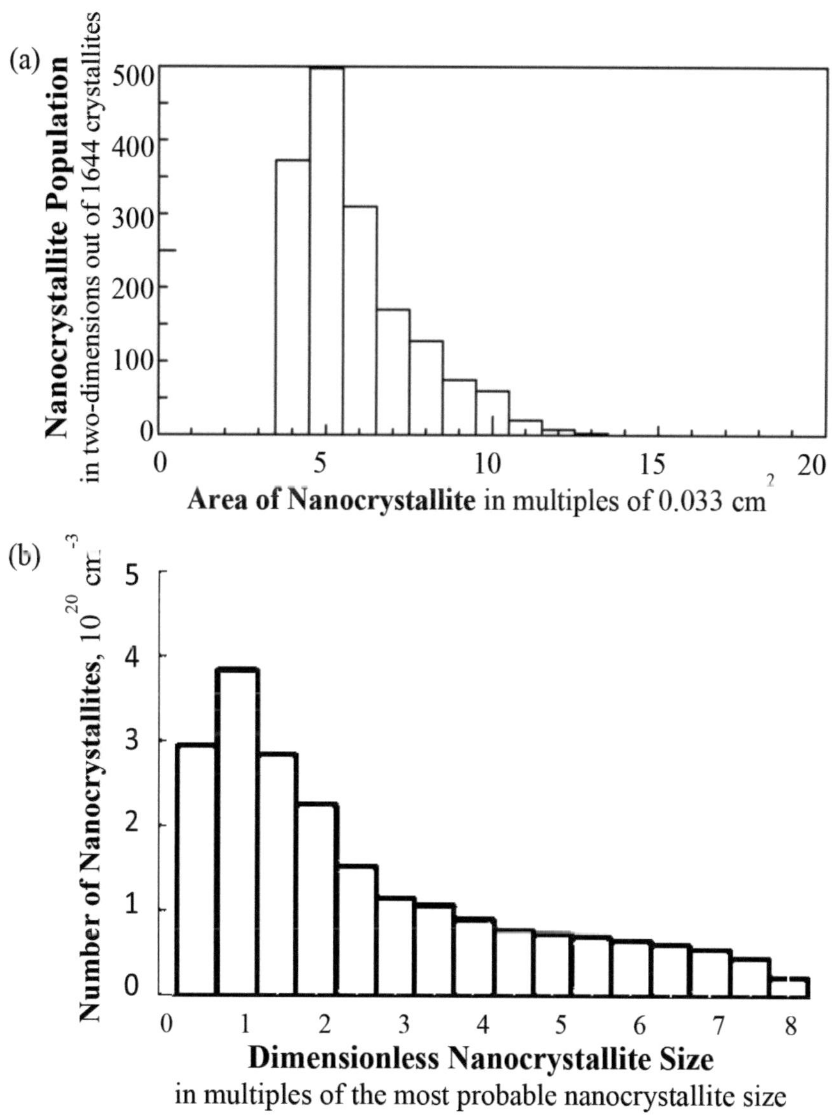

Figure 9. Nanocrystallite size distribution for alloy $AuCu_3$ at room temperature. The measured distribution function at alloy composition of 25W%Au:75W%Cu in two dimensions is shown in (a). The distribution after the transformation into three dimensions is shown in (b).

The initial nanocrystallite size distribution for alloy $AuCu_3$ is needed to carry out the law of mass action equation calculations for the degrees of dissociation at elevated temperature, and we make use of the measurement from the simulated alloy medium in two dimensions. The size of the nanocrystallite is represented in our modeling in terms of the area of a circle, and therefore the corresponding size of the nanocrystallite in three dimensions would be given by the volume of a sphere having the same radius of the circle in two dimensions. The 2D size distribution is transformed to three dimensions according to the reasoning that the size of a nanocrystallite in three dimensions would scale as the size in 2D raised to 3/2 power. Both the nanocrystallite size distribution in 2D and in 3D are shown in **Figure 9**.

In order to explore the morphology of the alloy as a function of the specimen temperature, it is necessary to carry out the law of mass action calculation at each given temperature to determine how the structure of the specimen evolves as a function of temperature. The value that is calculated through the set of equations is the degrees of dissociation or the amount of nanocrystallites of a certain size changing due to dissociation. Each numerical value of the degree of dissociation that is above zero means a thermal dissociation reaction for the nanocrystallite of particular size. The result of any such reaction means a change in the nanocrystallite size distribution as well as a change in the total number of atoms in the glassy state matter. The population at each nanocrystallite size is adjusted after the full set of degrees of dissociation values has been calculated. The number of *i*-sized nanocrystallites is adjusted according to the formula below:

$$N_i = N_{i0}(1-\alpha_{i,1}) + \sum_{j>i}^{j=max} N_j(1-\alpha_{j,j-i+1})\left(\prod_{k=1}^{k=j-i} \alpha_{j,k}\right) \tag{17}$$

N_{i0} is the initial number of *i*-sized nanocrystallites before the temperature has been raised. The first term in the formula accounts for any of the nanocrystallites that are lost out of the group due to dissociation, which will in turn reduce the number of the nanocrystallites of the particular size. The second term in the formula accounts for larger nanocrystallites that would become *i*-atom nanocrystallites after going through stages of dissociation. This term accounts for an increase in the number of *i*-sized crystallites.

Figure 10 displays the result from calculations of the degrees of dissociation for a series of temperatures for the alloy $AuCu_3$. As can be seen, the crystallite size distribution evolves as the temperature is raised, and **Figure 10** shows how the populations of nanocrystallites of different sizes are affected by the temperature change. As the nanocrystallites dissociate, the total number of atoms tied up in nanocrystallites decreases, and this causes a change in the degree of crystallinity. The numerical value of the degree of crystallinity at each temperature step is shown in the caption of **Figure 10**.

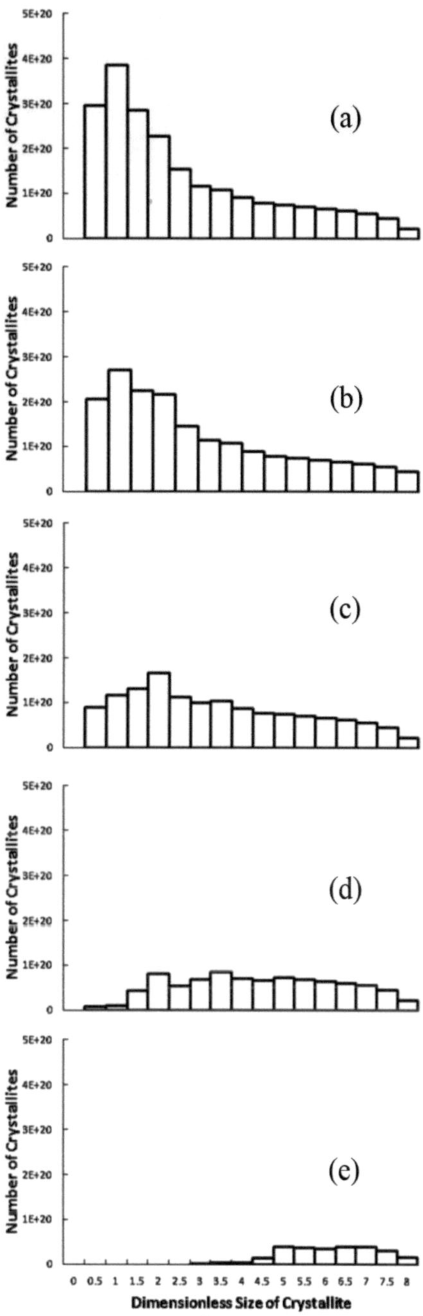

Figure 10. The nanocrystallite size distribution of the $AuCu_3$ alloy at elevated temperatures is computed using the law of mass action reaction equilibrium equations. Thermal dissociation of the distribution of nanocrystallites at room temperature is carried forward as the alloy temperature is increased. The vertical axis displays the number of nanocrystallites per cm^3. The temperature and the computed degree of crystallinity of each graph are as follows: (a) 300 K, γ =0.316; (b) 1000 K, $\gamma = 0.306$; (c) 1100 K, $\gamma = 0.276$; (d) 1200 K, $\gamma = 0.203$; and (e) 1400 K, $\gamma = 0.051$.

As the nanocrystallites in the alloy specimen undergo thermal dissociation, the number of glassy state atoms increases due to changing degrees of dissociation as follows:

$$N_A = N_{A0} + \sum_{i=2}^{i=max} N_{i0} \prod_{j=1}^{j=i-1} \alpha_{i,j} + \sum_{i=2}^{i=max} (N_i - N_{i0}) \tag{18}$$

N_{A0} is the number of glassy state atoms prior to heating, and there will be no decrease in the number because we are concerned with increases in temperature. The second term accounts for all of the fully dissociated nanocrystallites. The third term accounts for all glassy state atoms from partial dissociation reactions associated with each temperature rise. Note that with each thermal dissociation event, an atom is added into the ranks of glassy state matter.

Figure 11 shows the number of glassy state atoms in the alloy specimen as a function of temperature for the alloy $AuCu_3$. The vertical broken line inserted in the figure shows the known melting point of this particular alloy. As can be seen, the approximate model of the disordered alloy specimen not only reveals a well-behaved functional form of thermal dissociation of nanocrystallites in the alloy as the temperature is increased but also provides an effective physical mechanism for identifying the melting temperature of the alloy. A definition of the melting point could be the temperature at which one half of all nanocrystallite atoms within the alloy at room temperature are dissociated into glassy state atoms. Another definition is possible based on the temperature at which the thermal dissociation rate maximizes; this may provide a way to fine tune the melting point.

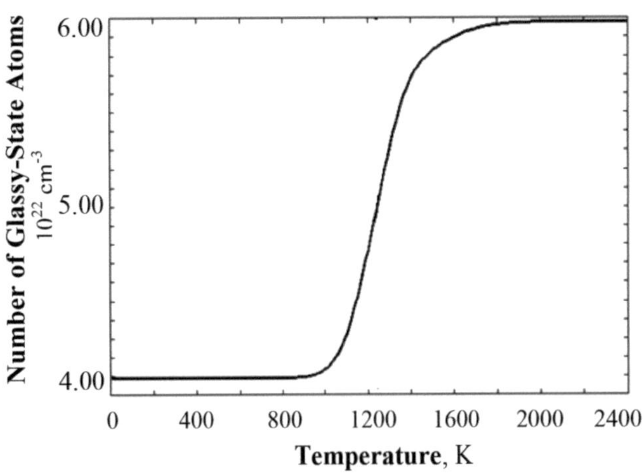

Figure 11. Number of glassy state atoms as a function of temperature in alloy $AuCu_3$. The dashed line shows the known melting temperature of 1240 K of the alloy in excellent agreement.

This technique was used to predict the melting temperature for a number of metallic elements and a group of six binary alloys. The results from the lowest order approximation of the

http://dx.doi.org/10.5772/64837

disorder in metallic alloys, as described above, are summarized in **Tables 1** and **2**. The good agreement of the computed melting points of the high temperature metals and alloys with the known experimental melting point values presents considerable confidence in the validity of our approach. The limited number of binary alloy calculations is due to the lack of available unit cell structure information for binary alloys, which is necessary for execution of our disorder modeling approach.

Element	Symbol	σ (Å)	ε (eV K)	Lattice constant (Å) [21]	Known melting point, K [21]	Predicted melting point (@ 50% dissociation), K
Rhodium	Rh	2.4622	0.6699	3.8	2233	2360
Iridium	Ir	2.4839	0.8077	3.84	2716	2730
Tantalum	Ta	2.6819	0.9809	3.31	3269	2905
Chromium	Cr	2.3357	0.4988	2.88	2163	2245
Tungsten	W	2.5618	1.0644	3.16	3653	3840
Hafnium	Hf	2.8917	0.750989	3.2	2423	2160
Ruthenium	Ru	2.4447	0.788343	2.7	2523	2410
Titanium	Ti	2.6841	0.568102	2.95	1948	1900
Zirconium	Zr	2.9318	0.738442	3.232	2125	2230

Table 1. Refractory metallic elements.

Alloy [22–26]	Lattice constant, Å	Known melting point, K	Predicted melting point (@ 50% dissociation), K
$AuCu_3$	3.74	1240.5	1320
AuCu	3.964 (2), 3.672	1183	1410
Au_3Cu	3.964	1151.5	1495
Al_3Ti	3.964	1340	1260
AlTi	4.00 (2), 4.075	1729	1935
$AlTi_3$	a =5.780, c=4.674	1473	1360

Table 2. Binary alloys.

8. Concluding remarks

The modeling technique presented above provides a first principle approach to modeling the morphology of a disordered binary alloy under a variety of equilibrium thermal forcing conditions. The success in predicting the accepted melting temperatures of the metallic elements and binary alloys demonstrates that the approximations made in the coarse-grained assessment of disorder in binary alloys are reasonable and by and large correct. By the same token, we note that there are many opportunities for improving the outcomes of the disordered alloy modeling.

The framework for modeling the morphology of disordered metallic binary alloys has been built on the basis of a randomly close-packed assembly of constituent atoms. The RCP structure is treated as a mixture of nanocrystallites and glassy state matter. The fraction of all atoms tied up in crystalline structures defines the degree of crystallinity. The distribution of nanocrystallites at given size of the structure has been measured in the form of the crystallite size distribution function in two dimensions. Both of these characterizations are found to be dependent on alloy composition. The histogram of the nanocrystallite size distribution provides the snapshot of the equilibrium structure of the alloy at room temperature when transformed into three dimensions. Our theoretical modeling takes off from the 2D RCP media in two dimensions at six different compositions. The morphological alloy data have been obtained from the simulated alloys by experiment, which have also provided the realistic measures of the degree of crystallinity.

The evolution of the morphology in a disordered alloy specimen is modeled as the case of thermal dissociation of nanocrystallites within the alloy specimen by means of the law of mass action. The theoretical model begins with the room temperature structure of the alloy specimen as summarized above. We assumed it to be a randomly close-packed medium of nanocrystallites and glassy state atoms. The temperature-dependent evolution of the size distribution of the nanocrystallites within an alloy specimen is treated as the case of reaction equilibrium according to the law of mass action, as imposed on each stage of dissociation of all nanocrystallites within the alloy specimen. The very large system of coupled law of mass action equations is solved for each possible dissociation step of the nanocrystallites of the specimen. To facilitate the computational procedure, we have introduced a set of degrees of dissociation, each as a measure of the fraction of nanocrystallites at given size that dissociate into smaller nanocrystallites while increasing the population of glassy-state atoms, at given temperature. Various alloy parameters are important in the calculation of each degree of dissociation equation, but the main determining parameter for the temperature at which dissociation happens is the dissociation potential, i.e., the energy needed to extract individual atoms from the surface of each nanocrystallite of given size. This quantity is calculated for an atom on the surface of a spherical crystallite of given size. The energy is found by considering all interactions with all other atoms of the nanocrystallite. This potential was calculated assuming interatomic interactions of the Lennard-Jones type. We have made use of the Lennard-Jones parameters that have been obtained by the method of quantum chemistry computation for the pairs of like atoms in the alloy [20]; for interactions between two dissimilar atoms, the Kong combination rules are used [21].

The success of the model is demonstrated by the accurate prediction of the melting temperatures for disordered high-temperature refractory metals as well as for some binary alloy specimens. The morphology of the disordered alloy structure can now be predicted analytically at arbitrarily high temperatures. This makes it possible to predict the thermal transport properties for disordered alloys as a function of temperature. The important features of the

http://dx.doi.org/10.5772/64837

alloy structure that pertain to calculation of thermal conductivity at elevated temperatures are the structure of the nanocrystallites as specified in terms of the nanocrystallite size distribution, the density of glassy state matter, and the interfaces between them. All necessary details of the structure become available from the computed solutions of the system of the law of mass action equations as applied to the specifics of the initial state of the alloy by means of the nanocrystallite size distribution function.

We have shown the full modeling sequences for thermal forcing of disordered binary alloys. The method is sufficiently general for applications to forcing of alloys by mechanical stress, neutron bombardment, or chemical reactions. Our modeling approach may also be fine-tuned for improved accuracy by adding more refined microscopic details that are specific to a given alloy system. For example, the spherical shape used for all nanocrystallites may be relaxed to allow for elliptical or rectangular shapes with variable aspect ratio or to highlight the prevailing crystalline symmetry properties of particular alloy composition. There are many other possibilities. We may broaden the modeling to include characterization of tertiary or even more complex alloys. Ultimately, it appears quite remarkable that one may effectively characterize the morphology of many disordered metallic alloys as a fluid-like mixture of nanocrystallites and glassy state atoms.

The authors gratefully acknowledge spirited contributions from a cadre of young physics students during the earlier stages of model development with measurements of nanocrystallite size distribution functions: Andrew Abrahams of Moravian College and Jerry Kim of UCLA. Nathan Tomer of Drake University, William Ferm of the University of Maine, Adam Hansell of Lehigh University, Daniel O'Driscoll of Trinity College Dublin of Ireland, and William Woodward of Lehigh University. Henry Koon of St. Olaf College has helped with fine-tuning of certain aspects of digital image analysis needed with the simulated experiment.

The authors gratefully acknowledge partial financial support for the research by the National Science Foundation and Lehigh University.

Author details

Ryan P. Cress and Yong W. Kim*

*Address all correspondence to: ywk0@lehigh.edu

Department of Physics, Lehigh University, Bethlehem, Pennsylvania, USA

References

[1] Kim, Y.W., "Routes to Development of Near-Surface Alloy Composition Anomaly," International Journal of Thermophysics 26, (2005): 1051.

[2] Kim, Y.W., "Development of Transport Property-Composition Relationship by Thermal Modification of Alloy Composition Profile," High Temperatures - High Pressures 38, (2009): 1.

[3] Kim, Y.W., "Surface Position-Resolved Thermophysical Properties for Metallic Alloys," International Journal of Thermophysics 28, (2007): 732.

[4] Kim, Y.W., "Simultaneous Multitemperature Measurements of Thermal Diffusivity and Composition," International Journal of Thermophysics 31, (2010): 926.

[5] Kim, Y.W. and R.P. Cress, "Effects of Thermal Forcing on Morphology of Disordered Binary Metallic Alloys: Local Equilibration and Modification of Near-Surface Elemental Composition," High Temperature – High Pressure 40, (2011): 335.

[6] Ho, C. Y., M. W. Ackerman, K. Y. Wu, S. G. Oh, and T. N. Havill. "Thermal Conductivity of Ten Selected Binary Alloy Systems," Journal of Physical and Chemical Reference Data 7, 3 (1978): 959.

[7] Van Swygenhoven, H., and J. Weertman. "Deformation in Nanocrystalline Metals," Materials Today 9, 5 (2006): 24-31.

[8] Tongjai, C., H. A. Murdoch, and C. A. Schuh. "Design of Stable Nanocrystalline Alloys," Science 337, 6097 (2012): 951-54.

[9] Barker, J. A., M. R. Hoare, and J. L. Finney. "Relaxation of the Bernal Model," Nature 257, (1975): 120.

[10] W. O. Smith, P. D. Foote, and P. F. Busang, Physical Review 34, (1929): 1271.

[11] J. D. Bernal, Proceedings of the Royal Society London A280, (1964): 299.

[12] C. H. Bennett, Journal of Applied Physics 43, (1972): 2727.

[13] C. A. Rogers, Packing and Covering, Cambridge University Press, 1964).

[14] J. L. Finney, Proceedings of the Royal Society of London A319, (1970): 479.

[15] G. D. Scott and D. M. Kilgour, British Journal of Applied Physics (Journal of Physics D) 2, (1969): 863.

[16] Berryman, J. G. "Random Close Packing of Hard Spheres and Disks." Physical Review A27, (1983): 1053.

[17] Rintoul, M. D., and S. Torquato, "Hard-sphere Statistics along the Metastable Amorphous Branch." Physical Review A58, (1998): 532.

[18] Anikeenko, A. V., and N. N. Medvedev, "Structural and Entropic Insights into the Nature of the Random-Close-Packing Limit," Physical Review E77, (2008): 031101.

[19] L.D. Landau and E.M. Lifshitz, Statistical Physics, Elsevier, 1980).

[20] S. Zhen and G. J. Davies, "Calculation of the Lennard-Jones N-m Potential Energy Parameters for Metals," Physica Status Solidi 78, (1983): 595.

[21] Kong, C. L., "Combining Rules for Intermolecular Potential Parameters. II. Rules for the Lennard-Jones (12–6) Potential and the Morse Potential," The Journal of Chemical Physics 59.5 (1973): 2464.

[22] Han, X. J., M. Chen, and Z. Y. Guo, "Thermophysical Properties of Undercooled Liquid Au–Cu Alloys from Molecular Dynamics Simulations," Journal of Physics: Condensed Matter 16, 6 (2004): 705-13.

[23] Lide, D. R., CRC Handbook of Chemistry and Physics, CRC Press, 1994)

[24] Okamoto, H., D. J. Chakrabarti, D. E. Laughlin, and T. B. Massalski, "The Au–Cu (Gold-Copper) System," Journal of Phase Equilibria 8, 5 (1987): 454-74.

[25] Batalu, D., G. Cosmeleata, and A. Aloman, "Critical Analysis of the Ti-Al Phase Diagrams." UPB Scientific Bulletin. B68.4 (2006).

[26] Djanarthany, S., J. C. Viala, and J. Bouix, "An Overview of Monolithic Titanium Aluminides Based on Ti_3Al and TiAl," Materials Chemistry and Physics 72, 3 (2001): 301-19.

Chapter 3

Amorphous and Nanocrystalline Metallic Alloys

Galina Abrosimova and Alexandr Aronin

Additional information is available at the end of the chapter

http://dx.doi.org/10.5772/64499

Abstract

In this chapter, evolution of structure and properties of amorphous alloys is discussed. Amorphous structure change before crystallization is analyzed for a lot of systems. Structure and property changes are discussed for both amorphous and amorphous-nanocrystalline materials. Nanocrystal formation in metallic glasses is considered at heating and deformation.

Keywords: metallic glasses, structure, properties, phase transformation, nanocrystalline structure

1. Introduction

Creation and development of modern technologies is based on fundamental research. Development of new materials is based on the knowledge of fundamentals of the processes that determine the formation of a material, as well as the correlation of structure and physical properties. Progress in the creation of new materials with certain properties (mechanical, electrical, magnetic) depends on the level of understanding of the processes that underlie the formation of a particular structure. The discovery and development of the materials with new physical properties, in turn, lead to the creation of new instruments and devices. Metallic glass and nanocrystalline metallic materials formed from metallic glasses, of course, apply to such materials.

Although many years passed after production of the first metallic glass, interest in them is only growing. In 1960, a group of researchers led by Professor Duwez first obtained metal alloy in a strange non-crystalline state [1]. X-ray diffraction patterns had no lines corresponding to whatever crystalline phases, and there was only a broad halo. This work [1] is considered to be the first publication, which refers to amorphous metal alloys or metallic glasses. However, a year earlier Miroshnichenko and Sally [2] in Dnepropetrovsk (USSR) have already demonstrated the ability to produce metal alloys in a non-crystalline state. Obviously, due to not very popular scientific "Factory laboratory" journal, this work went unnoticed. Whatever it was, work [1] is considered to be the first publication devoted to the study of the amorphous phase in metallic systems (as opposed to, e.g., already investigated oxide systems). After the first publication, an avalanche of the works went; an amorphous phase could be obtained in a growing number of systems, and within few years the number of amorphous alloys was already counted in the hundreds. Interest in the metallic glass was due to their unusual structure, quite untypical of alloys, and a whole set of outstanding physical and chemical properties. Amorphous alloys may be high-strength, magnetically hard and soft magnetic, corrosion-resistant, and others. Thus, microhardness H_V of metallic glasses based on transition metals (Fe, Co, Ni) may exceed 1000, tensile strength may be greater than 4.0 GN/m^2. These values exceed the maximum values of strength and hardness of usual metals and alloys used in the industry. For example, the strength of the wire from iron-based metallic glasses is higher than the strength of piano wire [3]. Iron-based alloys have very good magnetic properties of low coercivity (0.5–1 A/m) and high saturation magnetization exceeding 1.4 T. Even higher hysteresis properties were obtained for $Co_{70}Fe_5Si_{15}B_{10}$ [4] and Co-Fe-P-B [5] alloys that have almost zero magnetostriction. In general, the main characteristics of the soft magnetic amorphous alloys based on iron, cobalt, and nickel are the high values of residual induction and low magnetic reversal losses; high values of magnetic permeability (high iron content) or close to zero values of magnetostriction (high content of cobalt). Magnetic properties can also be increased with an addition of the alloying elements, the values of the magnetic permeability can be as high as 120,000 [6]. With a slight change in the composition, the properties of metallic glasses can vary quite significantly.

Most of the physical properties of solids are structurally sensitive. This dependence is typical for metallic glasses. For example, the soft magnetic properties of amorphous alloys can be improved by relaxation annealing, annealing in a magnetic field; mechanical properties naturally depend on the presence of residual stresses, corrosion properties depend on the chemical composition and state of the surface layer. Partially crystallized metallic glasses (peculiar composite consisting of amorphous and crystalline phases) have a number of very good properties that differ from the properties of both amorphous and crystalline materials. Importantly, the amorphous state is unstable; when heating or aging amorphous phase may decompose with a natural degradation of properties. Therefore, from the standpoint of basic science, and from the perspective of the industrial use of new materials, it is extremely important to study both the actual structure of the amorphous phase in metal alloys and its stability, the transition to the partially crystalline or fully crystalline state, as well as the correlation of structure and properties of the material.

The purpose of this review is to analyze the current state of research of structure evolution in amorphous and nanocrystalline metallic alloys. Particular attention is given to pre-crystallization processes and specific features of heterogeneous amorphous phase formation. The decomposition of homogeneous amorphous phase and the formation of regions with different

http://dx.doi.org/10.5772/64499

chemical compositions and/or different short-range order are considered for different types of metallic glasses. Formation of a nanocrystalline structure from homogeneous and heterogeneous amorphous phase is studied. Structure-property correlations are described.

2. Production of metallic glasses

The most common method of producing metallic glass or amorphous alloy is a melt quenching onto a moving substrate. When the melt is cooled, the cooling rate is about 10^6–10^9 K/s (depending on the method of quenching). With such a large cooling rate at room temperature, the structure of liquid is frozen and the sample is non-crystalline.

Not all alloys can be obtained in an amorphous state. The most commonly considered three main criteria that determine diffusionless solidification [7]: thermodynamic criterion, morphological criterion and heat criterion. Usually kinetic and structural criteria are also discussed [8–10]. Some alloys may be prepared as amorphous will have less cooling rate [11–24].

3. Metallic glass structure

3.1. Main methods of structure investigation

The most common method for studying the structure of amorphous alloys is a method of large-angle X-ray scattering or as it is often simply called X-ray method [25, 26]. For detailed information about the structure of the amorphous phase, it is necessary to build the partial radial distribution functions [27–31]. At present, methods of construction of total and partial radial distribution functions are not used very often, although it should be noted the study of Mattern et al. [32] or the study cited in [33]. In analyzing the structure of metallic glasses, as a rule, with the help of Ehrenfest equation, the radius of the first coordination sphere as well as its changes at all types of influences are estimated; and distortion of diffuse maxima (the appearance of an additional shoulder, splitting of the peak, and others.) is determined. In combination with other methods of research, such an approach appears to be more productive.

Another commonly used method for analyzing the structure of the amorphous phase is the method of small-angle X-ray (SAXS) and neutron scattering [34]. The most important feature of this method is the possibility of studying heterogeneities in the structure of disordered systems as are metallic glass. In an absolutely homogeneous medium, there is no small-angle scattering, and the scattering pattern varies considerably with the appearance of any heterogeneities of the electron density in the structure.

In addition to the X-ray methods for studying the structure, the most important method is the transmission electron microscopy (TEM). Without dwelling on the specifics of the method, it is important to note that the analysis of the structure of amorphous phase is based on both electron diffraction and analysis of bright-field and dark-field TEM images. The high-resolution electron microscopy and micro- or nanobeam diffraction methods are used to study the structural features of the early stages of crystallization of the amorphous phase, and the structure of the nanocrystals. The last method allows determining atomic arrangement in amorphous structure, strain in crystalline materials with a high spatial resolution, etc. The development of the nanobeam electron diffraction with a coherent electron beam lesser than 1 nm in diameter has made it possible to obtain two-dimensional diffraction patterns from a nanoscale region to detect local atomic structure. Combination of the nanobeam diffraction experimental method with ab initio molecular dynamics simulation allows obtaining and developing new knowledge about amorphous structure [35, 36]. The newly developed Cs-corrected TEM technique offers a great advantage to probe the local atomic structure of disordered metallic glasses since it allows achieving a coherent electron beam as small as ~3 Å in diameter, which cannot be obtained before by conventional TEM [37].

In studying the structure of the amorphous phase, indirect methods may be used such as the measurement of certain properties, whose changes reflect changes in the structure. For example, structure change can be fixed by measuring the temperature dependence of the magnetic properties. Thus, during heating of the ferromagnetic $Fe_{27}Ni_{63}P_{14}B_6$ amorphous alloy, there was an increase in the Curie temperature [38]. Since the Curie temperature is determined by the nearest environment of atoms, its change, of course, indicates a change in the structure of amorphous phase. In [39], it was shown that after the heat treatment, the amorphous Fe-P-B alloy is not uniform, and it is characterized by two different Curie temperatures. Numerous studies of amorphous alloys by Mössbauer spectroscopy showed that depending on the conditions of heat treatment, the parameters of the Mössbauer spectra may vary considerably [40]. Application of NMR spectroscopy revealed that the temperature change can lead to changing type of short-range order in the amorphous phase [41].

3.2. Structure models

A number of researchers have tried to describe the structure of the amorphous phase in metallic glasses by different models. The first model was the model of a chaotic close packing of hard spheres [42], which allowed successfully describing the distribution function of the atoms in the amorphous structure. Model chaotic close packing of hard spheres was initially designed to one-component systems, and later was adapted to the binary system [43]. Later this model has been improved by the introduction of interatomic potentials (soft-sphere model), which allowed to build atomic pair distribution function [44] with the more realistic values of the position and intensity of the main part and the shoulder of the second maximum of the distribution function. The next step in the modeling of the structure of metallic glasses was attempts to build a structure made up not of individual atoms and of coordination polyhedra [45, 46]. Another group of models of the metallic glass structure is based on the idea that elements of the crystal structure retained in liquid and amorphous alloys [47]. The review [33] lists all the main issues to date models of the structure of amorphous alloys: model of chaotic packing of hard spheres; model of polyhedra packing; the stereochemical model, model of efficient packing of quasi-equivalence clusters [47–51] and the model of middle-range order or fractal packaging [52].

http://dx.doi.org/10.5772/64499

3.3. Heterogeneities in amorphous structure

Any approach to the description of the amorphous structure suggests that it is a homogeneous isotropic structure. In fact, it was turned out that the structure of the amorphous phase in alloys cannot always be uniform and isotropic. **Figure 1** shows a typical X-ray scattering curve of amorphous alloy. The figure shows intensive first diffuse peak, determining the shortest distance between atoms and the subsequent weak halo. A feature of the amorphous phase in the metal-metalloid alloys is significantly weaker scattering from metalloid atoms compared to metallic atoms (e.g., an amorphous phase in alloys of Fe-B); therefore, the scattering is determined primarily by the metal atom. Other situation occurs in the case when the amorphous phase contains two or more metals with comparable scattering amplitude (e.g., Fe-Zr). In such systems, the appearance of inhomogeneity areas is much more pronounced, since the formation of regions with different chemical compositions (e.g., enriched in iron or zirconium) [53] leads to the appearance of at least two types of the shortest distances between atoms, which naturally affect the X-ray diffraction pattern. Such changes have been found in a number of systems [54–60].

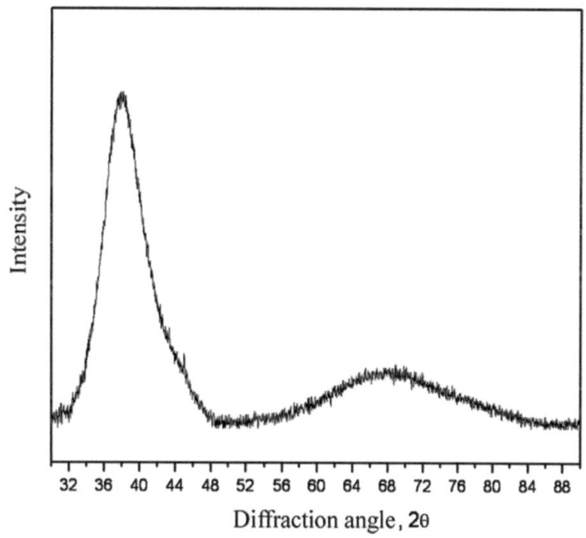

Figure 1. X-ray diffraction pattern of amorphous alloy.

It should be noted that X-ray diffraction pattern of the sample shown in **Figure 1** does not indicate that the amorphous phase is uniform. The appearance of a shoulder or a second peak occurs when the radii of the coordination spheres in different regions of the sample are different. If the areas, even with different chemical compositions, have the same radius of the first coordination sphere, no features on the X-ray diffraction patterns will be seen.

Heterogeneities in the structure of the amorphous phase were observed not only in as-prepared state. In some cases, the appearance of heterogeneities in the structure was found after various external influences: heat, irradiation, and others [61–65]. A lot of information about heterogeneities in the amorphous phase was obtained using the method of small-angle X-ray scattering [66–70]. It is interesting to note that the occurrence of heterogeneities may be related to the composition of the amorphous phase [56]. The impact of external influences on the structure of the amorphous phase will be discussed in more detail in Section 6.

4. Structure relaxation

Nucleation and growth of the crystals are suppressed at rapid cooling of the melt, and below the melting point metastable supercooled liquid is formed. With the temperature decrease the viscosity increases and the diffusion slows down, that is why at room temperature the solidified structure has the same atomic arrangement as it is in the melt prior to cooling. This non-equilibrium state is usually called the glass and the temperature at which solidification occurs—a glass transition temperature T_g. The T_g is generally defined as the temperature when the viscosity of the supercooled liquid reaches 10^{12} poises [6]. At the glass transition temperature, a lot of properties change drastically: specific heat, the temperature dependence of the specific volume, enthalpy, and others. Since amorphous alloys are obtained by melt quenching with cooling rates of 10^6–10^9 K/s, the structure of the amorphous phase after quenching is non-equilibrium. When heated, the amorphous phase relaxes to an equilibrium state. There is reversible and irreversible structural relaxation. The volume, viscosity, diffusion mobility, and embrittlement change irreversibly during relaxation; induced magnetic anisotropy may change reversibly, a decrease of internal stress (e.g., during annealing), Young's modulus, specific heat, coercive force, Curie temperature, the internal friction can vary and reversibly and irreversibly depending on the alloy composition [71, 72]. Relaxation processes were widely studied [73–81].

5. Deformation and mechanical properties of amorphous alloys

The main features of the deformation of metallic glasses are a great amount of elastic deformation (up to 4% in the bulk amorphous alloys), high strength, and fracture toughness [7]. Elementary carriers of deformation in metallic glasses are groups of atoms [82]. In contrast to crystalline materials in which the acts of deformation can be estimated from the behavior of dislocations by transmission electron microscopy, it is virtually impossible to observe such a process in metallic glasses. It is usually assumed that the reason for plastic flow and fracture is the formation of shear bands that develop over time. Deformation of metallic glasses proceeds differently in the two different modes of loading and temperature. At high stresses and relatively low temperatures, deformation occurs heterogeneously and it is much localized. As a result, narrow shear bands formed, the orientation of the shear bands is close to the orientation of the maximum shear stress [83, 84]. The time of formation of shear band is estimated as $\tau < 3$ ms. Shear bands and slip steps were observed for a number of amorphous alloys. Shear bands are oriented at an angle of $55 \pm 5°$ to the direction of deformation. Upon annealing, the difference in the structure of shear bands and the surrounding matrix can fully

http://dx.doi.org/10.5772/64499

or partially disappear. At low stresses, the deformation is carried out by a homogeneous creep mechanism.

It is important to note one more feature of the deformation of metallic glasses. When stress is applied initially, in accordance with Hooke's law, elastic deformation occurs in which the elongation linearly depends on the stress. At higher loads, this dependence deviates from the linear law. If so-called mechanical hysteresis loop is observed after the load removing, and the sample does not return to its original shape, one talks about the inelastic deformation of the material. The energy corresponding to the area of the loop, connected with the displacement of atoms in a stable position. The quantity of such displacement in the amorphous alloys is generally about an order of magnitude greater than in crystalline alloys. It is believed that the inelasticity of amorphous alloys is related to the free volume in their structure: if the free space is small, the inelastic deformation is also small. A further plastic deformation of amorphous alloys occurs by the formation and propagation of shear bands. It has been found that the velocity of propagation of shear bands does not depend on the strain rate in the range 2×10^{-4}–10^{-2} s^{-1} [84]. For such different metallic glasses as ductile $Pd_{40}Ni_{40}P_{20}$ alloy (value of the plastic strain of about 3%) and fragile $Mg_{58}Cu_{31}Y_6Nd_5$ (value of plastic deformation of 0%), the average velocity of propagation of shear bands varies slightly and is 317 and 366 μm/s, respectively [85].

Another feature of plastic deformation of amorphous alloys is the increase in the concentration of free volume in the shear bands, that is, increase in the average distance between the atoms. In general, the features of the shear band structure and, in particular, the reasons for the accelerated mass transfer in these areas are being actively discussed recently. A number of studies assumed that the severe plastic deformation leads to temperature increase up to the melting of the material in shear bands [86–87]. According to another view, shear bands are regions with a disordered amorphous structure of reduced density, so that the diffusion mass transfer in these areas can be facilitated. Probably, mass transfer processes depend both on the temperature rise and on the less dense structure, but these issues are still subject to a more detailed study.

The mechanical properties of amorphous alloys have aroused great interest from the moment when the metallic glasses were first obtained as a ribbon, and it was possible to carry out systematic research. One of the most complete early reviews devoted to the study of the mechanical properties was K. Pampillo's review [83]. Metallic glasses are high-strength materials, some of them can be cut, rolled, and even stamp, and it makes metallic glass attractive for technical applications [88–91]. The properties of metallic glasses are highly dependent on the prehistory of the samples and their chemical composition. Since metallic glasses contain fluctuations, atomic structure may vary slightly from place to place. Structural heterogeneities inevitably lead to irregularities in mechanical properties. Areas with less local viscosity are less stable and can be advantageous in some places of localization of inelastic deformation. These areas are places that facilitated the formation of shear bands [92–94]. Johnson and Samwer [95] derived a universal dependence of the yield strength on the temperature $(T/T_g)^{2/3}$. A number of studies suggest that this relationship may be more complicated [96, 97], but the overall trend has been maintained in different systems.

6. Effect of heat treatment and deformation on amorphous structure

6.1. Effect of heat treatment

When heated, amorphous alloys crystallize, but the crystallization is often preceded by separation of the amorphous phase into regions of different chemical compositions and different short-range order, that is, the formation of two or more amorphous phases. These phases do not separate with sharp interface, and the transformation may exhibit spinodal decomposition characteristics [54, 61]. Changes in the structure of the amorphous phase during annealing were studied for the alloys of various compositions. **Figure 2** presents the X-ray diffraction pattern of $Al_{87}Ni_8La_5$ sample annealed at 150°C [98]. After annealing, the samples were amorphous. The structure of the amorphous phase was found to change during annealing, which is revealed in the change of the diffuse maximum in the X-ray diffraction patterns. The first diffuse maximum in the X-ray diffraction pattern of the as-prepared amorphous ribbon was symmetrical, and after annealing it shows a shoulder on the large-angle side, the degree of the maximum distortion increasing with annealing time. **Figure 2** presents the X-ray diffraction pattern of the sample annealed for 25 h (first diffuse peak region). It is seen that subsequent to heat treatment the diffuse maximum is a superposition of two maxima. The positions of the scattering maxima are known to determine the radius of the first coordination sphere R_1,

$$R_1 = 7.73/(S_1)_{max} = 14.06/(S_2)_{max} = 20.46/(S_3)_{max} \quad (1)$$

where $(S_1)_{max} = 4\pi(\sin\theta\ /\ \lambda$ is the wave vector corresponding to the first (second, third...) maximum of the intensity curve, θ the scattering angle, and λ the radiation wavelength [25]. Thus, the two diffuse maxima point to the presence of amorphous matrix regions with different radii of the first coordination sphere. The difference in the angular positions of the diffuse maxima is indicative of the formation of amorphous regions with different chemical compositions of the two amorphous phases. The maximum located at the smaller angles corresponds to the amorphous phase with a large radius of the first coordination sphere (or the largest interatomic distance in the amorphous phase). Since in the system in question the largest atoms are those of lanthanum (radii $R_{Ni} = 0.124$ nm, $R_{Al} = 0.143$ nm, $R_{La} = 0.188$ nm), the amorphous phase is lanthanum-enriched. Thus, the isothermal annealing of amorphous $Al_{87}Ni_8La_5$ alloy leads to decomposition of the amorphous phase: formation of regions of different chemical compositions and with different types of short-range order.

Similar results were obtained in studies of the effect of prolonged low-temperature annealing on the structure of amorphous Fe-Zr alloy [53]. The annealing of $Fe_{90}Zr_{10}$ sample results in a shoulder on the smaller angle side of the first maximum in the X-ray diffraction pattern: the isothermal annealing leads to amorphous phase decomposition and formation of at least two new amorphous phases with different component concentrations. Such a structural change was also mentioned by the authors of [97] in the studies of metallic Fe-Zr glass by the methods

http://dx.doi.org/10.5772/64499

of anomalous X-ray scattering, nuclear gamma resonance spectroscopy, and magnetic property measurement.

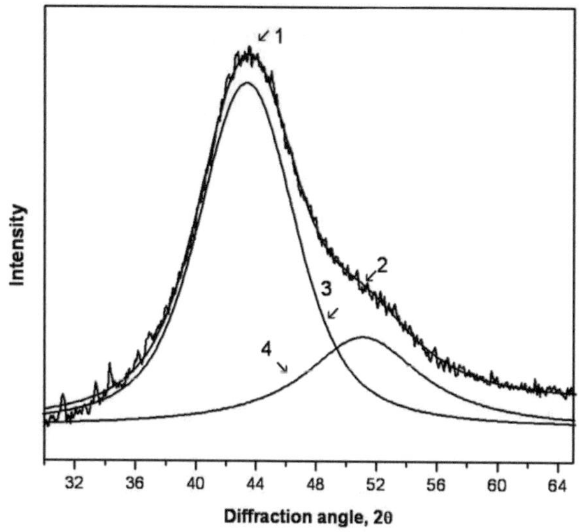

Figure 2. X-ray diffraction pattern of as-prepared amorphous $Al_{87}Ni_8La_5$ alloy (a) and sample annealed at 1500°C for 25 h (b); 1 denotes the experimental spectrum, 2 the summary curve, 3 and 4 lines the diffusion halos from the first and second amorphous phases.

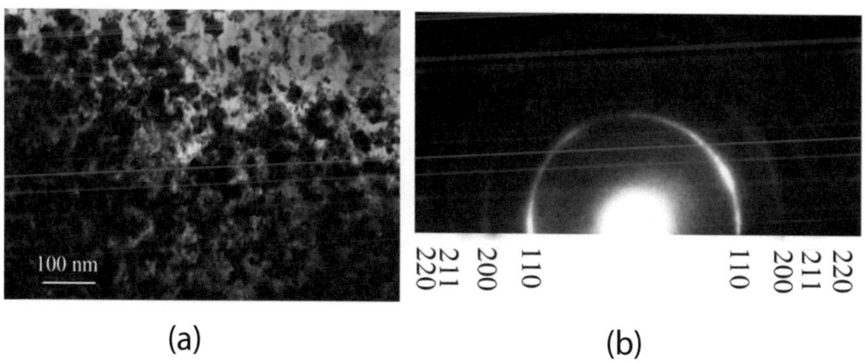

Figure 3. Bright field image (a) and electron diffraction pattern (b) of the sample annealed at 100°C for 6850 h.

Figure 3a presents the alloy structure after annealing at 100°C for 6850 h, and **Figure 3b** presents the corresponding electron diffraction pattern. After this annealing, the sample contains 10–30 nm size crystals. The nanocrystals have non-sharp boundaries in the transmission electron microscopy images. The structure is distinguished by diffuse grain boundaries in the microstructure image and persisting blurred rings in the microdiffraction pattern. Such diffusivity cannot be explained by the small grain size since the diffraction patterns of usual nanostructures with crystal sizes of 5–10 nm exhibit much more pronounced ring reflections [99]. With increasing annealing time, the structure becomes more distinct and the boundaries between nanocrystals become sharper, and rings in the electron diffraction patterns begin to split. **Figure 4** shows an electron diffraction pattern exhibiting splitting of ring reflection, its more pronounced areas indicated by arrows. It is important to note that these changes in structure occur without appreciable change in scale—the size of nanocrystals is not appreciably changed. The electron diffraction and X-ray diffraction patterns enabled to establish that two solid solutions of Zr in Fe form in the sample, their lattices having close but different parameters and, naturally, different component ratios. The above characteristic features of transformation indicate a progressive and continuous transition from single-phase to two-phase amorphous structure, then to crystalline structure with diffuse grain boundaries, and, finally, to fully crystalline structure with a constant size the regions of composition changes, point to the spinodal nature of transformation.

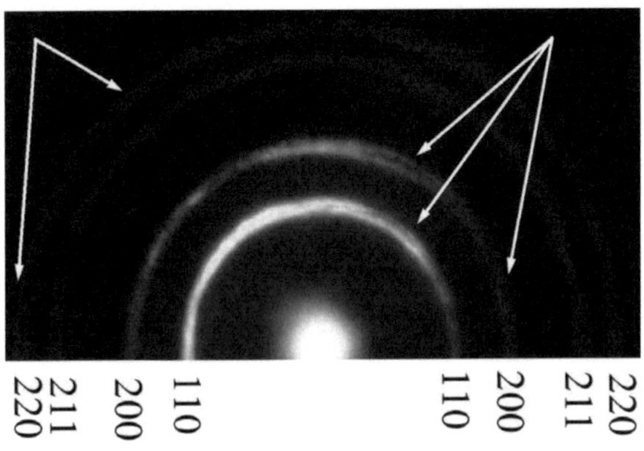

Figure 4. Electron diffraction pattern of the sample annealed at 100°C for 7000 h.

Evolution of the amorphous phase depends on temperature. As it was mentioned above, heating and annealing of the metallic glasses can be carried out in two different temperature ranges: above and below the glass transition temperature. Above the glass transition temperature, the amorphous phase is in a supercooled liquid state, below glass transition temperature it is in fact the amorphous state. It is known that when passing through the glass transition temperature T_g, material properties change sharply (viscosity, enthalpy, heat capacity, specific volume, and others.). This leads to significant differences in the process of diffusive mass transfer in these temperature ranges, and in turn, to changes in the structure. For comparison, let us consider the changes in the structure of the same alloy that occur above and below the glass transition temperature in the two nickel-based glasses: Ni-Mo-P and Ni-Mo-B.

1. *Amorphous* $Ni_{70}Mo_{10}P_{20}$ *alloy*. The glass transition temperature (T_g = 430°C for a heating rate of 20K/ min) of the alloy is below the crystallization temperature (T_x = 457°C) [100]. This allows to investigate the changes in the structure both in the supercooled liquid state (above T_g) and in the amorphous state (below T_g). Heating above the glass transition

temperature leads to a noticeable change in the structure of the amorphous phase. **Figures 5** and **6** show the TEM images of the as-prepared sample (**Figure 5**) and annealed above T_g sample (**Figure 6**) [100]. Under the same conditions of preparation of electron microscopy foils, in the annealed sample, in contrast to the original, there is a pronounced spotted contrast. **Figure 7** shows the X-ray diffraction patterns of as-prepared and annealed above T_g samples. After annealing, the shape of diffuse maximum distorted. When analyzing the structure of the amorphous alloy by micro-micro-diffraction method it has been found that the scattering vector corresponding to the maximum intensity of diffuse halo of an amorphous phase varies from one area to another along the sample. Observed area of spotted contrast and matrix is described by the different scattering vectors, indicating that the difference in short-range order in the light and dark areas. With further heating, the crystallization begins in places that appear lighter in the images. In these places, crystals of Ni(Mo) FCC solid solution form; the size of the crystals is 20–30 nm, and crystals are in direct contact with each other (**Figure 8**). If the heat treatment was carried out below the glass transition temperature, changes in the structure of amorphous phase were not observed. The structure of the sample annealed at 400°C for 1 hour (below T_g) contains the amorphous phase and crystalline eutectic colonies. Thus, the processes leading to separation of amorphous phase occur at the temperatures above T_g; above and below the glass transition temperature, amorphous structure varies differently, later leading to the formation of different crystal structures and, of course, different physical properties.

2. *Amorphous $Ni_{70}Mo_{10}B_{20}$ alloy* [101]. In this alloy, the glass transition temperature is also below the crystallization temperature. Upon heating the amorphous alloy above the glass transition temperature, the X-ray diffraction pattern also changes due to the appearance of areas with different chemical compositions. So before the crystallization the amorphous phase is heterogeneous and contains areas with higher and lower molybdenum concentration (and possibly boron). Crystallization of the alloy above the glass transition temperature leads to the formation of three crystalline phases simultaneously: two face-centered crystalline phases and an orthorhombic phase Ni_3B. FCC phases are Ni and Ni(Mo) solid solution. All three phases form substantially simultaneously. The crystal size is less than 50 nm. **Figures 9** and **10** show the images of Ni (**Figure 9**) and Ni(Mo) solid solution (**Figure 10**) nanocrystals formed above the glass transition temperature. Each nanocrystal is surrounded by the amorphous matrix. Studies of the structure by transmission and high resolution electron microscopy methods showed that the formation of the phases is independently from each other. The formation of each of the crystalline phases occurs in "their" concentration region, that is, the formation of crystalline phases in each concentration region can occur by the polymorphic mechanism without changing the chemical composition or by primary crystallization mechanism with slight change in concentration. These types of the crystallization (primary or polymorphic instead of eutectic crystallization) are due to changes in the amorphous structure before the crystallization which ensures the formation of amorphous regions to "fit" the compositional ordering. Below the glass transition temperature, phase separation is not observed, the crystallization occurs by the eutectic mechanism.

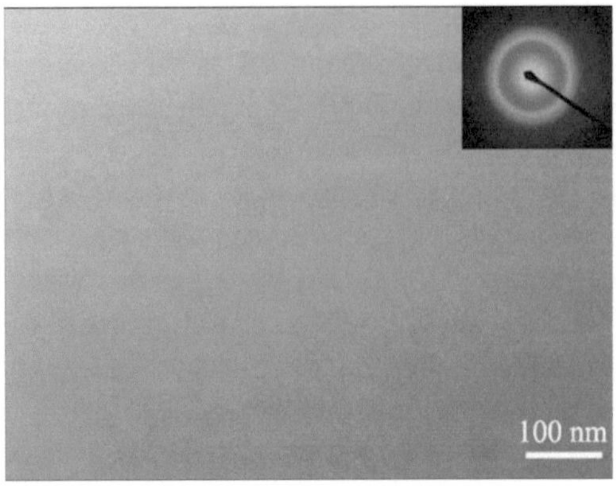

Figure 5. TEM image of the structure of as-prepared amorphous $Ni_{70}Mo_{10}P_{20}$ alloy.

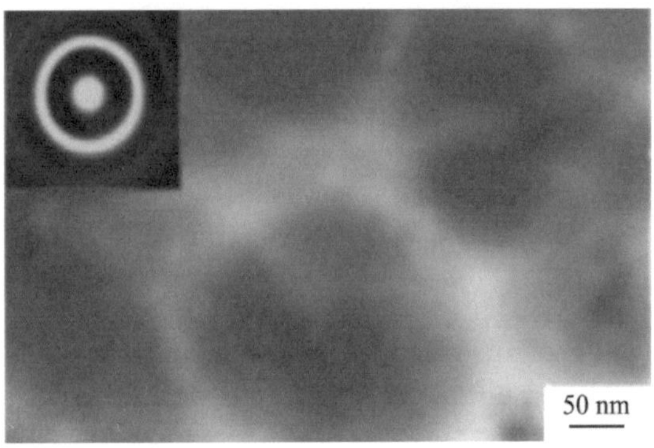

Figure 6. TEM image of the structure of $Ni_{70}Mo_{10}P_{20}$ alloy annealed above T_g.

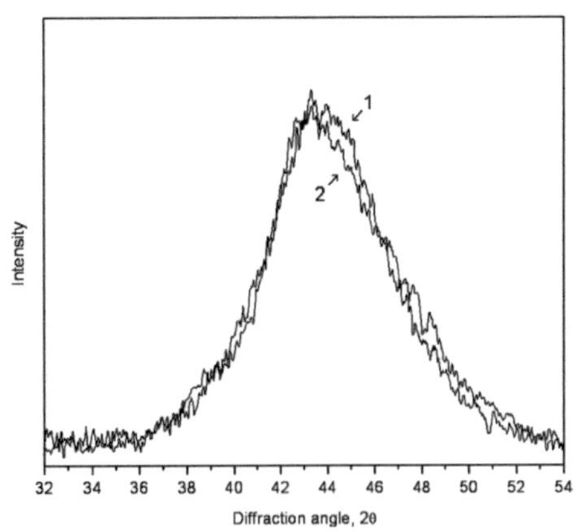

Figure 7. X-ray diffraction patterns of as-prepared and annealed above T_g $Ni_{70}Mo_{10}P_{20}$ sample.

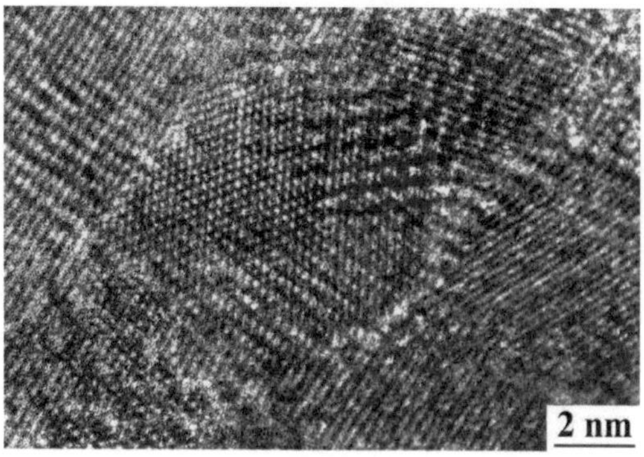

Figure 8. High-resolution electron microscopy image of alloy $Ni_{70}Mo_{10}P_{20}$ crystallized above T_g.

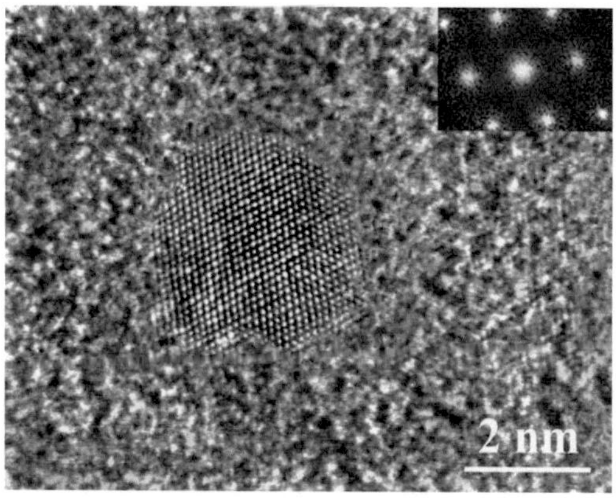

Figure 9. HREM image of Ni nanocrystals formed above the glass transition temperature.

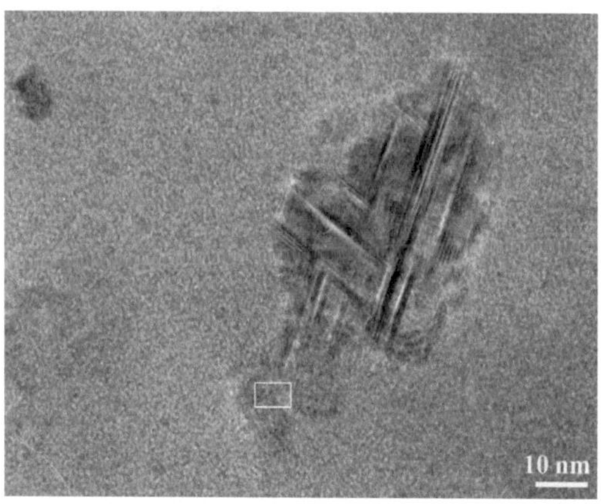

Figure 10. HREM image of Ni(Mo) nanocrystals formed above the glass transition temperature.

Thus, structure formed above and below the glass transition temperature varies markedly. It should be noted that the scale of the chemical composition change may be different: it is about (15–20 nanometers) in (Fe-Zr system) [102] or more than 50 nm (Ni-Mo-B system) [99]. This difference is connected with opportunity of implementing a diffusion mass transport in different temperature conditions. As noted above, the results, showing the development of heterogeneities in an amorphous matrix, were obtained using small-angle X-ray scattering method as well [63, 66, 99, 102–104]. Changes in the structure of the amorphous phase do not necessarily have the character of a separation at heat treatment. The type of short-range order can vary with the temperature [41, 105]. Concentration redistribution of the components in the amorphous phase can occur not only before the onset of crystallization. In some cases, the amorphous alloy separation may accompany the first stage of crystallization [99, 106].

6.2. Effect of deformation

The structure of amorphous phase can vary considerably, not only by heating but also by deformation. The question of a possible change in the structure under the influence of the deformation came into sight from the very beginning of amorphous alloy research. As was mentioned problems of homogeneous deformation and inhomogeneous flow, the fracture processes, their dependence on the temperature and composition was discussed in [83, 107]. In the last few years, new works were published that confirm the change in the structure of the amorphous phase at different types of influences which do not lead to crystallization [108]. A group of studies on the structure evolution carried out directly in the process of deformation of Zr-based amorphous alloys aroused great interest. The samples were deformed by tension, and synchrotron source was used for structure investigation in-situ. It allowed detecting the structure occurring at the elastic deformation [109–111]. It has been found that in the absence of plastic deformation, the tensile results in a change of the distance between atoms in an amorphous structure, and these changes depend on the orientation of the applied stress. It was shown that the initial symmetric with respect to zero diffuse peak position becomes asymmetric in the process of deformation. The results indicate that the first coordination sphere, characterizing the arrangement of atoms in the amorphous structure, transformed into an ellipsoid in the deformation process. In the study of elastic deformation of $Zr_{62}Al_8Ni_{13}Cu_{17}$ and $La_{62}Al_{14}(Cu_{5/6}Ag_{1/6})_{14}Ce_5Ni_5$ metallic glasses the deformation was shown [108] to be indeed an anisotropic. The Zr-based amorphous alloys are brittle, and they practically do not undergo plastic deformation, so only the region of elastic deformation was considered in the above-mentioned paper. At the same time, it would be interesting to determine whether the observed changes in the structure can be stored when the load is removed and whether they exist at plastic deformation. Such studies have been carried out for the amorphous $Pd_{40}Ni_{40}P_{20}$ alloy [112]. The deformation by multiple rolling was found to cause formation of anisotropic amorphous structure. The distance between atoms along the rolling direction was found to increase at the deformation, whereas it does not change in the perpendicular direction. Changes in the structure during deformation have been observed using a small-angle X-ray scattering method [113, 114].

In addition to the occurrence of the anisotropy in the structure, plastic deformation may lead to separation of the amorphous phase. **Figure 11** shows the X-ray diffraction patterns of amorphous $Al_{88}Ni_2Y_{10}$ and $Al_{88}Ni_{10}Y_2$ alloys after deformation by multiple rolling [115]. It is seen that the first diffuse maximum for $Al_{88}Ni_{10}Y_2$ alloy is asymmetric and, as in the case of heat treatment, it is a superposition of at least two maxima. Note that changing the structure under the influence of plastic deformation depends on the alloy composition. It is evident that the first diffuse maximum from the $Al_{88}Ni_{10}Y_2$ sample (curve 1 in the X-ray diffraction pattern) is asymmetric and, obviously, it is a superposition of at least two maxima. Two amorphous halos indicate the appearance of regions, which differ in composition. So, rolling of the $Al_{88}Ni_{10}Y_2$ alloy leads to separation of the amorphous phase into regions with different chemical compositions. Deformation of $Al_{88}Ni_2Y_{10}$ alloy (curve 2, **Figure 11**) leads to formation of a small amount of nanocrystals.

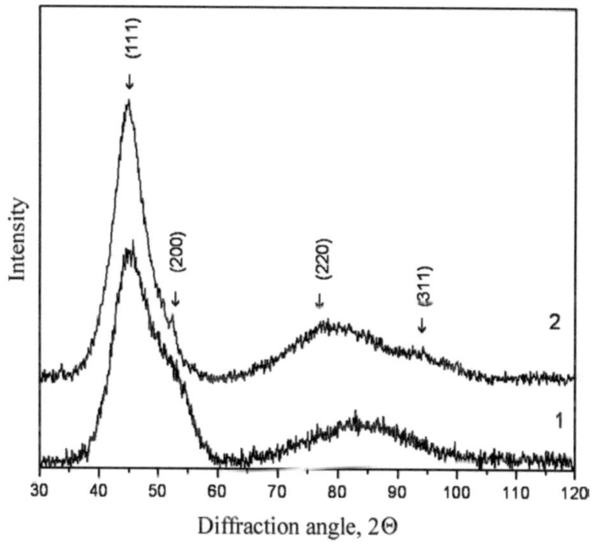

Figure 11. X-ray diffraction patterns of deformed $Al_{88}Ni_{10}Y_2$ (1) and $Al_{88}Ni_2Y_{10}$ (2) alloys.

7. Property change during amorphous phase evolution

The physical properties of amorphous alloys are very sensitive to structure evolution; they depend on external influences. Amorphous phase is not thermally stable, so the subsequent annealing, not leading to crystallization, causes the structure change:

- *The microhardness change.* Changes in the structure during annealing are accompanied by changes in microhardness of amorphous phase. For example, microhardness marked increases at low-temperature annealing of amorphous $Fe_{80}B_{20}$ alloy [116]. Since the microhardness of the material characterizes the strength of the bonds between atoms in the structure, change of microhardness shows evident amorphous structure evolution. The increase of the microhardness during annealing is obviously caused by both decreasing free volume and ordering processes.

- *The Curie temperature change.* Study of the amorphous $Fe_{27}Ni_{63}P_{14}B_6$ alloy [38] showed rising Curie temperature at heating in the temperature range below the crystallization temperature. Later, these changes were observed in a number of other alloys [39, 117]. Heat treatment of amorphous Fe-B-P alloy was found to lead to separation of the amorphous phase into region differing in composition and/or short-range order, which can be characterized by different Curie temperatures. The changes of the properties with heat treatment may be reversible [118, 119].

- *Changes in plasticity.* The ductility of amorphous alloy decreases during heating. This decrease can start at a sufficiently low temperature compared to the temperature of crystallization depending on the chemical composition. The embrittlement of amorphous alloys was first detected in the $Fe_{40}Ni_{40}P_{14}B$ alloy, and it was believed that it is caused by the presence of phosphorus, since it was found that the fracture surface is enriched with phosphorus [119]. However, Masumoto et al. [29] showed that the Fe-Si-B alloys without phosphorus are also fragile.

8. Crystallization processes

Since amorphous state is metastable, the amorphous alloys transform into a more stable crystalline state at heating. In most cases, the crystallization of the amorphous phase occurs by the mechanism of nucleation and growth. The most detailed crystallization reactions were studied by U. Koester in Bochum University [120]. The transition of the amorphous phase into crystalline phases depending upon the alloy composition can occur by one of the following reactions.

- primary (or preferential) crystallization;
- eutectic crystallization;
- polymorphic crystallization.

When the primary crystallization occurs, concentration gradient arises in front of the growing crystal. The growth rate decreases with time because the reaction involves the atoms diffusing on the long distances. It has been found that the spherical radius R of the growing crystal parabolically depends on the annealing time t in many metallic glasses which crystallize in the primary crystallization reaction [120, 121]. This means that the crystal growth is controlled by bulk diffusion,

$$R = \alpha \sqrt{(Dt)} \tag{2}$$

where α—dimensionless parameter depending on the composition at the interface particle/matrix, and D—the coefficient of volume diffusion.

The polymorphic and eutectic crystallization mechanisms are studied in [122, 123].

The crystal nucleation can be homogeneous and heterogeneous, in the last case the formation of crystals will be facilitated on the surfaces or so-called "frozen" crystallization centers [124, 125]. The ratio of nucleation and growth rates is a very important factor in the formation of structure. At low nucleation rate and high growth rate, small number of crystals forms and they can grow to considerable size. In the case of high nucleation rate and a small growth rate, very large number of crystals forms in the amorphous matrix.

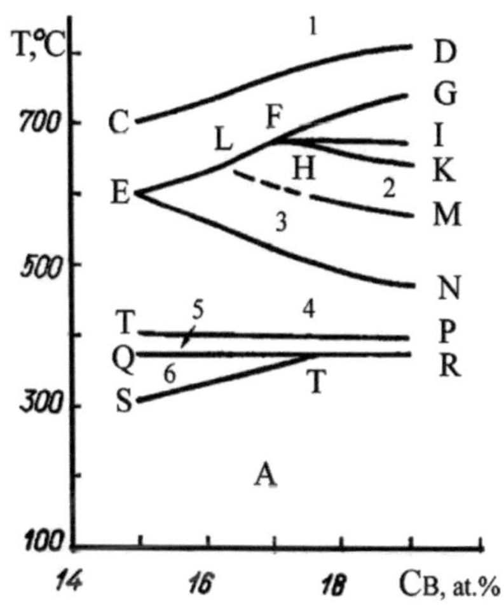

Figure 12. A diagram of phase transformation: I - α-Fe + Fe_2B, II - α-Fe + Fe_3B(P42/n), III - α-Fe + Fe_3B($P4_2$/n) + Fe_3B($I\bar{}4$), IV - α-Fe + Fe_3B($I\bar{}4$), V – A + α-Fe + Fe_3B($I\bar{}4$), VI – A + α-Fe, IHK - α-Fe + Fe_3B($P4_2$/n) + Fe_3B(P), GFI - α-Fe + Fe_3B(P) + Fe_3B(Pnma).

In addition to the above reaction of nucleation and growth crystallization, the mechanism of spinodal decomposition may occur at the metallic glass crystallization [126]. The possibility of such a separation in metallic glasses was discussed in a number of papers [127, 128–130]. The crystal structure formed during the crystallization of the amorphous phase can vary significantly in phase composition, morphology, crystal size, mutual arrangement of structural components [131]. In the last decades, much attention has been paid to nanocrystalline structure. The nanocrystalline structure in most cases is formed by the primary crystallization reaction. The Fe-Cu-Nb-Si-B alloy, called Finemet, was the first alloy obtained in the nanocrystalline structure [132].

An important feature of the crystallization of metallic glasses is the fact that in most cases, crystallization begins with the formation of metastable phases. As an example, crystallization of most extensively studied Fe-B metal glasses may be considered [133–135]. In accordance with the phase diagram of the system [106], the crystallization of compositions close to the eutectic must occur by α-Fe and tetragonal Fe_2B boride formation. However, the crystallization leads to the formation of α-Fe is formed of tetragonal Fe_3B boride. This boride may form different structures (crystalline lattice of different space group), which undergoes several changes before equilibrium phase turns into Fe_2B at heating or annealing. Specific features of metastable Fe_3B boride were studied in a lot of work [93, 134, 136, 137]. **Figure 12** illustrates a diagram of Fe-B metallic glass transformation; the regions of existence of different borides are shown.

Another example is the crystallization of the amorphous $Ni_{34}Zr_{66}$ alloy [138]. This alloy belongs to a small group of metallic glasses, which, in principle, can crystallize by polymorphic crystallization mechanism with formation of equilibrium $NiZr_2$ crystalline phase (tetragonal lattice of I4/mcm space group). However, $NiZr_2$ phase was not observed at the crystallization; a metastable phase with an orthorhombic lattice was formed. In contrast to the equilibrium phase, the structure of the metastable phase has a lower symmetry.

The formation of metastable phases at the crystallization of the amorphous phase is often observed for multi-component systems. Crystallization of multi-component Zr-based alloys and changes in the structure and properties at treatment were examined for the alloys of different composition [139–145]. The crystalline structure forming at amorphous phase decomposition is strongly dependent on the chemical composition. For example, the amorphous $Zr_{65}Cu_{17.5}Ni_{10}Al_{7.5}$ alloy crystallizes with simultaneous formation of three phases: $NiZr_2$-type phase with cubic lattice, $NiZr_2$ tetragonal phase and a metastable hexagonal phase, the crystals of all phases are randomly distributed in the amorphous matrix. When reducing the zirconium concentration, the phase composition varies. At the first stage of crystallization of the amorphous $Zr_{50}Ti_{16}Cu_{15}Ni_{19}$ alloy, three phases are formed, two of them are crystalline and the third phase is quasicrystalline. Quasicrystalline phase was observed in a number of crystallized metallic glasses of different composition [145–149].

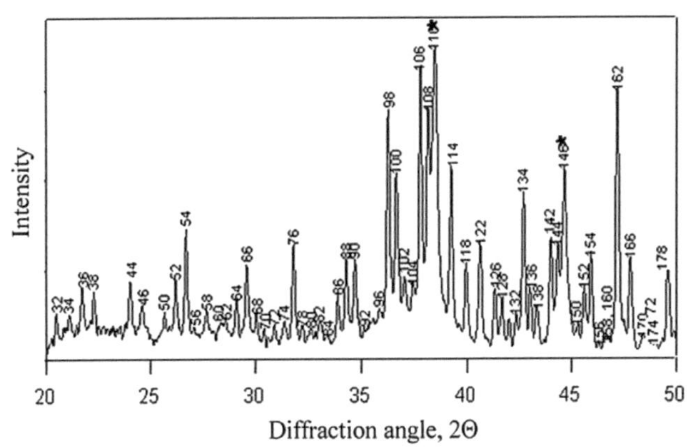

Figure 13. Part of X-ray diffraction pattern of $Al_{85}Ni_{10}Ce_5$ after the first stage of crystallization.

One more example: decomposition of the amorphous phase in the $Al_{85}Ni_{10}Ce_5$ metallic glass [150, 151]. The alloy crystallization begins with the formation of metastable phase. **Figure 13**

http://dx.doi.org/10.5772/64499

shows the X-ray diffraction pattern of the alloy after the first crystallization stage. After finishing the first stage, the sample is predominantly crystalline, but it retains a certain amount of amorphous phase. All the observed diffraction lines correspond to one crystalline phase with a body-centered cubic lattice. **Figure 13** shows only a part of the investigated spectrum; the sum of the squares of the indices $\sum = h^2 + k^2 + l^2$ is shown near each line. Usually the pure aluminum crystals form at the first stage of crystallization of Al-Ni-RE (RE = Y, Yb, Ce) alloys (nickel and rare earth components do not dissolve in FCC-Al). Since the ratio of the cell parameters of metastable phase and fcc-Al $a_{MS}/a_{Al} = 6.05$, it was assumed that the metastable phase forms in a situation when the concentration redistribution of the components of the "uniform" amorphous phase has no time to occur. As a result, metastable solid solution of the alloy components in Al lattice form, it has a much larger lattice parameter, wherein the concentration of nickel and cerium is noticeably greater than their equilibrium solubility in the All lattice.

Changes in the structure during the phase transformation from an amorphous phase to the crystalline equilibrium phase are accompanied by a change in the physical properties [118, 152]. Structure formed at the crystallization of metallic glasses varies also during deformation or pressure increase [153–155], icosahedral quasicrystals have been found at the crystallization of a number of metallic glasses [155–158].

9. Nanocrystalline structure formation

Under certain conditions, crystallization leads to the formation of a nanocrystalline structure. Nanocrystalline materials may be single-phase and multiphase polycrystalline samples with a grain size of up to 100 nm (at least in one direction). Owing to the extremely small size of the grain, the structure of nanocrystalline materials is characterized by a large volume fraction of grain boundaries and interfaces, which can largely determine a variety of physical and chemical properties of the material. Indeed, it has been found that many properties of nanocrystalline material are fundamentally different from those of conventional polycrystalline and amorphous alloys [159, 160]. For instance, nanocrystalline materials may have high strength and hardness, good ductility and toughness, reduced elastic modules, higher diffusion rates, large heat capacity and thermal expansion coefficient, and higher magnetic properties as compared to conventional materials [161–164]. The nanocrystalline materials not only provide an excellent opportunity to study the nature of structure, properties, and structure/properties correlation of solids down to the nanometer range, but they are also very attractive in terms of possible industrial use of new materials [165]. Changes in the structure and properties of amorphous and amorphous-nanocrystalline materials have been studied in a number of papers [166–169]. Small grain size makes a great development and length of the grain boundaries. The grain volume fraction of the grain boundaries in the total amount of the nanocrystalline particles in the material can be estimated in spherical shape approximation as in [170]:

$$\Delta V / V = [\pi d^3 / 6 - \pi(d - 2\delta)^3 / 6] / (\pi d^3 / 6) \approx 6\delta / d \tag{3}$$

where δ is the average thickness of the interface and d is an average grain diameter [170]. Thus, one can estimate the volume of the grain boundary. For example, when the thickness of the interface δ is 3–4 atomic monolayers (0.5–1.5 nm) and the average grain size is 10–20 nm, the fraction of surface layer is about 50%.

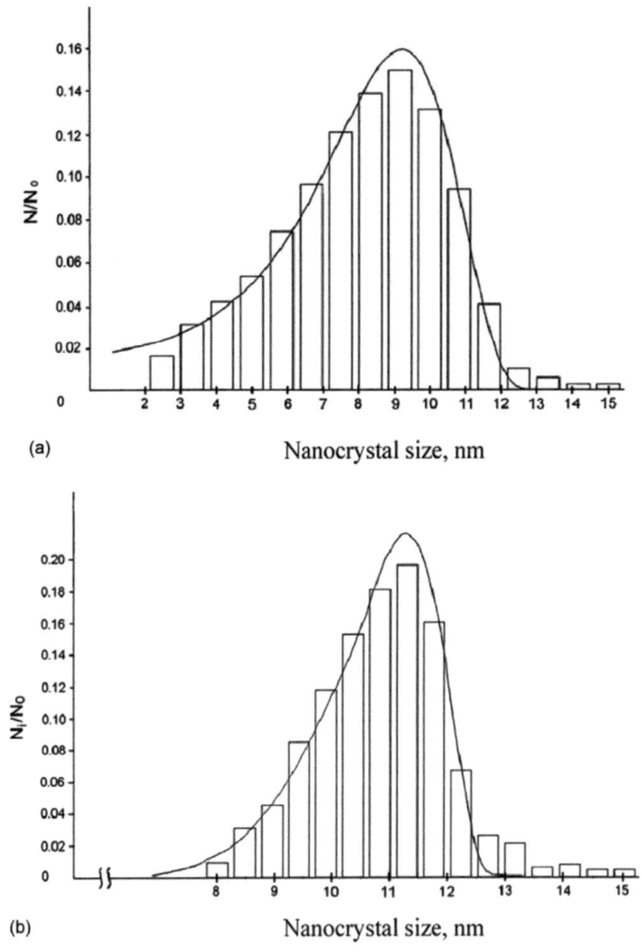

Figure 14. The crystal distribution by size after annealing at 473 K for 5 (a) and 15 (b) min: experimental (columns) and calculated (line) data.

Crystallization of amorphous alloys usually takes place under conditions of existence of the large driving forces and low diffusion mobility. These conditions are favorable rather for nucleation and not for crystal growth. Formation and specific features of nanocrystalline structure were studied for different materials [171–175]. Specific nanocrystalline/amorphous composite structure has excellent magnetic or mechanical properties in a number of alloys [176–178].

To establish the mechanism of nanocrystallization, one can use a method of analysis of crystal size distributions. For example, controlled crystallization of the amorphous $Al_{86}Ni_{11}Yb_3$ alloy [179] was found to lead to the formation of nanocrystalline structure with nanocrystal size of 5–12 nm. The nanocrystals are formed according to the mechanism of the non-stationary heterogeneous nucleation and from frozen-in crystal nuclei. For the conclusion of the nucleation mechanism, the distributions of nanocrystals by size were plotted for different durations of isothermal annealing; the distributions were analyzed. The size distributions obtained upon treatment for 5 and 15 min are depicted in **Figure 14a** and **b**, respectively. It is apparent that the heterogeneous nucleation with a latent period proceeds in our case. This is confirmed by the following facts. (1) Small-sized crystals are absent in the distribution obtained upon treatment for 15 min. (2) A sharply descending portion in the small-size range is observed in the distribution at a treatment time of 5 min, which is impossible in the case of homogeneous nucleation. (3) The fraction of large-sized particles (the right-hand branch of the distribution) decreases gradually, which is characteristic of the non-stationary rate of nanocrystal nucleation (with the latent period). Therefore, at the very early stage of metallic glass annealing, there exists a certain time interval during which the stationary size distribution of subcritical nuclei (corresponding to the classical theory) is reached.

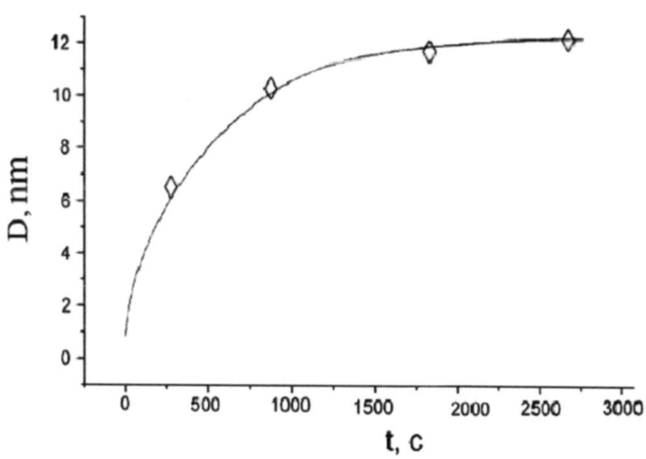

Figure 15. The dependence of the average crystal size in the $Al_{86}Ni_{11}Yb_{31}$ alloy on the annealing time time.

Computer simulations were carried out for heterogeneous nucleation and diffusion controlled growth. The calculated crystal distributions by size for annealed $Al_{86}Ni_{11}Yb_3$ alloy (473 K for 5 and 15 min) are shown by lines in **Figure 14**; the experimental size distributions are represented by columns. Good agreement with the experimental data was obtained for the following crystallization parameters: $Q_N = 90 \times 10^3$ J/mol and $\tau = 150$ s. The effective diffusion coefficient of Ni and Yb in amorphous $Al_{86}Ni_{11}Yb_3$ alloy is 1.4×10^{-19} m^2s^{-1} for 5 min annealing and 6×10^{-20} m^2s^{-1} for 15 min annealing. Since the diffusion coefficient of nickel, as a rule, is considerably larger than that of ytterbium, it is assumed that the nanocrystal growth is limited by the diffusion rate of ytterbium. As seen in **Figure 14**, the calculated and experimental data are very similar. However, the difference between the calculated and experimental distributions is in the presence of a tail of large-sized particles in the experimental histograms. Most likely, the occurrence of large-sized particles can be explained in terms of the presence of a small number of so-called "frozen-in crystal nuclei" in the initial alloy. The formation of crystals from these nuclei is facilitated. As a result, particles begin to grow earlier (prior to the completion of the latent period of attaining the stationary size distribution of subcritical nuclei) and reach larger sizes. **Figure 15** shows the dependence of the average crystal size in the $Al_{86}Ni_{11}Yb_3$ alloy on the annealing time. The similar results were obtained for the crystallization of amorphous Al-Ni-Y alloys [180]. It has been found that the nanocrystals are nucleated by heterogeneous mechanism in the aluminum-based alloys; and the nucleation begins in the regions depleted in the rare earth component [99, 174]. Specific features of the nanocrystal formation are studied in detail in [180–183].

10. Specific features in nanocrystalline structure

Since metallic glasses are often used as the starting material for producing nanostructures, and problem of deformation of the composite amorphous-nanocrystalline material is very important. Naturally, an important question is what component (nanocrystal or amorphous phase) determines the mechanical properties of the material and, in particular, if the nanocrystals can be deformed. Questions of the deformation of nanostructures are discussed in detail in [170]. Usually single unloaded volume unit contains a small amount of induced defects or does not contain them at all. The smaller the size of the nano volume, the smaller the number of induced defects it contains. A critical particle size with a free surface d_c was assumed to can be estimated if believe that below the probability of the existence of the defect sharply decreases [171, 172]. It was suggested that d_c is about 10–100 nm. Such estimates have been made for free nanoparticles. It was assumed that I also believe that the mechanisms of plastic deformation and fracture of nanocrystalline materials are determined by the size effect.

In this way, the solution to this problem largely rests on the question whether the nanocrystals contain dislocations, which are carriers of plastic deformation. In the early stages of research, it has been suggested that there is a critical size below which all the crystals are perfect; and the crystals contain stacking faults, twins, etc. when exceeding this size. In fact, at the initial moment when the crystals nucleate and grow in an isotropic amorphous phase, they really are perfect. When the crystal growing, when the volume per one atom in the crystal is differ then the volume per atom in the amorphous phase, there are tension stresses around the crystals. The value of these stresses increases as the crystals grow and conditions for the formation of linear defects arise. Is there, indeed, a certain critical size of crystal separating perfect and imperfect crystals? The answer is quite important in particular for determining the strength and ductility of the material. For example, it was supposed in [170] that such a critical crystallite size determines the boundaries of the regions where mechanisms of dislocation strengthening and plasticity cease to operate.

http://dx.doi.org/10.5772/64499

Specific features of the nanocrystalline structure were studied in detail in [174–178] for Al- and Ni-based alloys. The results showed that the Al nanocrystals and Ni nanocrystals formed in eutectic Ni-Mo-B systems were defect-free. These nanocrystals have equiaxial shape, small size, and they do not contain linear defects. The nanocrystals of Ni (Mo) solid solution in all studied alloys were defect-free at the beginning of crystallization, while the size of the nanocrystals is not more than about 5 nm. However, in the fully formed nanostructure the average size is much larger (20–50 nm), and the nanocrystals contain a significant number of defects. In analyzing the reasons of defect formation in the nanocrystals, two factors: the size factor and the energy of stacking fault formation, were considered. This size factor is associated with the inability to generate dislocations in nanocrystals smaller than 100 Burgers vector (operation conditions of dislocation sources, such as the Frank-Read [184]). It is also obvious that the probability of appearance of defects (e.g., stacking faults) depends on the energy of their formation, and of course, this factor must be taken into account when comparing the degree of perfection of different nanocrystals. In terms of the magnitude of the Burgers vector b, the average size of the nanocrystals is:

Ni in the alloy with 20 at.% B 8–20 ***b***,

Al ~40–45 ***b***

(the Burgers vector is $\boldsymbol{b}_{Ni} = a_{Ni}/2\,[110] = 0.250$ nm for Ni nanocrystals and $\boldsymbol{b}_{Al} = a_{Al}/2\,[110] =$ 0.286 nm for Al nanocrystals).

As was mentioned, when analyzing the degree of nanocrystals perfection in the Al- and Ni-based systems, the influence of the energy of stacking fault formation was considered [174]. The value of the energy of stacking fault formation is 135 erg/cm^2 for Al and 240 erg/cm^2 for Ni. However, it is known that the energy of stacking fault formation decreases with increasing electron concentration when doping [185]. For example, stacking fault energy in Cu-Ag alloys decreases ~10 times with increasing electron concentration to 20% [186, 187]. Unlike aluminum nanocrystals that are one-component, nickel nanocrystals contain about 17 at.% Mo, which leads to an increase in the electron concentration to about 20%. It is therefore natural to expect a significant reduction (several times) the stacking fault energy. This reduction of the energy of stacking fault formation can explain more perfect structure of Al nanocrystals as compared to Ni(Mo) nanocrystals.

However, what is a predominant reason of defect formation in the nanocrystals (the size factor or the energy of formation of stacking faults) is not clear.

What can cause the formation of defects in the nanocrystals? As shown above, a significant number of defects were observed in nanocrystals of Ni (Mo) solid solution. Since the nucleation and nanocrystal growth occur by the primary mechanism of crystallization, it is natural to expect that the concentration of molybdenum in nickel lattice varies from the center to the periphery of the nanocrystal. It is known that [187, 188] peculiar ordered structure can be formed in the crystals of two-component systems. It can be assumed that the formation of twins in the nanocrystals of Ni(Mo) solid solution can occur in this way, that is, the formation be facilitated due to the uneven distribution of molybdenum atoms in solid solution. Naturally, formation of stacking faults by this mechanism is not possible in the case of single-component nanocrystals.

It should be noted that the shape of Ni(Mo) nanocrystals with size of 30–50 nm differs from the equilibrium shape, and their growth leads to accumulation of the stresses associated with a crystallographic anisotropy and a difference in the density of the amorphous and crystalline phases. When increasing the size of the nanocrystals, the concentration of molybdenum in Ni (Mo) solid solution reduces, which, as previously noted, leads to increasing energy of stacking fault formation and, hence, decreasing the probability of twin formation by fluctuations of composition. On the other hand, molybdenum concentration in solid solution decreases during the growth of the nanocrystals; and its concentration increases at the reaction front (in the amorphous matrix in front of the crystal growing boundary). Details of the experimental observation of the distribution in front of the growing particle are given in [189].

In the processes of the primary crystallization the amorphous phase of Ni-based alloys is enriching in molybdenum. Naturally, the Mo concentration is increased near the growing nanocrystal and this increased concentration diminishes with the distance from the nanocrystal. Heterogeneities in the composition of the surrounded regions may be important for the origin of the twins or stacking faults near the nanocrystal boundary and its intergrowth into the nanocrystal. It should also be noted that by increasing the Mo concentration in the nickel lattice, stacking fault energy decreases, that is, formation of defects is facilitated. If these reasons, indeed, define the appearance of defects in the nanocrystals, you should expect the formation of twins and stacking faults in the near-boundary areas. The studies have shown that nanocrystals with defects in the near-boundary areas, in fact, present in the nanocrystalline structure. **Figure 16** shows a nanocrystal Ni(Mo) solid solution with defect-free central part and twins and stacking faults in the near-boundary regions.

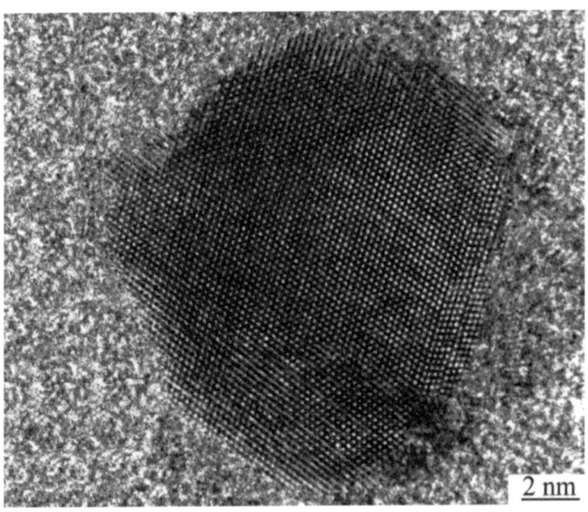

Figure 16. High-resolution electron microscopy image of the Ni(Mo) nanocrystal in the amorphous matrix.

Obviously, chemical heterogeneity is not only reason enough for the defects formation; small nanocrystals (up to ~ 5 nm) are always defect-free. Formation of the crystals in the amorphous matrix occurs in the field of stresses caused by the difference in specific volume per atom in the amorphous phase and in crystalline structure. Naturally, these stresses increase with the crystal sizes increasing. At the initial stages of crystal growth in isotropic amorphous phase these stresses are too small to induce the formation of defects in the nanocrystals. In the process of growth the nanocrystals reach the size when there is a growing stresses may cause the formation of defects.

It should be noted that the formation of microtwins and stacking faults cannot be determined by only size factor. One way of determining the degree of structure perfection of the nanocrystals formed in the amorphous phase was based on the following analysis of the diffraction line broadening:

- broadening of the diffraction line caused by the small grain size is proportional to 1/cos θ or sec θ (θ is the angle of the reflection);

- broadening of the diffraction line caused by randomly distributed dislocation is proportional to tg θ. If the nanocrystals do not contain defects, the half-widths of the diffraction reflections depend linearly on the secant of the diffraction angle. **Figure 17** shows such dependencies for Al-based alloy ($Al_{82}Ni_{11}Ce_3Si_4$) and Ni-based alloy ($(Ni_{70}Mo_{30})_{90}B_{10}$). These alloys contain the nanocrystals with practically the same size: 12 nm nanocrystals in $Al_{82}Ni_{11}Ce_3Si_4$ and 13 nm in $(Ni_{70}Mo_{30})_{90}B_{10}$. Therefore, curves 1 (for $Al_{82}Ni_{11}Ce_3Si_4$ alloy) and 2 (for $(Ni_{70}Mo_{30})_{90}B_{10}$ alloy) correspond to two nanocrystalline alloys containing nanocrystals with approximately the same size (~13 nm). (It should be noted that the lattice parameter of the aluminum and nickel varies markedly, which determines the different quantities of points in the curves.) It is evident that a linear relationship is indeed characteristic of nanocrystals in an aluminum alloy, while linear dependence was not observed for nickel-based alloy. This means that the nanocrystals of the same size may be defect-free or contain linear defects in the different systems.

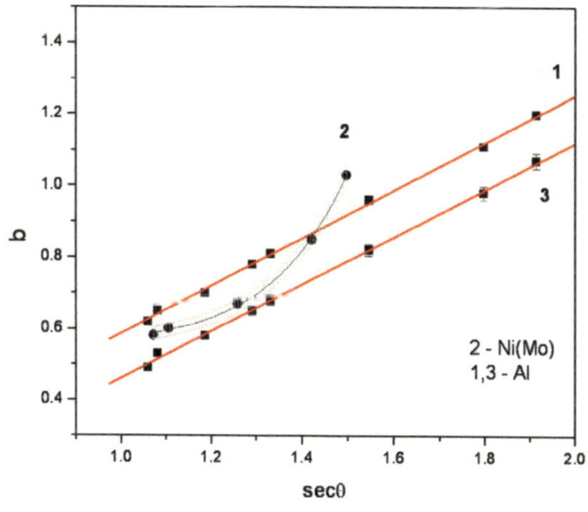

Figure 17. The angular dependence b of the diffraction line half-width (1 – $Al_{82}Ni_{11}Ce_3Si_4$, 2 – $(Ni_{70}Mo_{30})_{90}B_{10}$, 3 – $Al_{88}Ni_{10}Y_2$).

However, it is more correct to compare the nanocrystals with the same size, expressed in the Burgers vector units (i.e., with the same number of interatomic distances); these values are 42**b** and 52**b** for curves 1 and 2, respectively. Therefore, for comparison, the figure also shows curve 3 corresponding to $Al_{88}Ni_{10}Y_2$ alloy with the nanocrystal size of ~50**b**. As seen from the figure, in this case, the dependence (curve 3) is also linear, indicating that the broadening of the diffraction lines is caused by size factor only, that is, nanocrystals are defect-free. Thus, the size factor is not the only reason that determines the occurrence of defects in the nanocrystals; heterogeneities in composition play an important role as well.

In connection with the above, one point should be noted. As already mentioned, some researchers have suggested that there is a certain critical size of the nanocrystal, below which it is defect-free. For each such material critical size of the nanovolume can be estimated when a probability of the existence of induced defect sharply decreases [170]. Since the yield strength and elastic modulus for nickel are much larger (about three times) than that for aluminum, one would expect that the critical size of dislocation-free Al nanocrystals will be lower than that for N nanocrystals, but the experimental results do not match. The results show that the size of the nanocrystal, whereby there occur defects, actually, is different for different systems. It depends both on the chemical composition and concentration of the components and on the conditions of nanocrystal formation; one would expect that the formation of the nanocrystal, for example, in thin sections the size of defect-free crystal should be reduced significantly because the proximity surface will contribute to the resulting stress.

Obviously, to estimate perfection of the nanocrystalline structure, one should consider both the size of the crystals and the possible chemical heterogeneity, facilitating the formation of defects. The second factor is rather significant for nanocrystals comprising two or more components, the first one is important for any type of nanocrystals.

11. Structure design

As noted above, the changes occurring in the amorphous phase structure before the onset of crystallization may significantly affect the morphology and parameters of the crystalline structure forming on subsequent heating [116, 190, 191]. The amorphous phase is the parent phase for nanocrystals, which are formed by thermal and/or deformation effects. By varying the type and effect parameters, one can create amorphous-nanocrystalline materials with different size and volume fraction of nanocrystals. Nanocrystal formation usually occurs by the primary crystallization mechanism. This reaction leads to change of the amorphous phase composition and to enrich the amorphous phase in refractory components and metalloids. This process can result in an increased stability of the amorphous phase. The crystallization process thus can stop, and the resulting amorphous-nanocrystalline composites have high thermal stability. Heterogeneities in the amorphous phase may lead to the formation of different crystal phases. The noted structural changes do not exhaust all the possibilities for transformation of the system toward the equilibrium state. This wide variety of structures defines a set of good physical and chemical properties. For example, [116] annealing of amorphous Fe-B alloy at

http://dx.doi.org/10.5772/64499

temperatures below the crystallization temperature was found to alter microhardness and stability of the amorphous phase and ten-fold increase of the crystal size after subsequent crystallization. It has been found that heat treatment at different temperature ranges leads not only to a change of the phase composition but also to the formation of the structure with substantially smaller crystals upon subsequent heating. Formation of smaller crystals was also observed is the case of crystallization of heterogeneous amorphous structure. This last point is particularly important because it indicates the possibility of structure design, based on the decomposition and separation inside amorphous phase. The separation of amorphous phase makes it possible to produce amorphous samples with crystalline surfaces (**Figure 18**) and crystalline samples with amorphous surface (**Figure 19**) in Fe-B-P alloys [190].

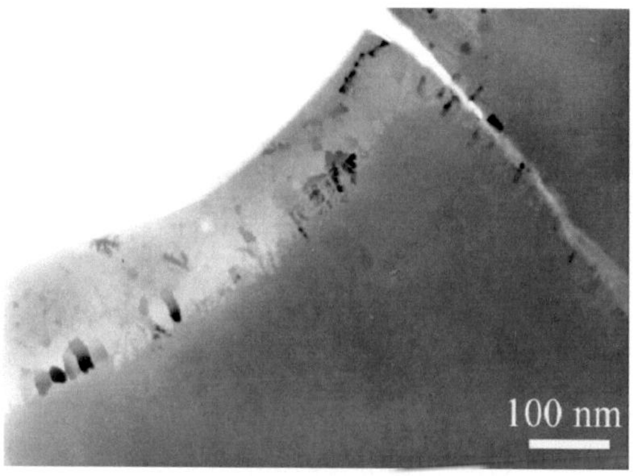

Figure 18. Microstructure of partially crystallized in-situ $Fe_{83}B_{10}P_7$ alloy.

Figure 19. Microstructure of annealed and partially crystallized in-situ $Fe_{83}B_{10}P_7$ alloy.

As most properties are structure-sensitive, the knowledge of the ways for structure design leads to the possibility of forming materials with the desired properties. Creating new materials with good properties, as already mentioned, is one of the major challenges of modern science.

Author details

Galina Abrosimova* and Alexandr Aronin

*Address all correspondence to: gea@issp.ac.ru

Institute of Solid State Physics RAS, Chernogolovka, Russia

References

[1] Willens RH, Klement W, Duwez P. Continuous series of metastable solid solutions in silver-copper alloys. J. Appl. Phys. 1960; 31: 1136-1137.

[2] Miroshnichenko IS, Sally IV. Device for crystallization of the alloys with high cooling rate. Factory Lab. 1959; 11: 1398-1399 (Rus.)

[3] Skakov YuA, Kraposhin VS. Solidification at rapid cooling and phase transfprmation at heating of metallic glasses. Metallovedenie i Termichaskaya Obrabotka. 1980; 13: 3-78

[4] He J, Zhou L, Zhao DL, Wang XL. Hysteresis loop shift behavior of CoFeSiB amorphous ribbons. J. Mater. Res. 2009; 24: 1607-1610

[5] Fukunaga T, Ichikawa T, Suzuki K. Amorphous Magnetism II. In: Levy RA, Hasegawa R, Editors. New York: Plenum Press, 1977. p. 52

[6] Suzuki K, Fujimori H, Hashimoto K. Amorphous Metals. Moscow: Metallurgy; 1987. 328 p.

[7] Cahn RW. Physical Metallurgy. 3rd ed. Amsterdam: North-Holland Physics Publishing. 1983. 2. 624 p.

[8] Uhlman BR. A kinetic treatment of glass formation. J. Non-Cryst. Sol. 1972; 7: 337-342

[9] Du XH, Huang JC, Liu CT, Lu ZP. New criterion of glass forming ability for bulk metallic glasses. J. Appl. Phys. 2007; 101: 086108

[10] Nagel SR, Tauc J. Correlations in binary liquid and glassy metals. Solid State Commun. 1977; 22: 129

[11] Drehman AJ, Greer AL, Turnbull D. Bulk formation of a metallic glasses: $Pd_{40}Ni_{40}P_{20}$. Appl. Phys. Lett. 1982; 41: 716

[12] Molokanov VV, Chebotnikov VN. Glass Forming Ability, Structure and Properties of Ti and Zr-Intermetallic Compound Based Alloys. Key. Eng. Mater. 1990; 40-41: 319

[13] Zhang T, Inoue A, Masumoto T. Amorphous Zr-Al-TM (Co=TM, Ni, Materials) Alloys with Significant Supercooled Liquid Region. Mater. Trans. JIM. 1991; 32: 1005

[14] Lad'yanov VI, Sterkhova IV, Kamaeva LV, Molokanov VV. On the solidification of the Fe50Cr15Mo14C15B6 bulk-amorphized alloy. J. Non-Cryst. Sol. 2010; 356:65-71

[15] Abrosimova GE, Aronin AS, Matveev DV. Formation and structure of nanocrystals in bulk Zr50Ti16Cu15Ni19 metallic glass. Phys. Sol. State. 2004; 46:2191-2195

[16] Molokanov V, Petrzhik MI, Mikhailova TN, et al. Formation of bulk(Zr, Ti)-based metallic glass. J. Non-Cryst. Sol. 1999; 250:560-565

[17] Abrosimova GE, Aronin AS, Kabanov YP, et al. Magnetic structure and properties of a bulk Fe72Al5P10Ga2C6B4Si1 alloy in the amorphous and nanocrystalline states. Phys. Sol. State. 2004; 46: 885-890

[18] Molokanov VV, Petrzhik MI, Mikhailova TN et al. Supercooled liquid region in Zr-Cu based bulk amorphous alloys. J. Non-Cryst. Sol. 1996; 207: 508-513

[19] Zhang Q-D, Wang L-F, Zhao Y. Remarkable improving plasticity of a brittle Zr-based bulk metallic glass by a high rheological rate forming method in centesimal seconds. Mater. Lett. 2016; 164: 348-352

[20] Ning Z, Liang W, Zhang M. High tensile plasticity and strength of a CuZr-based bulk metallic glass composite. Mater. Des. 2016; 90: 145-150

[21] Chen YH, Huang JC, Du XH, et al. Time-dependent creep behavior of amorphous ZrCu and nanocrystalline Zr thin films—a comparison. Intermetallics 2016; 68: 101-106

[22] Abrosimova GE, Aronin AS, Kabanov YP, et al. Magnetic structure and properties of a bulk Fe72Al5P10Ga2C6B4Si1 alloy in the amorphous and nanocrystalline states. Phys. Sol. State. 2004; 46: 885-890

[23] Abrosimova GE, Aronin AS, Matveev DV, Molokanov VV. Formation and structure of nanocrystals in bulk Zr50Ti16Cu15Ni19 metallic glass. Phys. Sol. State. 2004; 46: 2191-2195

[24] Halation S, Haddad-Sabzevar M. Thermal stability and non-isothermal crystallization kinetics of Ti41.5Cu42.5Ni7.5Zr2.5Hf5Si1 bulk metallic glass. J. Non-Cryst. Sol. 2015; 429: 164-170

[25] Skryshevskii AF, Structure Analysis of Liquids and Amorphous Solids. Moscow: High School; 1980

[26] Romanova AV, Illinskii AG. Amorphous Metallica Alloys. Kiev: Naukova Dumka; 1987. 248 p. (Rus.)

[27] Glassy Metals II. Atomic Structure and Dynamics, Electronic Structure, Magnetic Properties. Ed. H. Beck, H.-J. Güntherodt. Berlin: Spring-Verlag; 1983. 454 p.

[28] Abrosimova GE, Matz W. Construction of partial structural functions of amorphous Fe84B16 alloys by means of neutron diffraction. Phys. Met. 1989; 11:36-43 (in Russian)

[29] Waseda Y. The Structure of Liquids, Amorphous Solids and Solid Fast Ion Conductors. Pergamon Press. 1981. 122 p.

[30] Matz W, Hermann H, Mattern N. On the structure of amorphous Fe75B25. J. Non-Cryst. Sol. 1987; 93: 217-229

[31] Bhatia AB, Thornton DE. Structural aspects of the electrical resistivity of binary alloys. Phys. Rev. B. 1970; 2: 3004-3012

[32] Mattern N, Kühn U, Hermann H, Ehrenberg H, Neuefeind J., Eckert J. Short range order of $Zr_{62-x}Ti_xAl_{10}Cu_{20}Ni_8$ bulk metallic glasses. Acta Mater. 2002; 50: 305-314

[33] Cheng YQ, Ma E. Atomic-level structure and structure–property relationship in metallic glasses. Prog. Mater. Sci. 2011; 56:379–473

[34] Svergun DI, Feigin LA. X-ray and Neutron Small Angle Scattering. Moscow: Nauka; 1986. 280 p. (Rus.)

[35] Sheng HW, Luo WK, Alamgir FM, Bai JM, Ma E. Atomic packing and short-to-medium range order in metallic glasses. Nature. 2006; 439:419-425

[36] Li Huang, Wang CZ, Hao SG, Kramer MJ, Ho KM. Atomic size and chemical effects on the local order of Zr_2M (M=Co, Ni, Cu, and Ag) binary liquids. Phys. Rev. B. 2010; 81: 014108

[37] Hirata A, Guan P, Fujita T, Hirotsu Y, Inoue A, Yavari AR, Sakurai T, Chen M. Direct observation of local atomic order in a metallic glass. Nat. Mater. 2011; 10:28-33

[38] Liebermann HH, Graham CD, Jr., Flanders PJ. Changes in Curie temperature, physical dimensions, and magnetic anisotropy during annealing of amorphous magnetic alloys. IEEE. Trans. Magn. 1977; 13:1541

[39] Abrosimova GE, Serebryakov AV, Sokolovskaya ZhD. Change of structure and properties of Fe-based metal-metalloid amorphous alloys at heat treatment. Fizika Metallov I Metallovedenie. 1988; 66:727

[40] Ray R, Giessen BC, Grant NJ. New non-crystalline phases in splat cooled transition metal alloys. Scr. Met. 1968; 2:359

[41] Pokatilov VS. NMR-study of atomic-structure in amorphous FeB-alloys. Dokl. Akad. Nauk. SSSR 275 (1984) 79-83

[42] Ichikawa T. The assembly of hard spheres as a structure model of amorphous iron. Phys. Stat. Sol. 1975; a29:293-302

[43] Polk DE. Structural model for amorphous metallic alloys. Scr. Met. 1970; 4:117

[44] Yamamoto R, Matsuoka H. Relaxed bernal structure as a model of amorphous iron. J. Phys. F: Met. Phys. 1978; 7: L243

[45] Bernal JD. The Bakerian Lecture. 1962. The Structure of Liquids. London: Proc. Roy. Soc. (London). 1964; 280A:299-322

[46] Wang R. Short-range structure for amorphous intertransition metal alloys. Nature. 1979; 278: 700

[47] Egami T et al. Structure of bulk amorphous Pd–Ni–P alloys determined by synchrotron radiation. Metall. Mater. Trans. A, Phys. Met. Mater. Sci. 1998; 29(7):1805–1809

[48] Miracle DB. The efficient cluster packing model—an atomic structural model for metallic glasses. Acta Mater. 2006; 54(16): 4317–4336

[49] Sheng HW, Cheng YQ, Lee PL, Shastri SD, Ma E. Atomic packing in multicomponent aluminum-based metallic glasses. Acta Mater. 2008; 56(20): 6264–6272

[50] Sheng HW et al. Atomic packing and short-to-medium-range order in metallic glasses. Nature. 2006; 439(7075): 419–425

[51] Lu BC et al. A model of atom dense packing for metallic glasses with high-solute concentration. Appl. Phys. Lett. 2009; 94(24):241913

[52] Ma D, Stoica AD, Wang XL. Power-law scaling and fractal nature of medium-range order in metallic glasses. Nat. Mater. 2009; 8(1):30–34

[53] Abrosimova GE, Aronin AS. Evolution of the amorphous-phase structure in metal-metal type metallic glasses. J. Surf. Investig. X-ray Synchtr Tech 2015; 9(5):887–893

[54] Chou C-P, Turnbull D. Transformation behavior of Pd-Au-Si metallic glasses. J. Non-Cryst. Sol. 1975; 17:169

[55] Chen HS. Mater Glass temperature, formation and stability of Fe, Co, Ni, Pd and Pt based glasses. Sci. Eng. 1976; 23:151-154

[56] Tanner L, Ray R. Physical properties of Ti50Be40Zr10 glass. Scr. Met. 1980; 11:783-789

[57] Inoue A, Yamamoto M, Kimura HM, Masumoto T. Ductile aluminum-base amorphous alloys with two separate phases. J. Mater. Sci. 1987; 6:194

[58] Yavari AR. On the structure of metallic glasses with double diffraction halos. Acta Met. 1988; 36:1863

[59] Marcus M. Phase separation and crystallization in amorphous Pd-Si-Sb. J. Non-Cryst. Sol. 1979; 30:317

[60] Cohen MH, Turnbull D. Molecular transport in liquids and glasses. J. Chem. Phys. 1959; 31:1164-1169

[61] Chen HS, Turnbull D. Formation, stability and structure of palladium-silicon based alloy glasses. Acta Met. 1969; 17:1021

[62] Mehra M, Schulz R, Johnson WL. Structural studies and relaxation behavior of $(Mo_{0.6}Ru_{0.4})_{100-x}B_x$ metallic glasses. J. Non-Cryst. Sol. 1984; 61-62:859

[63] Nagarajan T, Asari UCh, Srinivasan S, Sridharan V, Narayanasamy A. Amorphous phase separation in METGLAS 2605CO. Hyperfine Interact. 1987; 34:491

[64] Mak A, Samwer K, Johnson WL. Evidence for two distinct amorphous phases in $(Zr_{0.667}Ni_{0.333})_{1-x}B_x$ alloys. Phys. Lett. 1983; 98A:353

[65] Doi K, Kayano H, Masumoto T. Small-angle scattering from neutron-irradiated amorphous $Pd_{80}Si_{20}$. J. Appl. Cryst. 1978; 11:605

[66] Terauchi H. Heterogeneous structure of amorphous materials. J. Phys. Soc. Jpn. 1983; 52:3454 -3459

[67] Osamura K. Structure and mechanical properties of a $Fe_{90}Zr_{10}$ amorphous alloy. J. Mater. Sci. 1984;19:1917

[68] Hermann H, Mattern N, Kuhn U, Heinemann A, Lazarev NP. Evolution of spatial heterogeneity in a Zr-based metallic glass. J. Non-Cryst. Sol. 2003; 317: 91-96

[69] Osamura K. SAXS study on the structure and crystallization of amorphous metallic alloys. Colloid. Polym Sci. 1981; 259:677

[70] Sheng HW et al. Polyamorphism in a metallic glass. Nat. Mater. 2007; 6(3):192–197

[71] Egami T, Vitek V, Srolovitz D. Microscopic model of structural relaxation in amorphous alloys. Proc. 4th Int. Conf. on Rapidly Quenched Metals (Sendai 1981) /ed. Masumoto T, Suzuki K. Sendai: Institute of Metals. 1982. p. 2

[72] Scott MG, Kursumovic A. Short-range ordering during structural relaxation of the metallic glass $Fe_{40}Ni_{40}B_{20}$. Acta Met. 1982; 30:853

[73] Shiga M. Magnetovolume: effects in ferromagnetic transition metals. J. Phys. Soc. Jpn. 1981; 50:2573 -2580

[74] Egami T. Structural relaxation in amorphous alloys— compositional short range ordering. Mater. Res. Bull. 1978; 13:557

[75] Masumoto T, Maddin R. Structural stability and mechanical properties of amorphous metals. Mater. Sci. Eng. 1975; 19:1 -24

[76] Chen HS. Glassy metals. Rep. Prog. Phys. 1980; 43:353

[77] Ultrarapid Quenching of Liquid Alloys. Ed, H. Hermann. New York: Academic Press; 1981. 375 p.

[78] Richardson MJ, Savill NG. Derivation of accurate glass transition temperatures by differential scanning calorimetry. Polymer. 1975;16:753-757

[79] Harmelin M, Calvayrac Y, Quivy A, Chevalier JP, Bigot J. On the effect of the initial quench on the thermal behaviour of amorphous $Cu_{60}Zr_{40}$ alloys. Scr. Met. 1982;16:707

[80] Greer AL. Crystallization kinetics of $Fe_{80}B_{20}$ glass. Acta Met. 1982; 30:171

[81] Chambron W, Chamberod A. Reversible enhancement of magnetic directional ordering rate associated with quenched-in defects, in an amorphous $Fe_{40}Ni_{38}Mo_4B_{18}$ alloy. Solid State Commun. 1980; 35:61

[82] Argon AS. Plastic-deformation in metallic glasses. Acta Met. 1979; 27:47–58

[83] Pampillo CA, Chen HS. Comprehensive plastic deformation of a bulk metallic glass. Mater. Sci. Eng. 1974; 13:181

[84] Neihauser G, Stossel RP. Homogeneous and Heterogeneous Deformation of metallic glasses. In: Rapid Quenching metallic alloys. Ed. Steeb S, Warkimont G. Moscow: Metallurgy; 1989 p. 247-252 (Rus.)

[85] Spaepen F. A microscopic mechanism for steady state inhomogeneous flow in metallic glasses. Acta Met. 1977; 25:407

[86] Chen H., Huang JG, Song SX, Nieh TG, Jang JSC. Flow serration and shear-band propagation in bulk metallic glasses. Appl. Phys. Lett. 2009; 94:141914

[87] Georgarakis K. Shear band melting and serrated flow in metallic glasses. App. Phys. Lett. 2008; 93:031907

[88] Handrich K, Kobe S. Amorphous Ferro- and Ferrimagnetics. Moscow: Mir; 1982. 106 c. (Rus.)

[89] Masumoto T. Mechanical characteristics of amorphous metals. Sci Repts. RITU. 1977; A26: 246-262.

[90] Zolotukhin IV, Khonik VA, Safonov IF. Strength, plasticity and stress relaxation in amorphous $La_{80}Al_{20}$ alloy. Fizika i chimiya stekla. 1983; 9: 67-73 (Rus).

[91] Ashby MF, Greer AL. Metallic glasses as structural materials. Scr. Mater. 2006; 54: 321-326.

[92] Ichitsubo T. Nanoscale elastic inhomogeneity of a Pd-based metallic glass: sound velocity from ultrasonic and inelastic X-ray scattering experiments. Phys. Rev. B. 2007; 76: 140201.

[93] Srolovitz D, Vitek V, Egami T. An atomistic study of deformation of amorphous metals. Acta Met. 1983; 31: 335–352

[94] Bulatov V.V., Argon A.S. A stochastic-model for continuum elastoplastic behavior. 1. Numerical approach and strain localization. Model Simul. Mater. Sci. Eng. 1994; 2: 167–184.

[95] Johnson WL, Samwer K. A universal criterion for plastic yielding of metallic glasses with a $(T/T\text{-}g)^{(2/3)}$ temperature dependence. Phys. Rev. Lett. 2005; 95: 195501.

[96] Cheng YQ, Sheng HW, Ma E. Relationship between structure, dynamics, and mechanical properties in metallic glassforming alloys. Phys. Rev. B. 2008; 78: 014207.

[97] Krebs HU, Biegel W, Bienenstock A. Phase separation in amorphous Zr-Fe alloys determined by anomalous X-ray scattering and magnetization measurements. Mater. Sci. Eng. 1988; 97; 163-7

[98] G. Abrosimova, A. Aronin, A. Budchenko. Amorphous phase decomposition in Al-Ni–RE system alloys. Mater. Lett. 2015; 139: 194-196

[99] Abrosimova GE, Aronin AS. Effect of the concentration of a rare earth component on the parameters of the nanocrystalline structure in aluminum-based alloys. Phys. Sol. State. 2009; 51: 1765-1771

[100] Abrosimova GE, Aronin AS, Ignat'eva EYu, Molokanov VV. Phase decomposition and nanocrystallization in amorphous $Ni_{70}Mo_{10}P_{20}$ alloy. JMMM. 1999; 203: 169-171

[101] Abrosimova G, Aronin A, Ignatieva E. Decomposition of amorphous phase in $Ni_{70}Mo_{10}B_{20}$ alloy above glass transition temperature. Mater. Sci. Eng. 2007; A449-451: 485-488

[102] Abrosimova GE, Aronin AS. Specific features of amorphous $Fe_{90}Zr_{10}$ alloy separation upon heating. Phys. Sol. State. 1998; 40: 1603-1606.

[103] Naudon A, Flank V. Relationship between space correlated fluctuations and initial alloy composition in some metallic glasses. J. Non-Cryst. Sol. 1984; 61/62: 355-360

[104] Abrosimova GE, Aronin AS, Asadchkov VE, Serebryakov AV. Structure evolution of Fe-B and Co-Fe-Si-B alloys during heating below crystallization temperature. Fizika metallov i metallovedenie. 1986; 62: 496-502

[105] Abrosimova GE, Aronin AS. Reversible structural changes on heat treatment of amorphous Fe-B alloys. Int. J. Rapid Sol. 1991; 6: 29-40

[106] Elliot PR. Constitution of Binary Alloys. First Supplement. NY: McGraw-Hill Book Company; 1970. 472 p.

[107] Pampillo CA. Flow and fracture in amorphous alloys. J. Mater. Sci. 1975; 10: 1194-1197.

[108] Zeng QS. Anomalous compression behavior in lanthanum/cerium-based metallic glass under high pressure. Proc. Natl. Acad. Sci. USA. 2007; 104(34): 13565-13568.

[109] Wang XD, Bednarcik J, Saksi K, Franz H, Cao QP, Jiang JZ. Tensile behavior of bulk metallic glasses by *in situ* X-ray diffraction. Appl. Phys. Lett. 2007; 91: 081913.

[110] Stoica M, Das J, Bednarcik J, Franz H, Mattern N, Wang WH, Eckert J. Strain distribution in $Zr_{64.13}Cu_{15.75}Ni_{10.12}Al_{10}$ bulk metallic glass investigated by in situ tensile tests under synchrotron radiation. J. Appl. Phys. 2008; 104: 013522.

[111] Wang XD, Bednarcik J, Franz H, Lou HB, He ZH, Cao QP, Jiang JA. Local strain behavior of bulk metallic glasses under tension studied by *in situ* X-ray diffraction. Appl. Phys. Lett. 2009; 94: 011911

[112] Abrosimova GE, Aronin AS, Afonikova NS, Kobelev NP. Effect of the deformation on structure change of amorphous $Pd_{40}Ni_{40}P_{20}$ phase. Phys. Sol. State.. 2010; 52: 1892

[113] Suzuki K. Small Angle and Very-Small Angle X-Ray Scattering from Amorphous $Pd_{80}Si_{20}$ before and after Cold Work. Jpn. J. Appl. Phys. 1981; 20: 271

[114] Walter J. J. Crystallization of the amorphous alloys $Fe_{50}Ni_{30}B_{20}$ and $Fe_{80}oB_{20}$. Mater. Sci. Eng. 1977; 29: 161

[115] Shirnina DP, G.E. Abrosimova, A.S. Aronin nanocrystal formation in Al-based alloys at the deformation. Des. Mater. Technol. 2011; 3(18): 80 (Rus.).

[116] Abrosimova GE, Aronin AS, Bezrukov AV, Pankratov SP, Serebryakov AV. Effect of heat treatment conditions on crystalllization of amorphous Fe-B alloys. Phys. Met. 1982; 4: 114

[117] Zhukov AP, Ponomarev BK. Dependence of the start field on the frequency and amplitude of remagnetizing field in Fe and Co based amorphous alloys. Fizika Tverdogo Tela. 1989; 31: 26-30

[118] Abrosimova GE, Zhukov AP, Ponomarev BK. Effect of heat treatment on some properties of Fe- and Co-based amorphous alloys. Phys. Stat. Sol. 1989; 111a: K237-K241

[119] Walter JL, Bacon F, Luborsky FE. Mater. Sci. Eng. 1976; 24: 239

[120] Herold U., Koster U. in Proceedings of the Third International Conference on Rapidly Quenched Metals. ed. by Cantor B., Brighton, Metals Society. London; 1978. V. 1. 281 p.

[121] Freed RL, VaderSande JB. The metallic glass $Cu_{56}Zr_{44}$: devitrification and the effects of devitrification on mechanical properties. Acta Met. 1980; 28: P. 103

[122] Rawson H. Inorganic Glass-Forming Systems. London: Academic Press. 1967.

[123] Fine ME. Phase Transformation in Condensed Systems. New York: MacMillan. 1964.

[124] Birac C, Lesueur D. Diffusion atomique du lithium dans l'alliage métallique amorphe Pd–Si. Phys. Status Solidi. 1976 ; 36a: 247.

[125] Köster U, Herold U. Metallic Glasses in Glassy Metals. Ed. H. Beck, H.-J. Güntherodt. Berlin: Spring-Verlag, 1983. 454 p.

[126] Martin JW, Doherty RD. Stability of Microsctructure in Metallic Systems. Cambridge: Cambridge University Press. 1976. 280 p.

[127] Ham RK, Kirkaldy JS, Plewes JW. The fatigue of Cu-Ni-Fe alloys. Acta Met. 1967; 15: 861-869.

[128] Chen HS, Turnbull D. Formation, stability and structure of palladium-silicon based alloy glasses. Acta Met. 1969; 17: 1021

[129] Stoering R, Conrad H, Metastable structures in liquid quenched and vapor quenched Ag-Cu alloys. Acta Met. 1969; 17: 933

[130] Mattern N, Goerigk G, Vainio U, Miller MK, Gemming T, Eckert J. Spinodal decomposition of Ni–Nb–Y metallic glasses. Acta Mater. 2009; 57: 903–908.

[131] Aronin AS, Serebryakov AV. Volume Discrepancy Compensation Upon Crystallization of Some Amorphous Alloys. Chernogolovka: ISSP. 1981. 9 p. (Rus).

[132] Yoshizawa Y, Oguma S, Yamauchi K. New Fe-based soft magnetic alloys composed of ultrafine grain structure. J. Appl. Phys. 1988; 64: 6044.

[133] Khan Y, Sostarich M. Dynamic temperature X-ray diffraction analysis of amorphous $Fe_{80}B_{20}$. Z. Met. 1981; 72: 266-268.

[134] Herold U., Koster U. Metastabile Phasen in extreme schnell erstarrten Eisen-Bor-Legierungen. Z. Met. 1978; 69: 326-332.

[135] Abrosimova GE, Serebryakov AV. Metastable phase decomposition in Fe-B alloys. Phys. Met. 1983; 5: 85-88

[136] Inal OT, Kekker L, Yost FG. High-temperature crystallization behavior of amorphous $Fe_{80}B_{20}$. J. Mater. Sci. 1980; 15: 1947-1961

[137] Khan Y. Sostarich M. Dynamic temperature X-ray diffraction analysis of amorphous $Fe_{80}B_{20}$. Z. Metallk. 1981; 72: 266-268

[138] Abrosimova GE, Aronin AS, Gantmakher VF, Levin YuB, Osherov MV. Variation of the electrical resistivity of the amorphous Ni-Zr alloy in the initial-stages of crystallization. Fizika Tverdogo Tela. 1988; 30: 1424-1430.

[139] Abrosimova GE, Aronin AS, Zverkova II, Kir'yanov YuV, Molokanov VV. Early stages of amorphous $Zr_{65}Cu_{17.5}Ni_{10}Al_{7.5}$ alloy crystallization. Fizika metallov I metallovedenie. 1998; 86: 91-96

[140] Abrosimova GE, Aronin AS, Matveev DV, Molokanov VV. Formation and structure of nanocrystals in bulk $Zr_{50}Ti_{16}Cu_{15}Ni_{19}$ metallic glass. Fizika Tverdogo Tela. 2004; 46: 2191-2195

[141] Abrosimova GE, Aronin AS, Kir'janov YuV, Matveev DV, Zver'kova II, Molokanov VV, Pan S, Slipenyuk A. The structure and mechanical properties of bulk $Zr_{50}Ti_{16.5}Cu_{14}Ni_{18.5}$ metallic glass. J. Mater. Sci. 2001; 36: 3933-3939,

[142] Abrosimova GE, Kobelev NP, Kolyvanov EV, Khonik VA. The influence of heat treatment on the ultrasonic velocity and elastic moduli of a Zr-Cu-Ni-Al-Ti bulk metallic glass. Phys. Sol. State. 2004; 46: 1859-1862

[143] Abrosimova GE, Kobelev NP, Kolyvanov EV, Khonik VA, Levin VM. Effect of heat treatment on elastic characteristics of a bulk amorphous Zr-Cu-Ni-Al-Ti alloy. Phys. Sol. State. 2006; 48: 2091-2094

[144] Frankwicz PS, Ram S, Fecht H-J. Enhanced microhardness in Zr65.0Al7.5Ni10.0Cu17.5 amorphous rods on coprecipitation of nanocrystallites through supersaturated intermediate solid phase particles. Appl. Phys. Lett. 1996; 68: 2825

[145] Köster U, Meinhardt J, Roos S, Liebertz H. Formation of quasicrystals in bulk glass forming Zr–Cu–Ni–Al alloys. Appl. Phys. Lett. 1996; 69: 179

[146] Xing LQ, Eckert J, Löser W, Schultz L. High-strength materials produced by precipitation of icosahedral quasicrystals in bulk Zr–Ti–Cu–Ni–Al amorphous alloys. Appl. Phys. Lett. 1999; 74: 664.

[147] Li CF Precipitation of icosahedral quasicrystalline phase in metallic glasses. Appl. Phys. Lett. 2000; 77; 528-530

[148] Abrosimova GE, Aronin AS, Serebraykov AV. Structure of rapidly quenched Al-Mn alloys. Suppl. Trams. JIM. 1988; 29: 485-488.

[149] He G, Löser W, Eckert J. Microstructure and mechanical properties of the $Zr_{66.4}Cu_{10.5}Ni_{8.7}Al_8Ti_{6.4}$. Scr. Mater. 2003; 48: 1531-1536

[150] Abrosimova GE, Aronin AS, Gurov AF. Formation and structure of light-weight nanocrystalline alloys of the aluminum-nickel-RE system. Phys. Met. Metallogr. 2000; 90: 180-185

[151] Abrosimova GE, Aronin AS, Zverkova II. Phase transformation upon crystallization of amorphous Al-Ni-RE alloys. Phys. Met. Metallogr. 2002; 94: 102-107

[152] Cremanshi V, Arcondo B, Sirkin H, Vazquez M, Asenjo F, Garcia JM, Abrosimova G, Aronin A. Huge magnetic hardening ascribed to metastable crystallites during first stages of devitrification of amorphous FeSiBNbSn alloys. J. Mater. Res. 2000; 15: 1936-1942,

[153] Wang WK, Iwasaki H, Suryanaryana C, Masumoto T. Crystallization characteristics of an amorphous $Ti_{80}Si_{20}$ alloy at high pressure. J. Mater. Sci. 1983; 18; 3765-3772.

[154] Ogawa Y, Nunogaki K, Endo S, Kiritani M, Fujita FF. Crystallization of amorphous Fe-B alloys under pressure. Sendai: Proc. 4th Int. Conf. on Rapidly Quenched Metals. 1981. P. 675-678.

[155] Steinhardt PJ. Solid-state physics – how does your quasicrystal grow? Nature. 2008; 452: 43

[156] Kelton KF. First X-ray scattering studies on electrostatically levitated metallic liquids: demonstrated influence of local icosahedral order on the nucleation barrier. Phys. Rev. Lett. 2003; 90: 195504.

[157] Shen YT. Icosahedral order, frustration, and the glass transition: evidence from time-dependent nucleation and supercooled liquid structure studies. Phys. Rev. Lett. 2009; 102: 057801.

[158] Saida J. Precipitation of icosahedral phase from a supercooled liquid region in $Zr_{65}Cu_{7.5}Al_{7.5}Ni_{10}Ag_{10}$ metallic glass. Appl. Phys. Lett. 1999; 75: 3497.

[159] Gleiter H. Nanocrystalline materials. Prog. Mater. Sci. 1989; 33: 223-315

[160] Birringer R. Nanocrystalline materials. Mater. Sci. Eng. 1989; A117: 33-43

[161] Turnbull D, Cohen MH. On free-volume model of liquid-glass transition. J. Chem. Phys. 1970; 52: 30308

[162] Zolotukhin IV. Physical properties of amorphous metallic materials. Moscow: Metallurgy. 1986. 176 p. (Rus)

[163] Ramachandrarao P, Cantor B, Cahn RW. Viscous behavior of undercooled metallic melts. J. Non-Cryst. Sol. 1977; 24: 109

[164] Ramachandrarao P, Cantor B, Cahn RW. Free volume theories of the glass transition and the special case of metallic glasses. J. Mater. Sci. 1977; 12: 2488

[165] Cahn RW. Rapid Solidification Processing: Principles and Technologies, eds. R. Mehrabian et al. Clattor's. LA: Baton Rouge. 1978. 129 p.

[166] Louzguine-Luzgin D, Inoue A. Nano-devitrification of glassy alloys. J. Nanosci. Nanotechnol. 2005; 5: 999-1014.

[167] Yavari AR, Osamura K, Okuda H, Amemia Y. Small-angle x-ray scattering study of phase separation in amorphous alloys during heating with use of synchrotron radiation. Phys. Rev. B. 1988; 37: 7759-7765.

[168] Inoue A, Yamamoto M, Kimura HM, Masumoto T. Ductile aluminium-base amorphous alloys with two separate phases. J. Mater. Sci. Lett. 1987; 6: 194.

[169] Aronin AS, Abrosimova GE, Gurov AF, Kir'janov YuV, Molokanov VV. Nanocrystallization of bulk Zr-Cu-Ti metallic glass. Mater. Sci. Eng. 2001; A304-306: 375-379.

[170] Gutkin MYu, Ovid'ko IA. Physical Mechanics of Deformed Nanostructures. S-Petersburg: Yanus, 2003. 194 p. (Rus).

[171] Gryaznov VG, Trusov LI. Size effect in micromechanics of nanocrystals. Prog. Mater. Sci. 1993; 37: 290-400.

[172] Gryaznov VG, Karepov AM, Romanov AE. Critical instability of dislocations in microcrystals. Pisma v zhurlal tekhnicheskoi fiziki. 1989. 15:39-44

[173] Wang N, Wang Z, Aust KT, Erb U. Effect of grain size in the mechanical properties of nanocrystalline materials. Acta Metall. Mater. 1995; 43: 519.

[174] Abrosimova GE, Aronin AS. Fine structure of FCC nanocrystals in Al- and Ni-based alloys. Phys. Sol. State. 2002; 44: 1003-1007

[175] Abrosimova GE, Aronin AS. Size effect on the structure of Al- and Ni-based nanocrystals. Phys. Sol. State. 2008; 50: 159-163

[176] Abrosimova GE, Aronin AS, Kir'janov YuV, Zver'kova II, Molokanov VV, Alves H, Köster U. The formation, structure and properties of nanocrystalline Ni-Mo-B alloys. J. Mater. Sci. 1999; 34: 1611-1618.

[177] Aronin AS, Abrosimova GE, Zver'kova II, Kir'janov YuV, Molokanov VV, Petrzhik MI. The structure of nanocrystalline Ni58.5Mo31.5B10 and structure evolution at heat treatment. Mater. Sci. Eng. 1997; A226-228: 536-540.

[178] Abrosimova GE, Aronin AS, Zver'kova II, Gurov AF, Kir'janov YuV, Formation, structure and microhardness of nanocrystalline Ni-No-B alloys. Phys. Sol. State. 1998; 40: 8-13.

[179] Aronin AS, Abrosimova GE, Kir'janov YuV. Formation and structure of nanocrystals in an Al86Ni11Yb3 alloy. Phys. Sol. State. 2001; 43: 2003–2011

[180] Abrosimova GE, Aronin AS, Barkalov OI, Dement'eva MM. Formation of the nanostructure in amorphous alloys of Al-Ni-Y system. Phys. Sol. State. 2013; 55: 1773-1778

[181] Köster U., Schunemann M., in: Rapidly Solidified Alloys, ed. H.H. Liebermann, New York: Marcel Dekker Inc. 1993, 303p.

[182] Ham FS, Theory of diffusion-limited precipitations. J. Phys. Chem. Sol. 1958; 6: 335 - 351

[183] Hebert RJ, Boucharat N, Perepezko JM, Rösher H, Wilde G. J. Alloys Compd. 2007; 434-435: 18

[184] Hirth JP, Lothe J, Theory of Dislocations. New York: McGraw-Hill Book Company. 1972

[185] Gallagher PCJ. Thermodynamic of metal interstitial solution. Trans. AIME. 1970; 1: 2429

[186] Howie A, Swann PR. Direct measurements of stacking-fault energies from observations of dislocation modes. Phil Mag. 1961; 6: 1215.

[187] Kelly A, Groves GW, Crystallography and Crystal Defects. London: Longman. 1970. 470 p.

[188] Bolling GF, Richman RH. Continual mechanical twinning: Part I: Formal description. Acta Met. 1965; 13: 709

[189] Hono K. Atom probe microanalysis and nanoscale microstructures in metallic materials. Acta Mater. 1999; 47: 3127-3145

[190] Aronin AS, Ivanov SA, Yakshin AE. Increase of the near-syrface crystallization temperature in a Fe-B-P amorphous alloy. Fizika Tverdogo Tela. 1991; 33: 2527-2532

[191] Abrosimova GE, Aronin AS. Effect of the concentration of a rare-earth component on the parameters of the nanocrystalline structure in aluminum-based alloys. Phys. Sol. State. 2009; 51: 1765-1771

Chapter 4

Assessment of Hardness Based on Phase Diagrams

Jose David Villegas Cárdenas,
Victor Manuel López Hirata,
Carlos Camacho Olguin,
Maribel L. Saucedo Muñoz and
Antonio de Ita de la Torre

Additional information is available at the end of the chapter

http://dx.doi.org/10.5772/64699

Abstract

This chapter summarizes the methodology and development of a general equation, in order to obtain a series of equations to assess the hardness of different Al-Cu-Zn alloys, based on their chemical composition. This methodology produces an assessment of hardness with a maximal deviation of 5%, in as-cast, homogenized and quenching alloys, for both alloys created in laboratories like commercials. This method entails the generation of linear equations by a linear regression method, obtained from a zone of the phase diagram, when the composition is changed from linear to planar form. Therefore, if the chemical composition of samples varies, the percentage of each phase will also vary, causing a change in mechanical properties in a linear manner. If the heat treatments are the same for all samples, then the changes in mechanical properties are proportional for each of them, maintaining the linear relationship in mechanical properties in accordance with chemical composition. This methodology is applicable for any ternary alloy along with its equilibrium diagram.

Keywords: quenching, Gibbs free energy, phase diagrams, alloys

1. Introduction

The cost of any product is primarily limited by the cost of raw materials. This determines 50% of the total product cost and that is why it is necessary to try and create products using cheaper materials, but that still possess the same physical, chemical and mechanical properties.

This phenomenon took place in China when the country started to create very cheap products, made possible thanks to the development of new, cheaper materials with the same mechanical properties as older, more expensive materials.

Even if cheaper materials are available, it is still necessary to assess their mechanical properties in order to decide whether the resultant product will be suitable.

In addition, in order to assess the mechanical properties of materials, it is necessary to know their phases, and this need is addressed by the phase diagram. The next section describes this in more detail.

2. Phase Diagrams

Phase diagrams in metallic alloys are a development of the relationship between the Gibbs free energy of the different phases that exist in a metallic mix. A phase is defined as "a portion of the system whose properties and composition are homogeneous and which is physically distinct from other parts of the system" [1].

Gibbs free energy measures the stability of a system according to certain temperatures and pressures. This energy generates a curve on the graphic representation of energy vs atomic percentage, and when the system in question has different phases, then each phase will generate a curve (**Figure 1(a)-(e)**).

The Gibbs free energy is defined by the equation:

$$G = E + PV - TS \tag{1}$$

Where:

E = Internal energy of the system

P = Pressure

V = Volume

T = Temperature

S = Entropy

In the case of binary alloys, the equation of Gibbs free energy is:

$$\Delta G_{mix} = N_a z\varepsilon X_A X_B + RT\left(X_A lnX_B + X_B lnX_A\right) \tag{2}$$

Where:

N_a is Avogadro's number

Z is the number of bonds per atom

ε is the difference between the A-B bond energy and the average of the A-A and B-B bond energies.

Equation (3) explain this further:

$$\varepsilon = \varepsilon_{AB} - \frac{1}{2}\left(\varepsilon_{AA} + \varepsilon_{BB}\right) \tag{3}$$

This can be simplified a little if $N_a z\varepsilon = \Omega$

The graphic of the binary alloy with respect to the Gibbs free energy at a constant pressure and temperature, is used to create the phase diagram. **Figure 1** represents the formation of a phase diagram by the Gibbs free energy.

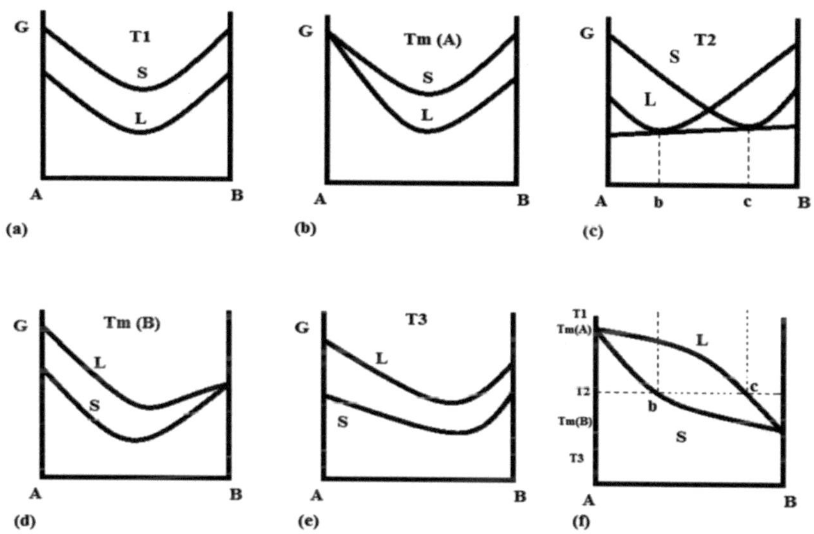

Figure 1. Phase diagram based on the Gibbs free energy in the liquid and solid phase.

Figure 1(a) shows both L and S phases, but in this case the L phase shows a lower energy in whole the combination percentage of the elements A and B. This indicates that the L phase is present in whole the combination of two elements at temperature T1, and this is shown in the phase diagram **Figure 1(f)**, where it can be observed that the temperature T1 is only observed the L phase in whole combination of elements A and B. If the curves generated is intercepted at a point in the temperature Tm, as is the case of **Figure 1(b)**, where an interception of the L and S phases can be observed. After this point, the L phase possesses less energy.

This is reflected in the phase diagram, where the temperature Tm (A) is observed in both phases at the initiation of the combination of A and B, and is later seen in the L phase only [2].

When the curves of Gibbs free energy intersect at certain temperatures, e.g. temperature T2, then the phases are mixed. This mixing occurs in the tangent line between the curves of the L and S phases (**Figure 1(c)**). At the point where element A begins, to the point of interception with the tangent line, only the S phase is present. In the points where the curves are intercepted by the tangent line, a mix of L and S phases is obtained. Finally, from point c to the start of element B, only the L phase is present.

As **Figure 1** shows, at a certain temperature it is possible to have a mix of phases. This is important because the proportion of the phases determines the mechanical, physical and chemical properties of the metallic alloys.

In the previous investigations, equations were obtained that link hardness with chemical percentage — that is to say with the tangent line of the curves of Gibbs free energy. It is therefore necessary to understand the mathematical form in which the tangent line is obtained.

In order to better understand this, the example below shows the creation of a tangent line to two parabolas represented by the equations:

$$y = 2x^2 + x \tag{4}$$

$$y = 3x^2 + 2x \tag{5}$$

The curves generated by equations (4) and (5), and the tangent line are shown below in **Figure 2**:

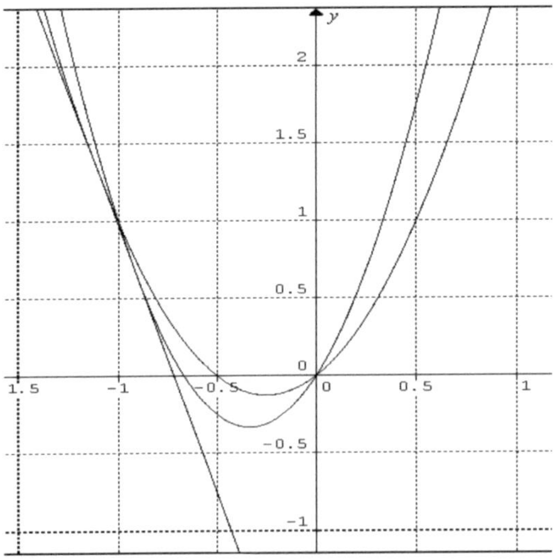

Figure 2. Graphic of equations (4) and (5) and associated tangent line.

The tangent line shown in **Figure 2** is represented by equation (6):

$$y = mx + b \tag{6}$$

However, at some points equation (6) is intercepted by equations (4) and (5) and then:

$$mx + b = 2x^2 + x \tag{7}$$

$$mx + b = 3x^2 + 2x \tag{8}$$

If we pass all of the left side of the equations, then:

$$2x^2 + x(1-m) - b = 0 \tag{9}$$

$$3x^2 + x(2-m) - b = 0 \tag{10}$$

The equation is solved using the quadratic equation:

$$x_{1,2} = \frac{-(1-m) \mp \sqrt{(1-m)^2 - 4(2)(-b)}}{2(2)} \tag{11}$$

$$x_{1,2} = \frac{-(2-m) \mp \sqrt{(2-m)^2 - 4(3)(-b)}}{2(3)} \tag{12}$$

However, equation (6) touch in only point to the equations (4) and (6), therefore when is applicate the quadratic equation the result into of square root.

$$(1-m)^2 - 4(2)(-b) = 0 \tag{13}$$

$$(2-m)^2 - 4(3)(-b) = 0 \tag{14}$$

By developing:

$$1 - 2m + m^2 + 8b = 0$$

$$4 - 4m + m^2 + 12b = 0$$

Then the quadratic equation is used:

$$m_{1,2} = \frac{2 \mp \sqrt{(-2)^2 - 4(1)(8b+1)}}{2(1)} \tag{15}$$

$$m_{1,2} = \frac{4 \mp \sqrt{(-4)^2 - 4(1)(12b+4)}}{2(1)} \tag{16}$$

As the slope is the same, it is possible to match equations (15) and (16).

$$\frac{2 \mp \sqrt{(-2)^2 - 4(1)(8b+1)}}{2(1)} = \frac{4 \mp \sqrt{(-4)^2 - 4(1)(12b+4)}}{2(1)}$$

$$2 + \sqrt{(-2)^2 - 4(1)(8b+1)} = 4 + \sqrt{(-4)^2 - 4(1)(12b+4)}$$

$$2 + \sqrt{4 - 32b - 4} = 4 + \sqrt{16 - 48b - 16}$$

$$2 + \sqrt{4 - 32b - 4} = 4 + \sqrt{16 - 48b - 16}$$

$$2 + \sqrt{-32b} = 4 + \sqrt{-48b}$$

Therefore, b is:

$$b = -\left[\frac{2}{\sqrt{32} - \sqrt{48}}\right]^2 = -2.4747$$

With this value, using equation (15) or (16) it is possible to obtain m:

$$2 + \sqrt{(-2)^2 - 4(1)(8(-2.4747+1)} = 4 + \sqrt{(-4)^2 - 4(1)(12(-2.4747+4)}$$

The equations are equal when m = −3.4495

The linear equation representing the tangent to each curve is:

http://dx.doi.org/10.5772/64699

$$y = -3.4495\times -2.4747 \tag{17}$$

Using this method we can obtain the linear equation generated by the curves of Gibbs free energy in a regular solution. For example:

$$\Delta G_{mix} = \Omega_1 X_A X_B + RT\left(X_A lnX_b + X_b lnX_A\right) \tag{18}$$

For phase 1.

$$\Delta G_{mix} = \Omega_2 X_A X_B + RT\left(X_A lnX_b + X_b lnX_A\right) \tag{19}$$

For phase 2.

So:

$$b = X_A X_B\left[\frac{X_A\Omega_1 - X_B\Omega_2}{X_A - X_B}\right] + RT\left(X_A lnX_B + X_B lnX_A\right) \tag{20}$$

And m is:

$$m = \frac{X_A X_b(\Omega_2 - \Omega_1)}{X_b - X_A} \tag{21}$$

In equations (20) and (21), it is possible to appreciate that the change in the phases depends principally on Ω, and this is also $N_a z\varepsilon$. Therefore, the mechanical properties of any material are easily transposable directly by the bond energy that exists between atoms.

When applying this method to a ternary alloy, the procedure is essentially the same, with the difference being that instead of having a tangent line, we have a tangent plane to the parabolas composed of each of the phases. **Figure 3** shows the formation of a ternary diagram from the curves of Gibbs free energy.

In the same fashion as the binary phase diagram, from the ternary diagram the percentage of each phase that exists for each composition can be obtained in such a way that it is possible to determine the mechanical, physical and chemical properties of each alloy. Therefore, it is possible —based on the composition and accordance with the phase diagram — to obtain the relationship with hardness. This is explained further in the next chapter.

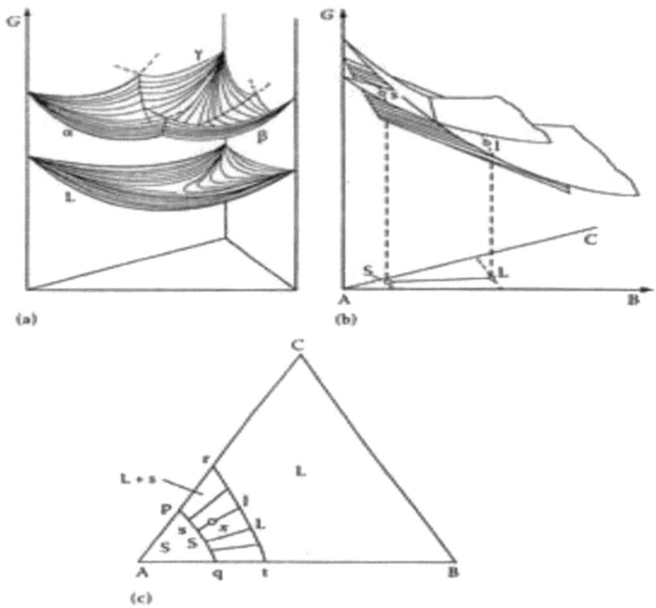

Figure 3. Representation of the ternary system curve developed with the Gibbs free energy [1].

3. Hardness

Hardness is defined as the capacity with which a body can contain the penetration of another body with a certain resistance. The deformation resistance is:

$$H = \frac{F}{A} \tag{22}$$

Where F is the force and A is the surface area.

The deformation of a sample during the indentation consist of two parts; the first part is elastic and the second is plastic.

Tabor's experimental studies suggest that hardness (H) is proportional to a representative tensile of contact σ_r;

$$H = 2.7\sigma_r \ldots \tag{23}$$

The answer of contact for materials of high hardness is often to the deformation elastic-plastic. These materials are governed by the equation:

http://dx.doi.org/10.5772/64699

$$\frac{H}{\sigma_r} = 1.44 + 0.264 \ln\left(\frac{E}{\sigma_r}\right) \tag{24}$$

where E is Young's modulus and equations (23) and (24) describe its relationship to the hardness. This is important because it is possible to obtain other mechanical properties only through the hardness.

In the last 10 years, new techniques have been developed in the measurement of hardness, such as micro and nanoindentation, in order to enable the complete characterization of mechanical properties in small areas the size of micrometers.

These techniques allow measurements of the curve force (P) and penetration (h). This curve describes the fundamental mechanical properties like Young's modulus and the stress-strain curve [3–7]. The micro and nanohardness allow for an assessment of the mechanical conduits to different levels microcstructurales. The difference between macro, micro and nanoindentation is the load force. For example, in microhardness, the minimum load is 200 mN while in nanohardness it is 0.01 mN. The problem with micro and nanoindentation, is that they only can characterize an extremely small region of the sample, speak in clear of metallic materials.

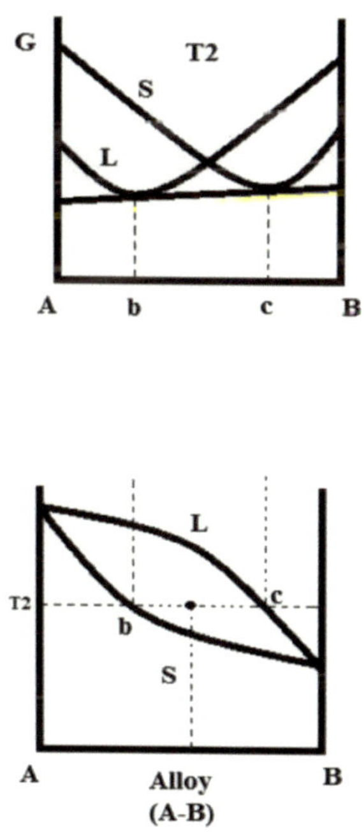

Figure 4. Alloy composed of L and S phases.

As previously mentioned, the points where a tangent line touches the two curves of the Gibbs free energy of each phase. **Figure 4** shows this:

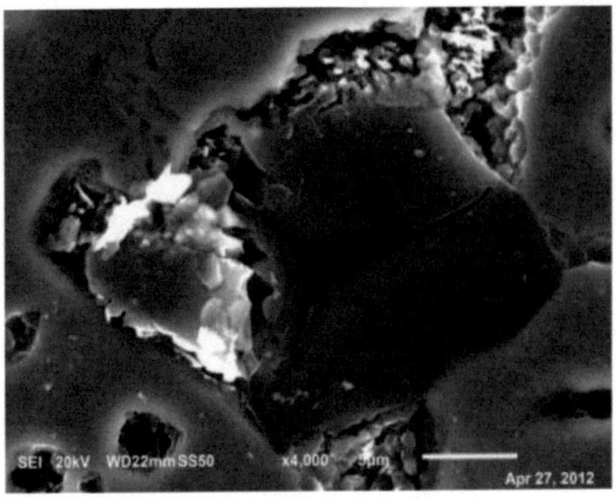

Figure 5. Nanoindentation seen by electronic microscopy.

The mechanical properties of any alloy are between point b and point c of **Figure 4**, and depend on the percentage of each phase present. For example, the alloy (a-B), consists of L and S phases, and to determine the percentage of each phase according to the chemical composition, it is necessary to use the lever rule. The alloy (A-B) can be measured with normal hardness and by the size of the track it generates, covering the two phases. As a result, the hardness is the average of the phases. However, when the hardness was measured with a nanohardness, only one phase was measured. For example, in **Figure 5** it is possible to appreciate the different phases present in the Al-Cu-Zn alloy. If we use conventional hardness, then the track is so big that it covers all the phases present, and nothing could observe the track to 4000 enhance.

The phase present in alloys, depends on temperature and pressure, and hence distinct phase diagrams can be seen according to these two variables.

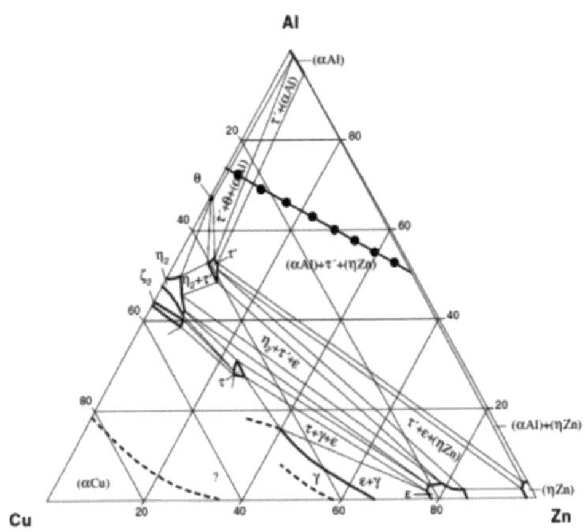

Figure 6. Ternary phase diagram of Al-Cu-Zn alloy. The line crosses the zone composed by the α, η and τ phases [8].

Figure 6 shows a ternary phase diagram of Al-Cu-Zn alloy to 200°C, at atmospheric pressure. A diagram of this kind does not exist in the literature at room temperature and all that's left is to think that keep the phases present of 200°C at room temperature. **Figure 6** shows a line that crosses the zone composed of the α, η y and τ phases. Each point on this line is one alloy, with a different percentage chemical composition, and of phases, each of these alloys will have certain mechanical properties.

The line that crosses the diagram can be modelled by a linear equation in such a way that this equation will have a relationship with the mechanical properties of the alloy. It is possible to assess the mechanical properties of each of the alloys that are on the line. The next section describes the methodology of the assessment of hardness, and the analysis of phases.

4. Assessment of hardness for Al-Cu-Zn alloys in as-cast, homogeneized and quenching conditions

4.1. Introduction

"The Al-Zn-Cu ternary system is one of the most important alloy systems. It is important for the 7xxx series Al-based alloys and Zn-Al based alloys, ZAMAK. For instance, Zn-Al-Cu alloys have been used in different tribological applications as a substitute to conventional bearing bronzes and cast irons. Nevertheless, these alloys have some limitations such as dimensional stability because of the transformation of the metastable ($CuZn_4$) with a hexagonal crystalline structure into the stable τ' (Al_4Cu_3Zn) phase with an ordered rhombohedric structure. It is formed by the four phase reaction $\alpha + \varepsilon \rightarrow \eta + \tau'$ which also involves the fcc Al-rich α and cph Zn-rich η phases. The low tensile strength of these alloys at temperatures above room temperature has been attributed to their low melting points and they also have low ductility and high cracking tendency caused by the presence of the Zn-rich and Cu-rich phases. In order to overcome these limitations, the change in chemical composition has been proposed as an alternative. For instance, it has recently been proposed to change the base from zinc to aluminium. Replacing zinc with aluminium in the alloy system causes the formation of the stable θ ($CuAl_2$) with a tetragonal crystalline structure instead of the metastable ε phase, promoting a considerably higher ductility. Therefore, the purpose of this work was to investigate the effect of change in chemical composition on the microstructure, mechanical properties and hardness, in order to establish a relationship which enables us to assess the hardness in the as-cast and homogenized alloys" [9].

4.2. Methods and materials

Two lines of work were developed, composed of 8 samples of each line, as shown in **Figure 7**. For line 1, the samples M1 to M8 were developed and for line 2 the samples M9 to M16 were developed. **Table 1** shows the chemical composition of each of the samples.

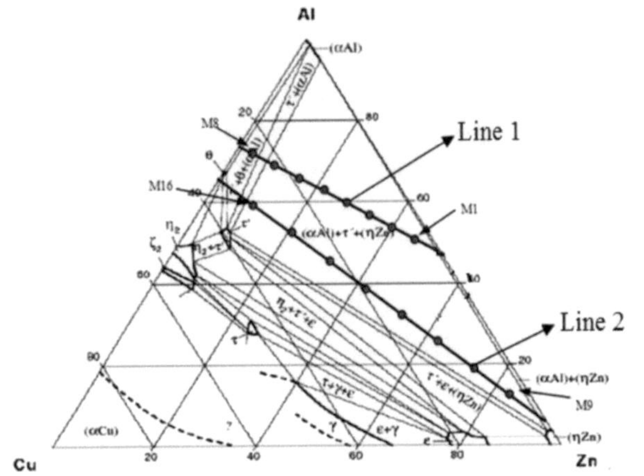

Figure 7. Ternary phase diagram of Al-Cu-Zn alloy, showing the two lines used. The points represent the alloys that were used in this work [8].

Every line is represented by an equation, so line one is represented by equation (25) and line two by equation (26).

$$X_{Zn1} = -1.9438(X_{Cu}) + 0.50334 \tag{25}$$

$$X_{Zn2} = -2.9823(X_{Cu}) + 0.97337 \tag{26}$$

where X_{Zn} and X_{Cu} are the atomic fractions of Zn and Cu respectively. The rest is Al, completing the 100%.

Sample	Cu		Zn		Al		Sample	Cu		Zn		Al	
	%p.	%at.	%p.	%at.	%p.	%at.		%p.	%at.	%p.	%at.	%p.	%at.
M1	5.00	3.54	63.18	43.44	31.82	53.02	M9	5.00	4.77	89.60	83.10	5.4	12.13
M2	10.00	6.87	55.31	36.96	34.70	56.17	M10	10.00	9.02	80.32	70.42	9.68	20.56
M3	15.00	10.03	47.43	30.82	37.57	59.15	M11	15.00	12.83	71.04	59.05	13.96	28.12
M4	20.00	13.01	39.56	25.02	40.44	61.97	M12	20.00	16.26	61.76	48.81	18.24	34.93
M5	25.00	15.84	31.69	19.52	43.31	64.64	M13	25.00	19.37	52.48	39.52	22.52	41.01
M6	30.00	18.53	23.82	14.30	46.19	67.17	M14	30.00	22.20	43.19	31.07	26.81	46.72
M7	35.00	21.08	15.94	9.33	49.06	69.59	M15	35.00	24.79	33.91	23.35	31.09	51.86
M8	40.00	23.51	8.07	4.61	51.93	71.88	M16	40.00	27.17	24.63	16.26	35.37	56.57

Table 1. Chemical percentage of each sample.

Each sample was prepared by the melting of pure elements at 750 °C under an argon atmosphere and then slow cooling. Alloys were homogenized at 350 °C for 180 h in order to eliminate the dendritic structure. All alloys were subsequently heated to 350 °C for 24 h and quenched

http://dx.doi.org/10.5772/64699

to 2 °C in ice water to retain the crystalline structure. Metallographic specimens were examined using optical and scanning electron microscopy (SEM). X-ray diffraction (XRD) analysis of samples was determined according to the standard procedure and hardness in Rockwell B.

4.3. Results and discussion

4.3.1. Phase diagrams using Thermo-Calc

Figure 8 shows the phase diagrams at different temperatures (20 °C to 400 °C) obtained using the Thermo-Calc program and shows each alloy. **Table 2** shows the phases obtained by this program for each alloy according to the temperature. When the temperature is between 25 °C to 200 °C, the phases are the same. Sample M11 shows the same phases (α, η y τ) between 20 °C to 400 °C. This explains the poor increase in hardness between different heat treatments. The rest of the samples present phase changes continuous with each heat treatment.

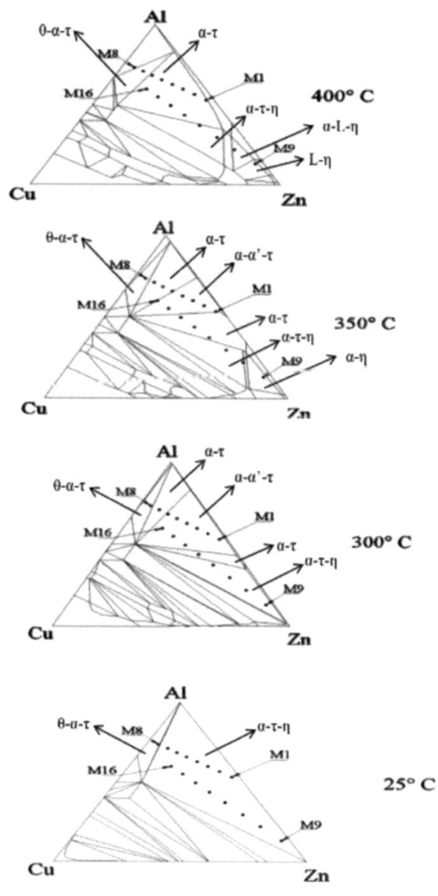

Figure 8. The results of simulations using Thermo-Calc [10].

Fases	Temperature (°C)	Muestra	Fases	Temperature (°C)	Muestra
τ, α	400	M1 – M6	L, η	400	M9
θ, α, τ	400	M7 – M8	L, α, η	400	M10
			τ, α, η	400	M11 – M12
			θ, α, τ	400	M13 – M16
τ, α, α′	350	M1 a M4	α, η	350	M9
τ, α	350	M5- M7	τ, α, η	350	M10 – M11
τ, α, θ	350	M8	τ, α	350	M12 – M14
			τ, α, α′	350	M15
			τ, α	350	M16
τ, α, α′	300	M1 – M5	τ, α, η	300	M9 – M12
τ, α	300	M6 – M7	τ, α	300	M13
τ, α, θ	300	M8	τ, α, α′	300	M14 – M16
τ, α, η	25	M1 – M7	τ, α, η	25	M9 – M16
τ, α, θ	25	M8			

Table 2. Phases present in each sample at specific temperatures according to Thermo-Calc.

Samples M1 to M8 show many changes in phase between 20 °C to 300 °C, compared to samples M10 and M11. Therefore, and as can be seen in the hardness results, samples M1 to M8 have an increase in the hardness of heat treatment to other.

The Thermo-Calc program also shows that the Börnstein's diagram is perfectly in line with these observations.

4.3.2. Structural and microstructural characterization of the alloys

The diffraction patterns of samples M1 to M8, and M9 to M16 are presented in **Figures 9** and **10** respectively.

Figure 10 shows that α, η, ε and τ phases are present in nearly all as-cast alloys. "The increase in Cu content was observed to be related to the increase in the intensity of X–ray diffraction peaks corresponding to the θ phase. This phase may be formed by the eutectic reaction $L \rightarrow \alpha + \theta$ located in the Al-Cu rich side. In contrast, the peak intensity of the Zn-rich η phase also increases with an increase in Zn content. This fact suggests the increase in the volume fraction of this phase. The amount of α and ε phases showed no clear tendency to increase or decrease the contents of either Al or Zn. A low volume fraction of the β phase is present in the as-cast M1 to M7, M9, and M14 to M16 alloys. This presence seems to indicate that the chemical compositions of alloys on line 1 are closer to the β phase field at high temperatures than the alloy compositions on line 2. Likewise, the τ′ phase is, in general, more stable as the Cu content increases. This suggests that the four-phase reaction, $\alpha + \varepsilon \rightarrow \eta + \tau'$, took place during the cooling of these alloy compositions"[11].

http://dx.doi.org/10.5772/64699

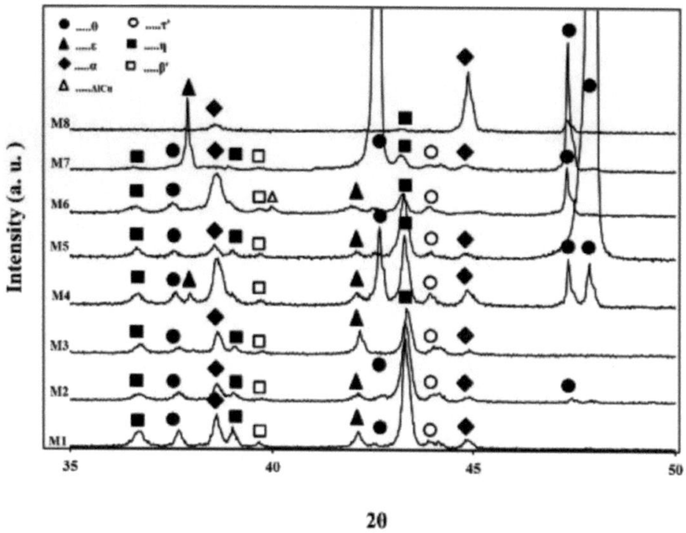

Figure 9. Diffraction pattern in as-cast alloys of samples M1 to M8.

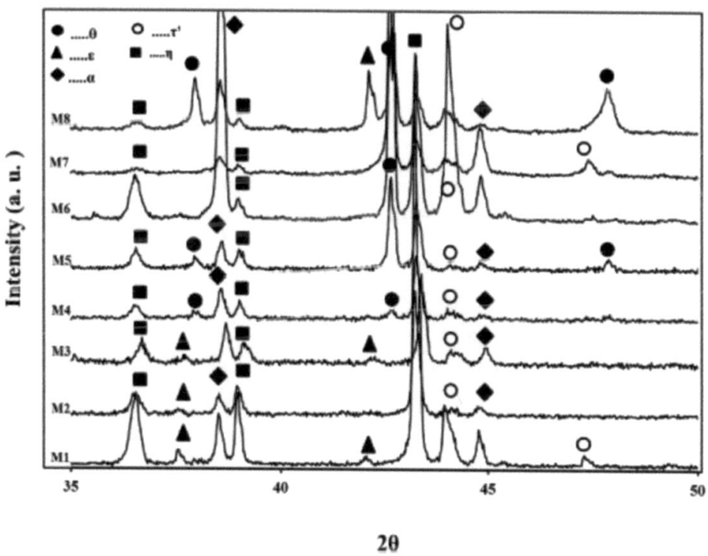

Figure 10. Diffraction pattern in as-cast alloys of samples M9 to M16.

Micrographs taken by SEM of samples M2, M7, M10 and M15 are shown in **Figure 11**.

The micrographs show that there is an increases in the percentage by volume of the θ phase with respect to an increase in the quantity of Cu and Al. All samples with a low Cu and Al content show a very heterogeneous microstructure, while samples with a high content of these elements show a uniform appearance.

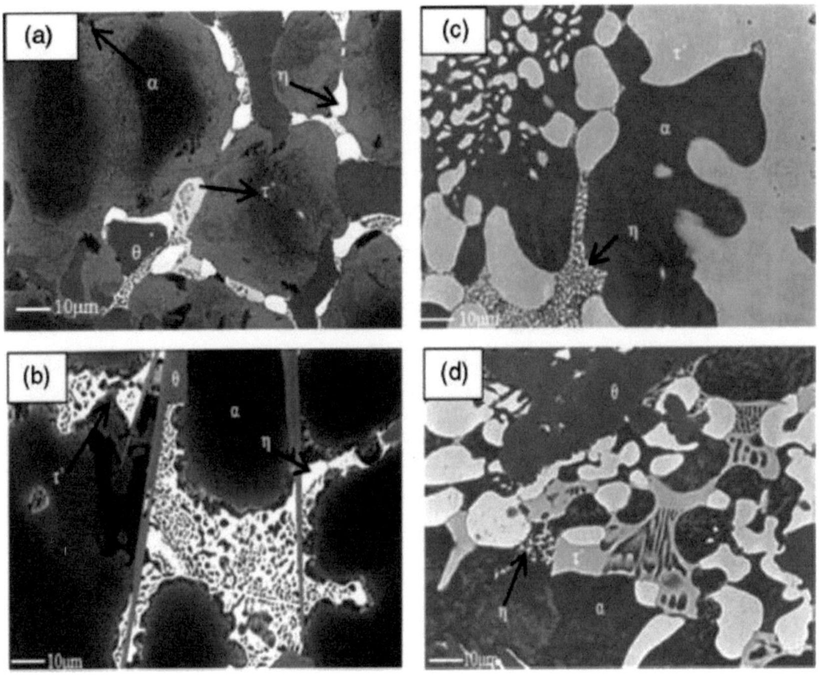

Figure 11. SEM Micrographs of different as-cast samples, (a) sample M2, (b) sample M7, (c) sample M10 and (d) sample M15.

Figures 12 and **13** show the diffraction patterns of the homogenized samples M1 to M16.

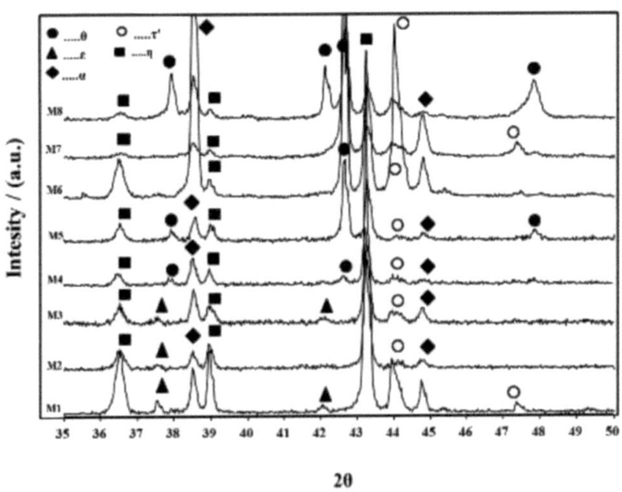

Figure 12. Diffraction pattern of the homogenized samples M1 to M8.

The ε phase appears in samples M1 to M3. This is important because the phase is present in the transformations of the four phases. Samples M4 to M8 do not have the ε phase, but they have the θ phase, which is more stable because it is not involved in the transformation of the four phases [12]. In the samples of series two, the ε phase is presented in samples M9 to M13, and the θ phase in samples M15 to M16 even though it should be emphasized that in these samples the β phase is present. It should be remembered that the β phase, in accordance with Aragon's investigations [13] occurs after quenching and then disappears after this to room temperature when Cu is present in small proportions. However, when the content of Cu is enhanced, the β phase is retained and is harder to eliminate.

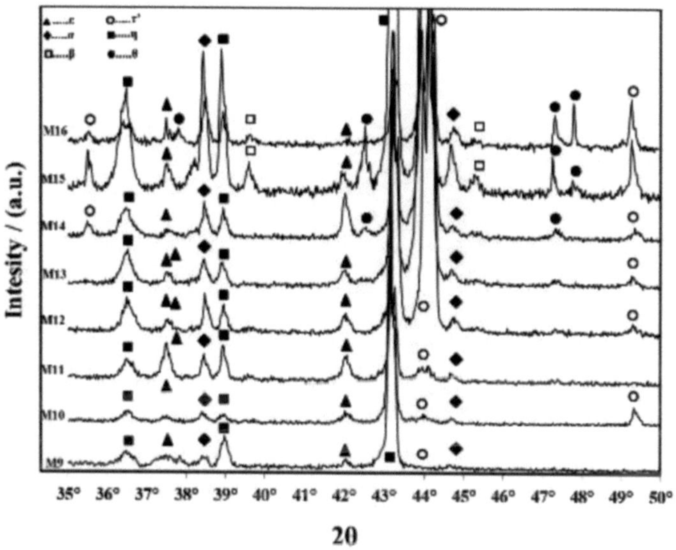

Figure 13. Diffraction pattern of the homogenized samples M9 to M16.

It can also be observed that an enhanced volumetric of the θ phase causes an increase in the hardness of the samples.

Figure 14 (a) and **(b)** show the diffraction patterns after quenching. Samples M1 to M4, and M9 to M11 show the same phases as follow the process of homogenization. However, when comparing the diffraction patterns of homogenized and quenching alloys of these samples, an increase in the intensity of both τ and α phases can be observed. It should be noted that the samples are solid and not in powder form, therefore there is texture, but even so it is possible to observe an increase in the phases.

Figure 15 shows micrographs of samples M2 (a–b) after homogenized, and sample M8 (c–d) after quenching. We can see that in sample M2 there is no dendritic structure. Moreover, when we compare samples M2 and M8, it can be seen that an increase in the quantity of Al and Cu enhances the percentage of the θ phase.

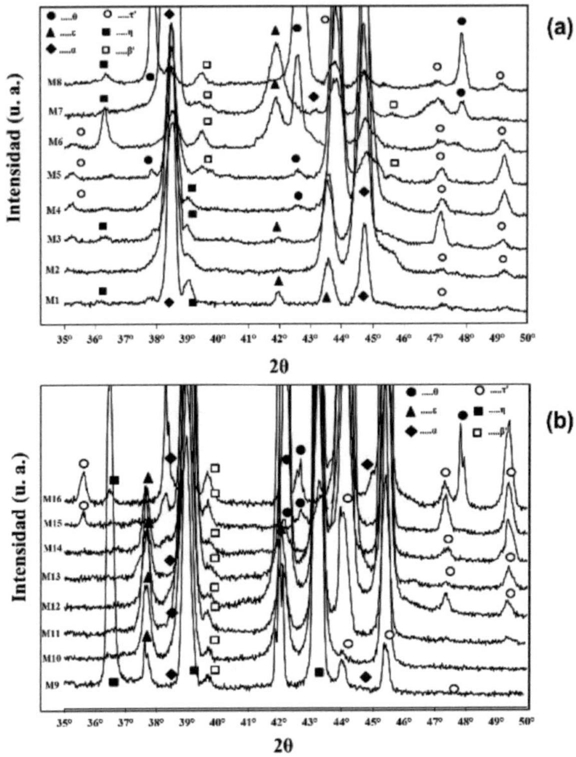

Figure 14. Diffraction patterns of samples M1 to M16 after quenching: (a) samples M1 to M8, (b) samples M9 to M16.

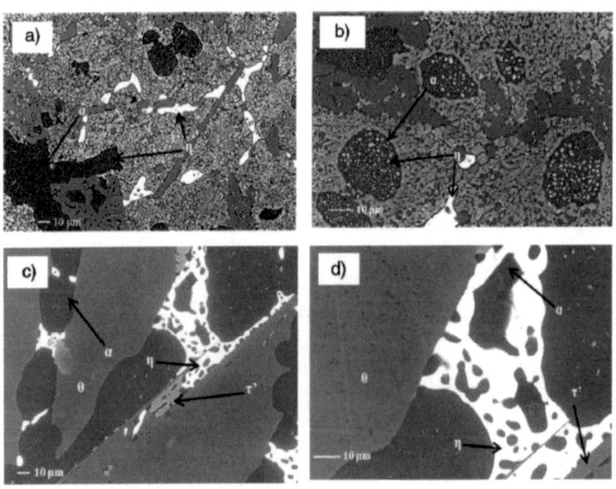

Figure 15. Micrographs of samples M2 and M8: (a) Sample M2 homogenized to 400X, (b) Sample M2 homogenized to 1000X, (c) Sample M8 quenching to 400X, (d) Sample M8 quenching to 1000X.

5. Hardness results

The hardness obtained from each of the samples, namely as-cast, homogenized and quenching, is shown in **Figure 16**. Each thermal process of the series 1 compound for samples M1 to M8, as well as the series 2 compound for samples M9 to M16, shows a linear trend.

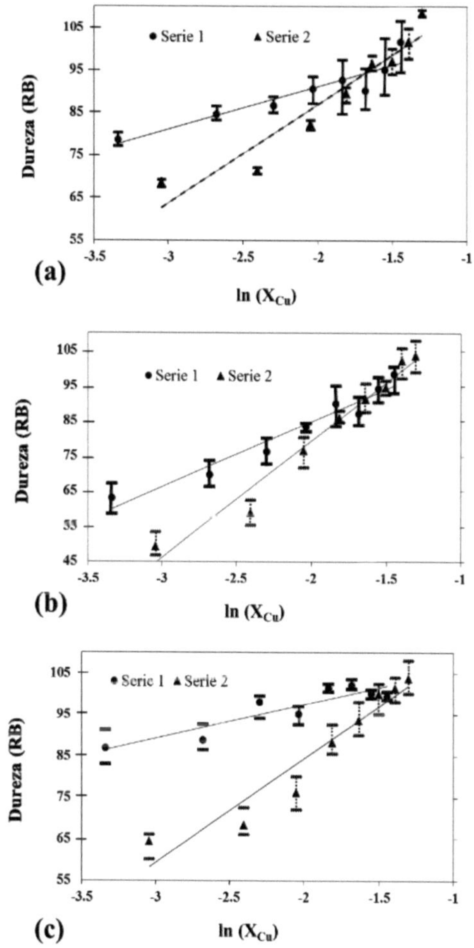

Figure 16. Graphic of the hardness of samples M1 to M16 subjected to different heat treatments, (a) as-cast; (b) homogenized and (c) quenching [10].

Using a least squares regression produces a correlation coefficient of between 0.82 and 0.97, indicating a good fit. The table shows the equations that relate the hardness to the chemical composition of each of the samples.

The equations in **Table 3** are used in order to provide an assessment of hardness according to chemical composition. In these equations, the value of only one of the three elements used in the alloy is present. As a result, it is necessary to use equations 27 and 28. If a certain hardness is desired in accordance to a certain heat treatment, then we use the equation relating to that heat treatment. For example, if you want to have 65 RB after a homogenized heat treatment, then the homogenized equation of **Table 3** should be used, but the series 1 and 2:

$$63 = 18.617 \ln X_{Cu} + 122.12 \quad \ldots \text{series 1}$$
$$63 = 33.326 \ln X_{Cu} + 146.05 \quad \ldots \text{series 2}$$

Treatment Thermic	Equation	R^2
Series 1		
As-Casting	$HB = 10.12 \ln X_{Cu} + 111.38$	0.8673
Homogenizad	$HB = 18.617 \ln X_{Cu} + 122.12$	0.9464
Quenching	$HB = 8.1666 \ln X_{Cu} + 113.81$	0.8197
Series 2		
As-Casting	$HB = 23.334 \ln X_{Cu} + 133.63$	0.9169
Homogenizad	$HB = 33.326 \ln X_{Cu} + 146.05$	0.9704
Quenching	$HB = 25.211 \ln X_{Cu} + 134.83$	0.9186

Table 3. Equations generated from linearly regressing each heat treatment with respective values of R^2.

To resolve X_{Cu}:

$$X_{Cu} = 0.04204 \quad \ldots \text{atomic fraction of series 1}$$
$$X_{Cu} = 0.08274 \quad \ldots \text{atomic fraction of series 2}$$

In order to know the atomic fractions of Zn and Al, equations (25) and (26) are used,

$$X_{Zu1} = -1.9438(0.04204) + 0.50334 = 0.4216 \quad \ldots \text{atomic fraction of Zn series 1}$$
$$X_{Zu2} = -2.9823(0.08274) + 0.97337 = 0.7266 \quad \ldots \text{atomic fraction of Zn series 2}$$

Therefore, in order to have an alloy with 63 RB, it is necessary to have 4.204 % at. Cu, 42.16% at. Zn and 56.63% of Al in the case of series 1.

However, it is also possible to obtain 63 RB by using 8.27% at. Cu, 72.66% at. Zn and 19.06% at. Al.

As can be seen, it is possible to obtain the same hardness, but with a different chemical composition. If we want to use less of the more expensive Cu, then there is a need to use an increased amount of Al, whereby we can have a product with the same mechanical properties but at a cheaper cost. For example, the international standard cost of Aluminum is 1.55 dollars/kg, the cost of Cu is 8.2 dollar/kg and that of Zn 2.07 dollars/kg. If we require an amount of 100 kg of the product, then the cost will be of 247.86 dollars using a larger quantity of Cu. If the same product is made but a larger quantity of Al and Zn is used, then the cost is 204.82 dollars, making a saving of 17.36% —almost a fifth of the total product.

6. Assessment of hardness

Using the equations it is possible to produce an assessment of the hardness of any alloy that is found in the line of either series 1 or 2. However, it is also possible to produce an assessment of the hardness of any alloy which is situated between the two series. It must be remembered that equations (25) and (26) provide the chemical composition of any alloys that are on either line. The difference between the lines of series 1 or 2 is the angle, and this is directly proportional to the slope of the hardness. For this reason, it is possible to relate the two. **Figure 17**

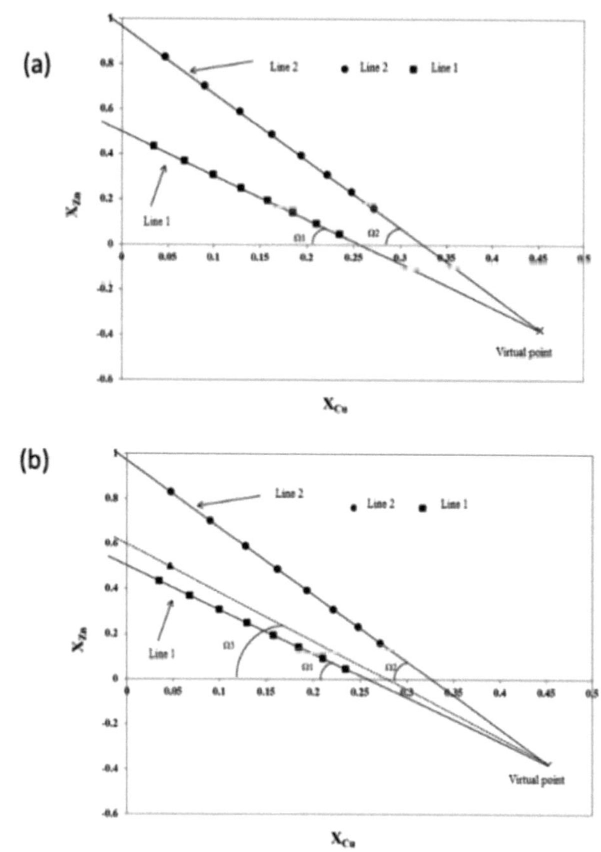

Figure 17. (a) Representation in cartesian coordinate of equations (27) and (28), (b) Representation of cartesian coordinate ZnAl27Cu3 [10].

shows a cartesian coordinate where "x" is X_{Cu} (the atomic fraction of Cu) and "y" is X_{Zn} (the atomic fraction of Zn) and the angle is represented by Ω_1 and Ω_2.

In order to determine the angle, it is necessary to use a virtual point that is the intersection of the two lines in **Figure 7**, or of equations (25) and (26). This point is virtual because the percentage of Zn is negative (-37.46 % at.), and for this reason it is impossible for this point to exist. However, the virtual point is useful in determining the angle of any alloy that exists between these two lines.

For example, we can consider the alloy ZnA127Cu3, which has 50.54% at. Zn; 2.23% at. Cu with the remaining percentage being Al [14]. The composition of ZnA127Cu3 is only a point in the Cartesian coordinate and this point exists between line 1 and 2. When this point is joined with the virtual point, there is an angle of 63.99° as seen in **Figure 17(b)**. As mentioned above, there is a direct relationship between the slopes, composition and hardness. This is because the percentage of each of the phases changes with the chemical composition. Therefore it is necessary to elucidate the relationship between the angle in the Cartesian coordinate in the chemical composition (**Figure 17**), and the angle in the graphic of the hardness vs $\ln X_{Cu}$. For example, the angle of line 1 in the cartesian coordinate of the composition has a value of 62.78°, but the angle in the hardness in as-cast, is 84.74° and the same is occurs with line 2. In the case of the alloy ZnA127Cu3, the angle of hardness is found by interpolation, as seen in **Table 4**.

Sample	**Angle (Percentage at. of each element)**	**Angle (Hardness in as-cast)**
Line 1	62.78°	84.74°
ZnA127Cu3	63.99°	85.13° by interpolation
Line 2	71.47°	87.55°

Table 4. Results of the angles of each of the series according to its composition and hardness in as-cast.

Once the slope of hardness in the ZnA127Cu3 alloy is present, the hardness intercept is then required. For this, is necessary to first obtain the hardness equation and then subsequently the hardness of the alloy according to the chemical composition. The intercept of the origin is obtained using the point of interception that exists between the equations of hardness. For example, the point of intersection in the equations in as-cast is 18.57% at. Cu and 94.34 RB. The point of intersection is shown in **Figure 16(a)**.

At this point, we have the slope, the hardness and the percentage Cu, therefore the equation is:

$$94.34 = \tan(85.13)\ln(0.1857) + b$$

Solve for b

$$b = 94.34 - \tan(85.13)\ln(0.1857) = 114.11$$

So the equations of the line that crosses for the ZnAl27Cu3 alloy is:

$$HB = \tan(85.13)\ln(X_{Cu}) + 114.11$$

If the atomic percentage of ZnAl27Cu3 alloy is 50.54% at. Zn; 2.23 % at. Cu and the remaining percentage being Al then the hardness is:

$$HB = \tan(85.13)\ln(0.023) + 114.11 = 69.45$$

The result obtained by Savaskan [14] was 67.15 RB with an error of only 5.43%. In this way, any alloy between the two lines can be assessed, as long as it has a heat treatment of as-cast, homogenized or quenching.

Tables 5 and **6** show the assessment obtained by this method for certain alloys made by Savaskan and Ciach [14, 15].

Sample	Savaskan et al. (2003)	Assessment	Error (%)
ZnAl27Cu1	64.74	59.94	7.41
ZnAl27Cu2	66.21	66.98	1.15
ZnAl27Cu3	67.15	71.04	5.48
ZnAl27Cu4	68.5	73.91	7.32
ZnAl27Cu5	70.75	76.12	7.05
		Error average =	5.68

Table 5. Comparative results of hardness as determined by Savaskan et al. (2003) and the assessment values.

Sample	Savaskan et al., (2003)	Assessment	Error (%)
AlZn78Cu1	73	66.99	7.41
AlZn78Cu2	78	73.12	1.15
AlZn78Cu3	81	76.53	5.48
		Error average =	6.67

Table 6. Comparative results of hardness as determined by Ciach et al. (1969) and the assessment values.

The same methodology can be used in any ternary alloy as long as there is a phase diagram, and this methodology can even be used with commercial alloys and the average error is only 5%. Therefore, it is possible to obtain alloys tailored to any ternary alloy, or to develop alloys cheaper or with better mechanical properties.

7. Conclusions

- The evidence from equations (20) and (21) shows a direct relationship between the bond energies of the atoms (Ω) and the mechanical properties in such a way that in future it will be possible to directly obtain the mechanical properties of bond energies.
- The θ phase increases the hardness, and this phase intensifies along with an increase of Al and Cu.
- The β phase along with ε phase causes a decrease in hardness.
- The β phase remains to room temperature with an increase of Cu.
- The simulation using Thermo-Calc enables a better understanding of the changes of phase shown in each of the samples.
- The methodology presented here has been shown to be very effective, and it is possible use this methodology with other alloys.
- It has also been shown that is possible to obtain alloys with similar mechanical properties, but at a cheaper cost.

Author details

Jose David Villegas Cárdenas[1*], Victor Manuel López Hirata[2], Carlos Camacho Olguin[1], Maribel L. Saucedo Muñoz[2] and Antonio de Ita de la Torre[3]

*Address all correspondence to: jdvc76@yahoo.com.mx

1 Universidad Politécnica del Valle de México

2 Instituto Politécnico Naciona

3 Universidad Autónoma Metropolitana

References

[1] Porter, D., Easterling, K., Sherif, M., (2009) "Phase transformations in metals and alloys", Ed. CRC Press, 3er edic. E.U., 1 – 105.

[2] Konrad, H. (2011), "Harness testing: principles and applications", Ed. ASM international, isbn 1615038329, 978-1-61503-832-9, 1er edic, E.U. 50 – 250.

[3] Alcala, J., Giannakopoulos, A. E., & Suresh, S. (1998). Continuous measurements of load-penetration curves with spherical microindenters and the estimation of mechanical properties. Journal of Materials Research, 13(05), 1390-1400. 1997.

[4] Alcalá, J., (2000), "Instrumented micro-indentation of zirconia ceramics", J. Am. Ceram. Soc.,No. 83, p. 1977-1984.

[5] Suresh, S. Giannakopoulos, A., Alcalá, J. (1997),"Spherical indentation of compositionally graded materials: theory and experiments", Acta Mater., No. 45, p. 1307-1321.

[6] Alcalá, J., Gaudette, F., Suresh, S., Sampath, S., (2001) "Instrumented spherical micro-indentation of plasma-sprayed coatings", Mater. Sci. & Engn. A, No. 316, p. 1-10.

[7] Dao, M., Chollacoop, N., Van Vliet, K., Venkatesh T., Suresh, S., (2001) "Computational modeling of the forward and reverse problems in instrumented sharp indentation", Acta Mater., No. 49, p. 3899-3918.

[8] Bornstein, L. (2005). Ternary Alloy System Phase Diagrams, Crystallographic and Thermodynamic Data, Ed. Springer, Vol. 1, 11, New York, USA, pp. 183–205.

[9] Villegas-Cardenas, J. D., Lopez-Hirata, V. M., Torre, A. D. I. D. L., & Saucedo-Munoz, M. L. (2011). Assessment of hardness in as-cast and homogenized Zn-Al-Cu alloys. Materials Transactions, 52(8), 1581-1584.

[10] Villegas J. D., Saucedo M. L., Lopez V. M., Dorantes H. J., "Predicción de la dureza de aleaciones Al-Cu-Zn en estado de colada y templado", Revista de Metalurgia, Vol. 50, No. 2, 2014. DOI: 10.3989/revmetalm.015

[11] Villegas J. D., Saucedo M. L., Lopez V. M., Dorantes H. J., "Effect of phase transformations on hardness in Zn-Al-Cu alloys". Materials Research, V17, No. 5. DOI:http://dx.doi.org/10.1590/1516-1439.228913.

[12] Savaskan, T., Heikimoğlu, P., PURcEK, G. (2004). Effect of copper content on the mechanical and sliding wear properties of monotectoid-based zinc-aluminium-copper alloys. Tribol. Int. 37 (1), 45–50. http://dx.doi.org/10.1016/S0301-679X(03) 00113-0.

[13] Aragon, J.A., Miranda, J.R., Garcia, A. (2007b). Obtencion de una microestructura nueva en la aleacion Zn -40% at. Al-1,5 % at. Cu. Rev. Mex. Fís. 53 (3), 149–158. http://rmf.smf.mx/pdf/rmf/53/3/53_3_149.pdf.

[14] Savaskan, T., Purcek, G., Heikimoğlu, P. (2003). Effect of copper on the mechanical and tribological properties of ZnAl27 based alloys. Tribol. Lett. 15 (3), 257–263.http://dx.doi.org/10.1023/A:1024817304351.

[15] Ciach, R., Krol, J., Wegryn, K. (1969). Studies on four phases transformation in AlZn78 alloy containing 1–3 per cent of copper. Bulletin de L'Academie Polonaise des Sciences 17 (4), 371–378.

Chapter 5

Differential Speed Rolling: A New Method for a Fabrication of Metallic Sheets with Enhanced Mechanical Properties

Wojciech Polkowski

Additional information is available at the end of the chapter

http://dx.doi.org/10.5772/64418

Abstract

This chapter presents an actual progress in the usage of a new processing method called differential speed rolling (DSR) to a fabrication of metallic sheet materials with enhanced mechanical properties. An introduction of the rolls speed differentiation to a rolling process results with qualitative and quantitative changes of imposed strain and thus to new structural effects as compared to a conventional rolling. The presence of additional high through-thickness shear strain in the DSR technique is utilized to a substantial grain refinement (to a fabrication of high-strength ultrafine-grained materials) and therefore is regarded as one of the severe plastic deformation (SPD) methods. In this chapter, mechanical properties of selected DSRed metals are compared to those of their counterparts processed by competitive hydrostatic SPD methods. Moreover, the imposed complex strain state in the DSR method significantly affects a crystallographic texture of fabricated sheet materials leading to more favorable anisotropic characteristics and to an improved formability that is especially important in the case of aluminum and magnesium alloys.

Keywords: Sheet forming, Ultrafine-grained materials, Severe plastic deformation, Differential speed rolling, Formability

1. Introduction

A fabrication of high-strength metals and alloys is an emerging field of materials science and engineering originating from continuously increased demands of industry, leading to a formation of linkage between various aspects of mechanical metallurgy, the physics or material's mechanics.

According to a well-known Hall-Petch relationship, a grain refinement through a properly conducted thermomechanical processing is one of the most efficient ways to improve a mechanical strength of metallic materials. It has been already established that the best compromise between a high-strength and an acceptable ductility is generally achieved when a grain size is in the submicron (ultrafine) regime of 100–1,000 nm. Furthermore, a large fraction of high-angle grain boundaries (HAGBs) in a material's volume is a crucial feature supporting a kinetics of diffusion-related phenomena and resulting with, e.g., an improved environmental resistance of ultrafine-grained (UFG) materials, as compared to their counterparts with a coarse-grained structure. Processing methods involving a severe plastic deformation (SPD) have been already recognized as the most efficient and thus industrially preferable techniques of the UFG material fabrication. In SPD processes, the material is subjected to a very large plastic deformation (true strain ε is even greater than 80) usually being conducted under a hydrostatic pressure and room temperature conditions. As a consequence of the imposed intense plastic straining, a mechanical fragmentation of material grains by introducing a number of mutually intersected shear bands or a formation of fine-grained structure due to an activation of structure restoration processes (i.e., a dynamic or post-dynamic recovery and recrystallization) takes place. So far, a number of hydrostatic SPD methods such as cyclic extrusion compression (CEC), equal-channel angular pressing (ECAP), and high-pressure torsion (HPT) have been developed and implemented. However, despite of a successful fabrication of various UFG materials, these SPD methods also exhibit some serious disadvantages, e.g., a poor process efficiency, small dimensions of produced (semi-) products, or a necessity of using specialized machines and tools. Therefore, a development of SPD methods that are based on highly efficient, continuous plastic deformation processes conducted on widely available conventional processing devices has generated a considerable research interest.

2. Differential speed rolling (DSR): a new continuous severe plastic deformation method

A differential speed rolling (DSR) is a modification of the rolling process which involves a deformation with different values of a rotational speed of the upper and the lower rolls. This kind of processing belongs to the group of asymmetric rolling processes that have been already introduced to a large-scale production of flat steel products [1]. The asymmetry is introduced to the rolling process by using different diameters, different materials (that generate a differentiation of friction conditions on upper and lower surface of a deformed sample), or different rotational speed of working rolls (**Figure 1**). From a standpoint of design simplicity, the easiest and the best solution is the differentiation of rolls speed. In the case of the DSR process, this modification gives an unequal rolling velocity imposed to upper and lower surface of the processed sample. The main characteristic of the DSR method is a value of a rolls speed differentiation coefficient R defined as a ratio of the upper to lower rolls speed.

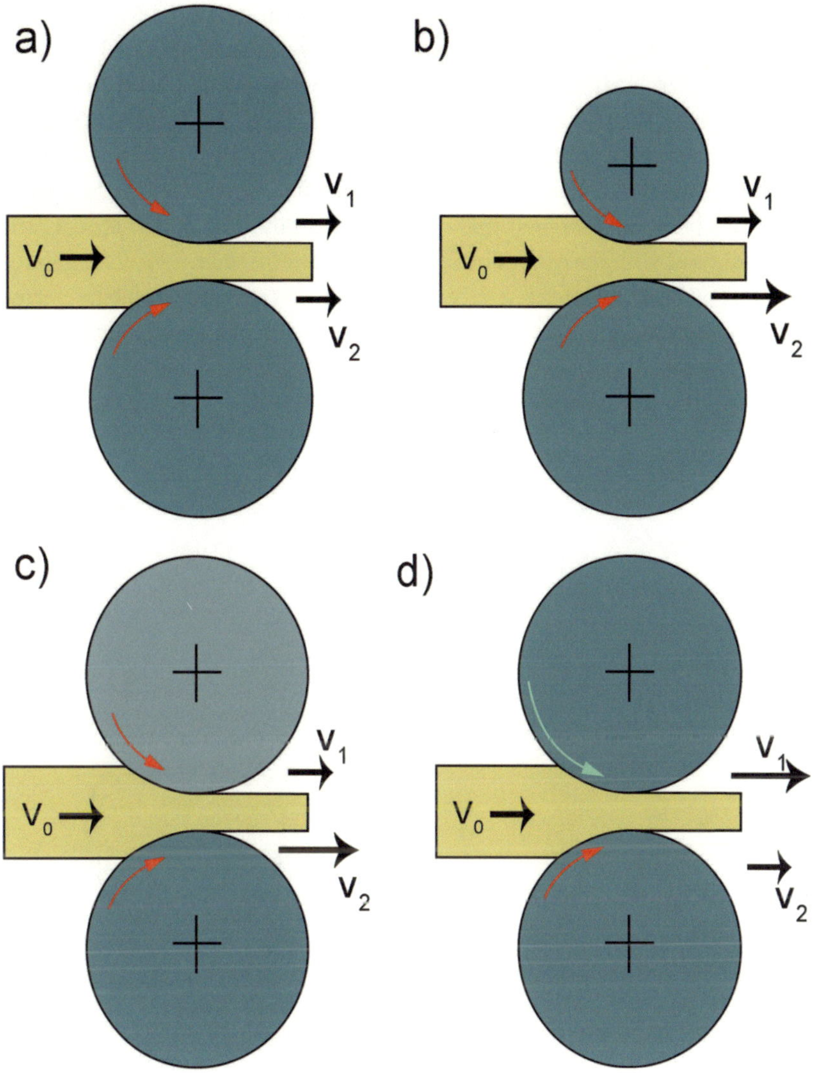

Figure 1. A schematic drawing of (a) a normal rolling process and different variants of the asymmetric rolling (b) with an unequal rolls diameter (differential diameter rolling), (c) with a different rolls materials (a differential friction rolling), and (d) with a differential speed rolling.

The first theoretical description of such a process was proposed in the 1940s of the last century [2], and a further development of this method was mainly devoted to the improvement of technological aspects of the rolling process. Between 1960s and 1980s, many experiments and theoretical calculations have been carried out on the roll forces, roll torques, and rolled product shape alteration due to the existed asymmetry of rolling gap [3–6]. It has been generally established that the asymmetry of the roll gap may be utilized as a factor improving work of the hydraulic gauge control in a plate-rolling mill leading to, e.g., a prominent decrease of the rolling force and torque and improvement of a rolled strip shape. Additionally, a lower rolling force and torque are not only beneficial in terms of providing a lower wearing of tools and a higher rigidity of the rolling cage but also give a better control and make a process more applicable to produce thin sheets or foils [7]. It was shown by Dyja et al. [8] on the real example of a plate mill in the steel plant, Częstochowa (Poland), that the introduction of rolls speed asymmetry to the rolling of eight different carbon-manganese steels leads to decreasing of the rolling force in the last finishing passes up to more than 70 %, depending on the applied process parameters. It has been documented that the implementation of asymmetrical rolling in the rolling mills equipped with the hydraulically controlled adjustment of rolling gap results with an improvement of plate geometry, e.g., a flatness and transverse profile or decrease of the thickness differentiation along the plate. These results are in line with those presented by Kawalek et al. [9] indicating that a lower value of the unit press decreases an elastic deflection of a rolling stand upon an asymmetric rolling pass.

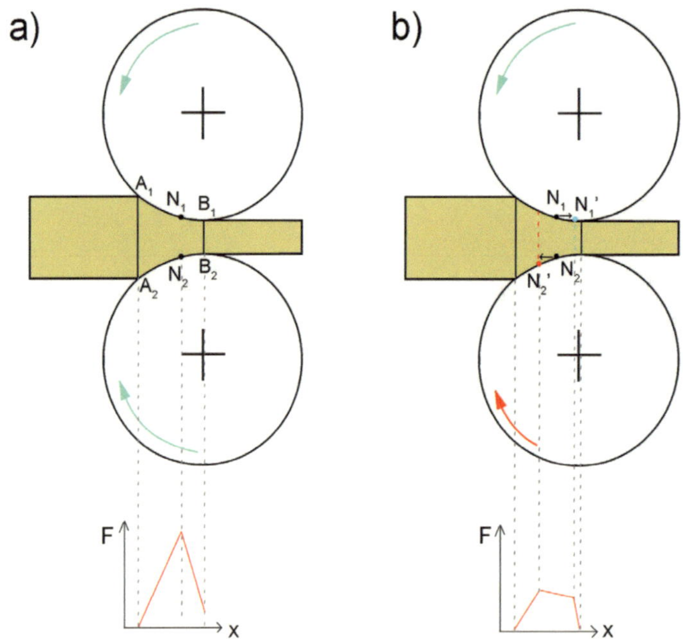

Figure 2. A rolling gap geometry in (a) the equal speed rolling and (b) the differential speed rolling. In the case of the DSR process, a shear zone is located between neutral points (N1':N2') shifted to different positions (based on [10]).

This positive effect of the DSR implementation on technological aspects of a rolling process is attributed to a change in the deformation geometry. Roumina and Sinclair [10] reported that the differentiation of rolls speed results with a shifting of so-called neutral points (the position where the sheet velocity equals the roll velocity) on upper and lower surfaces of the sample. The neutral point associated with the slow roll is shifted toward the entrance of the roll gap, while the neutral point associated with the fast roll is moved toward the exit of the roll gap (**Figure 2**). This situation leads to both a different distribution of rolling pressure (and thus lowering of the rolling force) and an imposition of a high through-thickness shear strain to the material. It was also confirmed by Tian et al. [11] that the extent of cross shear region increases with the increase of the speed ratio, whereas the rolling force decreases.

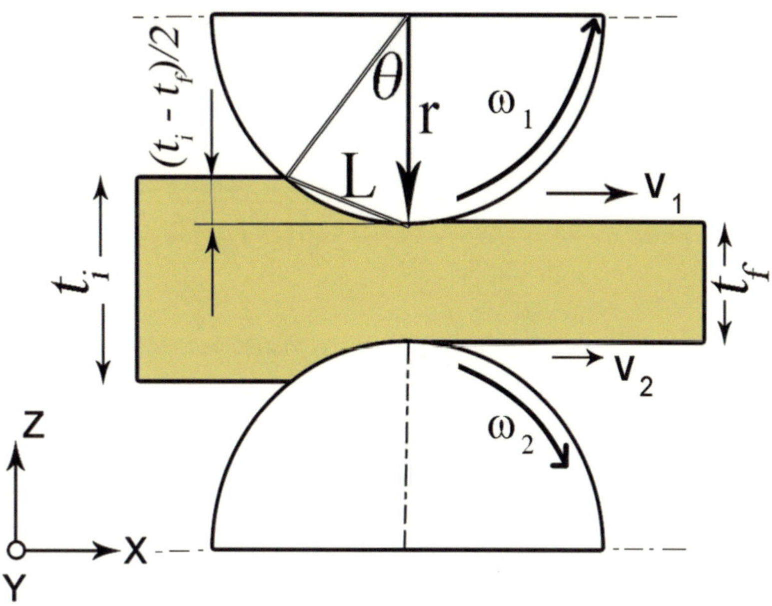

Figure 3. A schematic drawing that is used for strain and strain rate assessment in the DSR process (based on [12]).

Based on the assumption that the deformation gradient in the DSR method is approximated by a superposition of a plane strain (which is specific for the normal rolling process) and a simple shear in the rolling direction (Eq. (1)), Ko et al. [12] proposed approximate equations for the strain and strain rate imposed by DSR (Eq. (6)). This evaluation is based on trigonometric relationships between a geometry of a sample and a rolling gap schematically shown in **Figure 3**. However, it is worth noted that in this attempt the friction between a sample and the rolls is neglected; thus, calculated strain values may be underestimated:

$$\begin{bmatrix} \varepsilon_{xx} & 0 & 0 \\ 0 & 0 & 0 \\ 0 & 0 & \varepsilon_{zz} = -\varepsilon_{xx} \end{bmatrix} + \begin{bmatrix} 0 & 0 & \varepsilon_{xz} \\ 0 & 0 & 0 \\ 0 & 0 & 0 \end{bmatrix} = \begin{bmatrix} \varepsilon_{xx} & 0 & \varepsilon_{xz} \\ 0 & 0 & 0 \\ 0 & 0 & \varepsilon_{zz} = -\varepsilon_{xx} \end{bmatrix} \tag{1}$$

By taking into assumption that θ is a central arc angle defined by positions of inlet and outlet points of the rolling gap, t_i and t_f are an initial and a final thickness of the sample, and r is the rolls radius, the following calculation procedure has been proposed.

The θ value may be calculated by using an assumption that for satisfied contact conditions the angle value is closed to its sinus, so

$$\sin\theta \approx \theta = \frac{\sqrt{r\left(t_i - t_f\right) - \left(\frac{t_i - t_f}{2}\right)^2}}{r} \tag{2}$$

Therefore, the contact arc length L is equal to

$$L = r \times \theta = r \times \frac{\sqrt{r\left(t_i - t_f\right) - \left(\frac{t_i - t_f}{2}\right)^2}}{r} = \sqrt{r\left(t_i - t_f\right) - \left(\frac{t_i - t_f}{2}\right)^2} \tag{3}$$

Subsequently, the average deformation time t_{ave} is calculated as follows:

$$t_{ave} = \frac{L}{v_{ave}} \tag{4}$$

where v_{ave} is an average rolls velocity. By having calculated the t_{ave} value, the simple shear component ε_{xz} may be defined as

$$\varepsilon_{xz} = \left(\frac{\left(v_G - v_D\right) \cdot t_{ave}}{\left(\frac{t_i + t_f}{2}\right)} \right) \cdot 1/2 \tag{5}$$

Finally, the total equivalent strain is expressed as

$$\varepsilon = \frac{2}{\sqrt{3}} \cdot \sqrt{\left(\varepsilon_{xx}^2 + \varepsilon_{xz}^2\right)} \tag{6}$$

where ε_{xx} is a true logarithmic strain equal to $\ln(t_f/t_i)$.

It should be underlined that in the "technological" attempt to the DSR process, the introduced asymmetry is rather low (usually the R value is not greater than 1.2) and thus is treated as a "correction coefficient," whereas, since the end of 1990s, a new growing direction in the research on the DSR process has been started. In these years, first works on a possible utilization of a shear deformation imposed upon the DSR process to control a microstructure and property evolution of various engineering materials were published [13, 14] starting a "material approach" to this subject. While the shear strain is found to be proportional to the rolls speed mismatch, it is assumed that extensive microstructure changes need both a large deformation and a high ratio of rolls speed asymmetry (the *R* values are usually higher than 2).

Generally, there are two main purposes for using the DSR method in a material processing. The first one is to produce high-strength materials by a grain refinement through a high-plastic strain accumulation—due to the presence of additional shear deformation, the DSR method is considered as one of the SPD techniques. The second one is to control a deformation texture that affects anisotropy of mechanical properties (and also determines a recrystallization texture formed upon a subsequent annealing). Obviously, these two purposes are in many cases simultaneously achieved.

Material	Processing		UTS [MPa]	Reference
Pure copper	ECAP (1 cycle)		344	[16]
	ECAP (4 cycles)		413	[17]
	ECAP (8 cycles)		378	[18]
	ECAP (8 cycles)		386	[19]
	HPT (5 cycles)		445	[20]
	ARB	(1 cycle)	290	[21]
		(2 cycles)	350	
		(3 cycles)	370	
		(4 cycles)	380	
		(5 cycles)	388	
		(6 cycles)	395	
	As-annealed		200	[15]
	Equal speed rolling (1-pass reduction of 65 %)		350	
	DSR (R = 3, 1-pass reduction of 65 %)		**470**	

Table 1. A comparison of ultimate tensile strength (UTS) of pure copper processed by various SPD methods, an equal speed rolling and the DSR technique.

The grain refinement effect by the DSR method was previously observed in numerous pure metals and alloys. Kim et al. [15] reported that in oxygen-free copper, submicron grain size of 820 nm is obtained after the 65 % thickness reduction in a single rolling pass by the DSR method (the *R* = 3). This processing also leads to a formation of a large fraction of high-angle grain boundaries (HAGBs) (~60 %) and maintaining a high electrical conductivity (that proves a low level of structure defects). Furthermore, reported results of tensile tests show that the DSR-deformed copper exhibits a superior ultimate tensile strength (UTS) than that of copper subjected to the equal speed rolling, accumulative roll bonding (ARB), or hydrostatic SPD methods (the HPT and an equal-channel angular pressing (ECAP)) (**Table 1**). These results clearly point toward a high efficiency of the DSR process—prominently higher strength than in the case of other competitive processing methods in a large bulk material is achieved in one simple operation.

Similar findings were shown by Jiang et al. [22] on pure aluminum subjected to the DSR process. The authors found that the cold rolling (with the $R = 3$) of commercially pure aluminum to 90 % of thickness reduction leads to a formation of microstructure composed of submicron-equiaxed grains and a high fraction of HAGBs (~50 %). As in the case of the copper, the DSR-processed pure aluminum exhibits a higher strength than that processed by a normal rolling and competitive SPD-based fabrication methods (**Table 2**), confirming a superiority of this process over hydrostatic techniques.

Material	Processing		UTS [MPa]	Reference
Commercially pure aluminum	ECAP	(1 cycle)	120	[23]
		(2 cycles)	130	
	ECAP	(1 cycle)	110	[24]
		(2 cycles)	135	
		(8 cycles)	165	
	Repetitive tube expansion and shrinking (RTES) (1 cycle)		140	[25]
	Rotatory swaging (1 cycle—true reduction of 3)		163	[26]
	Constrained groove pressing (CGP) (2 cycles)		105	[27]
	HPT (2 cycles)		145 (estimated from reported hardness)	[28]
	Equal speed rolling (total thickness reduction of 90 %)		150	[22]
	DSR (R = 3, total thickness reduction of 90 %)		**250**	

Table 2. A comparison of ultimate tensile strength (UTS) of commercially pure aluminum processed by various SPD methods, an equal speed rolling and the DSR technique.

Some works were also devoted to a fabrication of high-strength sheets made of ultrafine-grained titanium via the DSR process. Kim et al. [29] documented that this purpose may be successfully achieved by an effective grain refinement (the grain size in the range of 100–300 nm) through the differential speed rolling (with the $R = 3$) to 63 % of the thickness reduction obtained in one single rolling pass at room temperature. Consequently, as-received titanium was characterized by the UTS of 895–915 MPa, which is a better result than that represented by pure Ti processed by other SPD methods (**Table 3**). Moreover, the same authors showed in the further work [30] that a microstructure state formed upon the proposed DSR processing ensures an enhanced corrosion resistance in acid environments (H_2SO_4 and HCl solutions) by altering growth kinetics of passivating oxide film. Since a formation of continuous, passive surface layer is a key factor in a biomedical usage of titanium, the DSR method shows a good usefulness in the field of various human healthcare applications.

Material	Processing		UTS [MPa]	Reference
Commercially pure titanium	ECAP (1 cycle)		780	[31]
	ECAP	1 cycle	571	[32]
		3 cycles	624	
		5 cycles	665	
	ARB (6 cycles)		892	[33]
	HPT (4 cycles)		870	[34]
	DSR (R = 3, 1-pass reduction of 63 %)		**895–915**	[29]

Table 3. A comparison of ultimate tensile strength (UTS) of commercially pure titanium processed by various SPD methods and the DSR technique.

The strengthening of metals upon the DSR deformation is related to the structure refinement by a formation of narrowly spaced shear bands distributed homogeneously over the entire section of the sheet and a high temperature rise during the rolling. It was presented by Kim et al. [15] that the temperature rise may exceed a value of 284 K upon the cold rolling (with the $R = 3$) of pure copper to 65 % under a non-lubricated condition. Recent results of a more detailed experimental study by Megantoro et al. [35] confirm that the temperature rise increases with increasing either the thickness reduction or the rolls speed ratio (showing a near-linear relationship). Consequently, these conditions allow activating dynamic structure restoration phenomena (namely, continuous or discontinuous dynamic recovery or recrystallization) leading to the formation of thermally stable (up to a certain temperature [36]) microstructure composed of dislocation-free volumes (grains, subgrains) with a size in the submicron range.

The second purpose for using the DSR processing—the control of a deformation texture that affects anisotropy of mechanical properties—is especially important in the case of aluminum and magnesium alloys. These materials mainly due to their very good strength to weight ratio are considered as candidates in many car and plane body applications. However, the main drawback of these materials is their lower formability (namely, a susceptibility to deep drawing) than conventional low carbon steels. Aluminum and aluminum alloys are well known for their high-mechanical property anisotropy (a so-called earing behavior) upon a deep drawing process. The presence of this kind of shape defects of processed components generates a necessity for using additional operations and a waste of large quantity of the material. It was recognized that the main determinant of such behavior is a {1 0 0}<1 0 0> cubic crystallographic texture formed in a fully annealed state. On the other hand, it was proposed by Lequeu and Jonas [37] that formation of the undesired {1 0 0}<1 0 0> recrystallization texture component may be prominently inhibited through an application of shear deformation prior a heat treatment. Therefore, a number of works have been devoted to a development of asymmetric rolling-based processing techniques that allow for the fabrication of aluminum alloy sheets with enhanced formability.

Engler et al. [38] reported that the most efficient method of the formability improvement is to introduce {1 1 1} textures (composed of crystallographic orientations that are characterized by {1 1 1} crystallographic planes parallel to a rolling plane). Since these orientations are normally found in bcc metals and alloys (and are responsible for an excellent drawability of low carbon steels), in the case of *fcc* metals, they may be produced only by shearing. Jin and Lloyd [39] proved that the recrystallization texture of AA5754 aluminum alloy is randomized (the {0 0 1}<1 0 0> component is prominently reduced), when a high-ratio (R = 1.5 and R = 2) asymmetric rolling is applied before the annealing treatment. The {1 1 1} shear deformation texture is maintained in the material after annealing that allows lowering a so-called planar anisotropy (that characterizes an alteration of mechanical properties in different directions lying in the rolling plane). Analogous results were also documented by Sakai et al. [40] who showed that 5052 aluminum alloy cold deformed to 75 % of thickness reduction in a two-pass asymmetric rolling process followed by recrystallization annealing at temperature of 310–460 °C exhibits almost perfectly isotropic mechanical behavior (values of the planar anisotropy coefficient were reduced to near zero). Similar findings on 5251 aluminum alloy were also presented by Polkowski and Jóźwik [41].

While magnesium alloys (especially these containing Al and Zn additions—a so-called AZ series) are in many fields superior to aluminum alloys (e.g., possess a lower density and thus a better specific strength), their problematic formability concerns even a greater attention. Over a last few years, a number of scientific works have been devoted to a fabrication of Mg alloy sheets with a good drawability. It was proposed that the main reason for a very poor cold formability and a high mechanical anisotropy is an induction of a strong {0 0 0 1} basal texture in conventional plastic-forming processing [42] due to limited number of slip systems in hexagonal close-packed (hcp) crystal structure [43]. Results of an extensive study on various Mg alloys, e.g., AZ31 [44–47], AZ91 [48], AM31 [49, 50], or ZK60 [51, 52] alloys, showed that the DSR has a great impact on the intensity of the basal texture and plasticity of these materials. Generally, it was established that increasing the shear deformation by raising either the rolls speed ratio or a rolling reduction leads to weakening of the basal texture through facilitating the activation of prismatic slip during deformation. The basal texture weakening effect at high speed ratios is attributed to extensive tension twinning that occurred in the basal-oriented matrix, which in turn is exceptionally found in a conventional rolling process. Consequently, the DSR-fabricated Mg alloy sheets are characterized not only by more isotropic properties but also by the enhanced plasticity combined with exceptionally high strength (that is related to the simultaneous structure refinement [53–55]). Therefore, the DSR process is considered to be one of the most efficient techniques for processing these materials.

A newly proposed interesting usage of the DSR process is a fabrication of composite materials via a powder compaction [56] or an improvement of properties of these materials in the additional step of a manufacturing process. It was recently reported by Yoo et al. [57] on the example of carbon nanotube/copper and by Kim et al. [58] in a study on TiC/aluminum metal matrix composites that the large amount of redundant shear strain induced during the DSR significantly facilitates the dispersion of the reinforcement (through breaking up their clusters), having also a positive impact on mechanical properties of processed materials. It is believed that this research direction will become more and more important in a near future.

By summarizing, it should be again underlined that the DSR process exhibits a great potential in a large-scale fabrication of bulk metal components with enhanced mechanical properties and formability. The high imposed shear strain leads to an extensive structural evolution involving a grain refinement, affecting a crystallographic texture and a distribution of "second-phase" particles. What is very important, resulting mechanical properties of DSRed materials are much better than those of conventionally cold-rolled materials and at least not worse than those of counterparts subjected to hydrostatic SPD methods while having an undeniable advantage of a better efficiency in terms of a larger quantity of processed material and a lower number of needed operations.

Author details

Wojciech Polkowski

Address all correspondence to: wpolkowski@wat.edu.pl

Department of Advanced Materials and Technologies, Faculty of Advanced Technologies and Chemistry, Military University of Technology, Warsaw, Poland

References

[1] A. Nilsson, I. Salvator, P.D. Putz. Using asymmetrical rolling for increased production and improved material properties. Report of Research Programme of the Research Fund for Coal and Steel, RFSR-CT-2006-00011, 2006–2009. 2012. pp. 5–27 DOI: 10.2777/14782

[2] E. Siebel. The theory of rolling process with separately driven rolls. Archive for the Iron and Steel Industry. 1941;15:125–128.

[3] W. Johnson, G. Needham. Further experiments in asymmetrical rolling. International Journal of Mechanical Sciences. 1966;8:443–455. DOI: 10.1016/0020-7403(66)90014-2

[4] P. Dewhurst, I.F. Collins, W. Johnson. A theoretical and experimental investigation into asymmetrical hot rolling. International Journal of Mechanical Sciences. 1974;16:389–397. DOI: 10.1016/0020-7403(74)90013-7

[5] I.F. Collins. A slipline field analysis of asymmetrical hot rolling. International Journal of Mechanical Sciences. 1975;17:643–651. DOI: 10.1016/0020-7403(75)90082-X

[6] D. Pan, D.H. Sansome. An experimental study of the effect of roll-speed mismatch on the rolling load during the cold rolling of thin strip. Journal of Mechanical Working Technology. 1982;6:361–377. DOI: 10.1016/0378-3804(82)90034-1

[7] Y. Zuo, X. Fu, J.-Z. Cui, X.Y. Tang, L. Mao, L. Li, Q.-F. Zhu. Shear deformation and plate shape control of hot-rolled aluminium alloy thick plate prepared by asymmetric rolling process. Transactions of Nonferrous Metals Society of China. 2014;24:2220–2225. DOI: 10.1016/S1003-6326(14)63336-7

[8] H. Dyja, J. Markowski, D. Stoiniski. Asymmetry of the roll gap as a factor improving work of the hydraulic gauge control in the plate rolling mill. Journal of Materials Processing Technology. 1996;60:73–80. DOI: 10.1016/0924-0136(96)02310-2

[9] A. Kawalek, H. Dyja, S. Mróz, M. Knapinski. Effect of plate asymmetric rolling parameters on the change of the total unit pressure of roll. Metalurgija. 2011;50:163–166.

[10] R. Roumina, C.W. Sinclair. Deformation geometry and through-thickness strain gradients in asymmetric rolling. Metallurgical and Materials Transactions A. 2008;39A: 2495–2503. DOI: 10.1007/s11661-008-9582-6

[11] Y. Tian, Y.Ch. Guo, Z.-D. Wang, G.-D. Wang. Analysis of rolling pressure in asymmetrical rolling process by slab method. International Journal of Iron and Steel Research. 2009;16:22–26. DOI: 10.1016/S1006-706X(09)60055-8

[12] Y. G. Ko, J. Suharto, J. S. Lee, B. H. Park, and D. H. Shin. Effect of roll speed ratio on deformation characteristics of if steel subjected to differential speed rolling. Metals and Materials International. 2013;19:603–609. DOI: 10.1007/s12540-013-3033-7

[13] Ch. H. Choi, K.H. Kim, D. N. Lee. The effect of shear texture development on the formability in rolled aluminum alloys sheets. Materials Science Forum. 1998;273–275:391–396. DOI: 10.4028/www.scientific.net/MSF.273-275.391

[14] Q. Cui, K. Ohori. Grain refinement of high purity aluminium by asymmetric rolling. Materials Science and Technology. 2000;16:1095–1101. DOI: 10.1179/026708300101507019

[15] W.J. Kim, K.E. Lee, S.-H. Choi. Mechanical properties and microstructure of ultra fine-grained copper prepared by a high-speed-ratio differential speed rolling. Materials Science and Engineering A. 2009;506:71–79. DOI: 10.1016/j.msea.2008.11.029

[16] M.Z. Bian, Y.L. Li, M. Mathesh, D. Abreud, N.D. Nam. Microstructure and texture evolutions and mechanical properties in pure copper by equal-channel angular pressing. Journal of Alloys and Compounds. 2013;578:369–372. DOI: 10.1016/j.jallcom.2013.06.024

[17] S. Z. Han, M. Goto, Ch. Lim, Ch. Joo Kim, S. Kim. Fatigue behavior of nano-grained copper prepared by ECAP. Journal of Alloys and Compounds. 2007;434–435:304–306. DOI: 10.1016/j.jallcom.2006.08.179

[18] C.F. Zhu, F.P. Du, Q.Y. Jiao, X.M. Wang, A.Y. Chen, F. Liu, D. Pan. Microstructure and strength of pure Cu with large grains processed by equal channel angular pressing. Materials and Design. 2013;52:23–29. DOI: 10.1016/j.matdes.2013.05.029

[19] F. Salimyanfard, M. Reza Toroghinejad, F. Ashrafizadeh, M. Hoseini, J. A. Szpunar. Investigation of texture and mechanical properties of copper processed by new route of equal channel angular pressing. Materials and Design. 2013;44:374–381. DOI: 10.1016/j.matdes.2012.08.002

[20] I.V. Alexandrov, R.Z. Valiev. Developing of SPD processing and enhanced properties in bulk nanostructured metals. Scripta Materialia. 2001;44:1605–1608. DOI: 10.1016/S1359-6462(01)00783-7

[21] C.Y. Lim, S.Z. Han, S.H. Lee. Formation of nano-sized grains in Cu and Cu-Fe-P alloys by accumulative roll bonding process . Metals and Materials International. 2006;12:225–231. DOI: 10.1007/BF03027535

[22] J. Jiang, Y. Ding, F. Zuo, A. Shan. Mechanical properties and microstructures of ultrafine-grained pure aluminum by asymmetric rolling. Scripta Materialia. 2009;60:905–908. DOI: 10.1016/j.scriptamat.2009.02.016

[23] J.W. Wang, Q.Q. Duan, C.X. Huang, S.D.Wu, Z.F. Zhang. Tensile and compressive deformation behaviors of commercially pure Al processed by equal-channel angular pressing with different dies. Materials Science and Engineering A. 2008;496:409–416. DOI: 10.1016/j.msea.2008.05.041

[24] M. Ibrahim Abd El Aal, H.Y. Um, E.Y. Yoon, H.S. Kim. Microstructure evolution and mechanical properties of pure aluminum deformed by equal channel angular pressing and direct extrusion in one step through an integrated die. Materials Science & Engineering A. 2015;625:252–263. DOI: 10.1016/j.msea.2014.11.089

[25] H. Jafarzadeh, K. Abrinia. Fabrication of ultra-fine grained aluminium tubes by RTES technique. Materials Characterization. 2015;102:1–8. DOI: 10.1016/j.matchar.2014.12.025

[26] S.S. Satheesh Kumar, T. Raghu. Structural and mechanical behaviour of severe plastically deformed high purity aluminium sheets processed by constrained groove pressing technique. Materials and Design. 2014;57:114–120. DOI: 10.1016/j.matdes.2013.12.053.

[27] A. Abdulstaar, E. A. El-Danaf, N. S. Waluyo, L. Wagner. Severe plastic deformation of commercial purity aluminum by rotary swaging: microstructure evolution and mechanical properties. Materials Science and Engineering A. 2013;565:351–358. DOI: 10.1016/j.msea.2012.12.046

[28] Y. Ito, Z. Horita. Microstructural evolution in pure aluminum processed by high-pressure torsion. Materials Science and Engineering A. 2009;503:32–36. DOI: 10.1016/j.msea.2008.03.055

[29] W.J. Kim, S.J. Yoo, J.B. Lee. Microstructure and mechanical properties of pure Ti processed by high-ratio differential speed rolling at room temperature. Scripta Materialia. 2010;62:451–454. DOI: 10.1016/j.scriptamat.2009.12.008

[30] H.S. Kim, S.J. Yoob, J. W. Ahn, D. H. Kim, W.J. Kim. Ultrafine grained titanium sheets with high strength and high corrosion resistance. Materials Science and Engineering A. 2011;528:8479–8485. DOI: 10.1016/j.msea.2011.07.074

[31] X. Zhao, W. Fu, X. Yang, T.G. Langdon. Microstructure and properties of pure titanium processed by equal-channel angular pressing at room temperature. Scripta Materialia. 2008;59:542–545. DOI: 10.1016/j.scriptamat.2008.05.001

[32] D.H. Kang, T.W. Kim. Mechanical behavior and microstructural evolution of commercially pure titanium in enhanced multi-pass equal channel angular pressing and cold extrusion. Materials and Design. 2010;31:S54–S60. DOI: 10.1016/j.matdes.2010.01.004

[33] D. Terada, S. Inoue, N. Tsuji. Microstructure and mechanical properties of commercial purity titanium severely deformed by ARB process. Journal of Materials Science. 2007;42:1673–1681. DOI: 10.1007/s10853-006-0909-7

[34] K. Edalati, E. Matsubara, Z. Horita. Processing pure Ti by high-pressure torsion in wide ranges of pressures and strain. Metallurgical and Materials Transactions A. 2009;40:2079–2086. DOI: 10.1007/s11661-009-9890-5

[35] R.B. Megantoro, Loorentz, Y.G. Ko. Temperature rise during differential speed rolling. Journal of Alloys and Compounds. 2014;586:S254–S257. DOI: 10.1016/j.jallcom.2012.11.173

[36] J. Suharto, Y.G. Ko. Annealing behavior of severely deformed IF steel via the differential speed rolling method. Materials Science and Engineering A. 2012;558:90–94. DOI: 10.1016/j.msea.2012.07.081

[37] Ph. Lequeu, J.J. Jonas. Modeling of the plastic anisotropy of textured sheet. Metallurgical and Materials Transactions. 1988;19:105–120. DOI: 10.1007/BF02669819

[38] O. Engler; H.C. Kim; M.Y. Huh. Formation of {1 1 1} fibre texture in recrystallised aluminium sheet. Materials Science and Technology. 2001;17:75–86. DOI: 10.1179/026708301101508990

[39] H. Jin, D.J. Lloyd. The reduction of planar anisotropy by texture modification through asymmetric rolling and annealing in AA5754. Materials Science and Engineering A. 2005;399:358–367. DOI: 10.1016/j.msea.2005.04.027

[40] T. Sakai, S. Hamada, Y. Saito. Improvement of the r-value in 5052 aluminum alloy sheets having through-thickness shear texture by 2-pass single-roll drive unidirectional shear rolling. Scripta Materialia. 2001;44:2569–2573. DOI: 10.1016/S1359-6462(01)00932-0

[41] W. Polkowski, P. Jóźwik. Effect of asymmetric rolling on mechanical properties of 5251 aluminum alloy (in Polish). In: J. Pacyna, editor. XXXIX Szkoła Inżynierii Materiałowej AGH 2011; 24–27. IX. 2011; Krynica-Kraków. AGH, Kraków: 2011. p. 450–453.

[42] J. Bohlen, M.R. Nürnberga, J.W. Senn, D. Letzig, S.R. Agnew. The texture and anisotropy of magnesium-zinc-rare earth alloy sheets. Acta Materialia. 2007;55:2101–2112. DOI: 10.1016/j.actamat.2006.11.013

[43] T. Obara, H. Yoshinga, S. Morozumi. {1 1 2 2}<1 1 2 3> slip system in magnesium. Acta Metallurgica. 1973;21:845–853. DOI: 10.1016/0001-6160(73)90141-7

[44] L. Huaqiang, T. Di, C. Qingwu, L. Zhen. Texture of AZ31B magnesium alloy sheets produced by differential speed rolling technologies. Rare Metals. 2012;31:415–419. DOI: 10.1007/s12598-012-0530-7

[45] J.B. Lee, T.J. Konno, H.G. Jeong. Effect of differential speed rolling on the anisotropy of mechanical properties and texture evolution of AZ31 Mg alloys. Journal of Alloys and Compounds. 2010;499:273–277. DOI: 10.1016/j.jallcom.2010.03.186

[46] W.J. Kim, B.G. Hwang, M.J. Lee, Y.B. Park. Effect of speed-ratio on microstructure, and mechanical properties of Mg-3Al-1Zn alloy in differential speed rolling. Journal of Alloys and Compounds. 2011;509:8510–8517. DOI: 10.1016/j.jallcom.2011.05.063

[47] X. Huang, K. Suzuki, A. Watazu, I. Shigematsu, N. Saito. Improvement of formability of Mg-Al-Zn alloy sheet at low temperatures using differential speed rolling. Journal of Alloys and Compounds. 2009;470:263–268. DOI: 10.1016/j.jallcom.2008.02.029

[48] W.J. Kim, J.D. Park, W.Y. Kim. Effect of differential speed rolling on microstructure and mechanical properties of an AZ91 magnesium alloy. Journal of Alloys and Compounds. 2008;460:289–293. DOI: 10.1016/j.jallcom.2007.06.050

[49] L.L. Chang, J.H. Cho, S.K. Kang. Microstructure and mechanical properties of twin roll cast AM31 magnesium alloy sheet processed by differential speed rolling. Materials and Design. 2012;34:746–752. DOI: 10.1016/j.matdes.2011.06.060

[50] L.L. Chang, S.B. Kang, J.H. Cho. Influence of strain path on the microstructure evolution and mechanical properties in AM31 magnesium alloy sheets processed by differential speed rolling. Materials and Design. 2013;44:144–148. DOI: 10.1016/j.matdes.2012.07.052

[51] W.J. Kim, M.J. Kim, J.Y. Wang. Superplastic behavior of a fine-grained ZK60 magnesium alloy processed by high-ratio differential speed rolling. Materials Science and Engineering A. 2009;527:322–327. DOI: 10.1016/j.msea.2009.08.064

[52] X.B. Gong, S.B. Kang, S.Y. Li, J.H. Cho. Enhanced plasticity of twin-roll cast ZK60 magnesium alloy through differential speed rolling. Materials and Design. 2009;30:3345–3350. DOI: 10.1016/j.matdes.2009.03.040

[53] W.J. Kim, Y.G. Lee, M.J. Lee, J.Y. Wang, Y.B. Park. Exceptionally high strength in Mg-3Al-1Zn alloy processed by high-ratio differential speed rolling. Scripta Materialia. 2011;65:1105–1108. DOI: 10.1016/j.scriptamat.2011.09.029

[54] W.Y. Kim, W.J. Kim. Fabrication of ultrafine-grained Mg–3Al–1Zn magnesium alloy sheets using a continuous high-ratio differential speed rolling technique. Materials Science and Engineering. 2014;594:189–192. DOI: 10.1016/j.msea.2013.11.066

[55] J.B. Lee, T.J. Konno, H.G. Jeong. Grain refinement and texture evolution in AZ31 Mg alloys sheet processed by differential speed rolling. Materials Science and Engineering B. 2009;161:166–169. DOI: 10.1016/j.mseb.2009.02.021

[56] T. Hirohata, S. Masaki, S. Shima. Experiment in metal powder compaction by differential speed rolling. Journal of Materials Processing Technology. 2001;111:113–117. DOI: 10.1016/S0924-0136(01)00492-7

[57] S.J. Yoo, S.H. Han, W.J. Kim. A combination of ball milling and high-ratio differential speed rolling for synthesizing carbon nanotube/copper composites. Carbon. 2013;61:487–500. DOI: 10.1016/j.carbon.2013.04.105

[58] W.J. Kim, S.I. Hong, J.M. Lee, S.H. Kim. Dispersion of TiC particles in an in situ aluminum matrix composite by shear plastic flow during high-ratio differential speed rolling. Materials Science and Engineering A. 2013;559:325–332. DOI: 10.1016/j.msea.2012.08.106

Section 2

New Alloys for Special Application

Chapter 6

The Superconducting Tape of Nb_3Al Compound

V.P. Korzhov

Additional information is available at the end of the chapter

http://dx.doi.org/10.5772/64500

Abstract

Annotation. Compound Nb_3Al with critical temperature $T_c \approx 18.5$ K has the upper critical magnetic field H_{c2} of 30 *T*, which is significantly higher than that of the well-known Nb_3Sn. Besides, aluminum is lighter and cheaper than tin and is widely abundant in nature. With such properties, Nb_3Al is a good alternative to Nb_3Sn. The next chapter proposed steps for the development of manufacturing technology of superconductor Nb_3Al tapes from its approbation on a three-layer composite Nb/Al/Nb tape to obtain experienced semi-finished pieces of multilayer Nb_3Al-tape with length of about 100 *m*. This is despite the fact that in order to achieve the above values, T_c and H_{c2} needed exposure for 1–2 seconds at a temperature of 1850°C.

Laboratory technology allowed the investigation of such unknown during the time of obtaining the compound Nb_3Al as: a joint deformation of niobium and aluminum; reactive diffusion in semi-infinite pair niobium/aluminum; the dependence of the superconducting parameters from regimes of high-temperature heating and low-temperature annealing; fastening of superconducting vortices at the grain boundaries of compound Nb_3AlGe. Usage of duralumin (alloy Al–Cu) in a multilayer tape, instead of pure aluminum, has caused the need to investigate the effect of copper on superconductivity Nb_3Al. Its positive impact at that time was quite not obvious.

Keywords: superconductivity, type Cr_3Si structure, Nb_3Al and $Nb_3Al_{0.8}Ge_{0.2}$ compounds, the critical temperature, the critical current density, the critical current, the upper critical magnetic field, pinning, the superconducting tape, reactive diffusion, package rolling, extrusion

1. Introduction

Immediately after the discovery by Kamerlingh Onnes of superconductivity in 1911, work began on the practical use of this phenomenon in order to obtain large constant magnetic fields. However, the concrete development of superconducting materials, with the assurance that such goal is really achievable, began only after 1961, when Kunzler discovered that the compound Nb_3Sn has high critical currents in magnetic fields of the order of 7 Tesla. Since then, they started a new department of science of solids – «Metal science of superconductors», dedicated to the development of new hard superconducting materials and technologies for their production and processing. In about the same year, intensive researches of the deformable alloys Nb–Ti with continuous solubility in the liquid and solid states began as subjects of superconductivity. It turned out that at a content 40–50 at. %Ti alloy with a critical temperature ~9.5 K had the second critical magnetic field of 15 T at 4.2 K.

In the early 1970s, the class of intermetallic compounds with a structure of the A15 was distinguished, except for Nb_3Sn, including Nb_3Al, $Nb_3Al_{0.8}Ge_{0.2}$ and others, having a temperature of superconducting transition T_c up to 20 K and able to withstand magnetic fields up to 20 T and more. From the day of the discovery of superconductivity in mercury at $T_c = 4.25$ K, maximum critical temperature was adjusted to 23 K only in 1973, when almost simultaneously J.R. Gavaler and Yu.F. Bychkov with employees discovered superconductivity in thin films Nb_3Ge. This level was kept in the next 13 years – until the discovery of high temperature superconductivity.

Of course, in the near future, high-temperature superconductivity will come into modern life in a much larger scale than the low-temperature superconductors. But, most likely, they will always find a worthy place for themselves to use. Direct evidence are the international mega-projects such as «The Large Hadron Collider» and «Experimental Thermonuclear Reactor» (International Thermonuclear Experimental Reactor – ITER), in magnetic systems, which were designed low-temperature superconducting materials. The magnetic system of the modern MR-tomographs is also made of superconducting cable of an alloy Nb–Ti and owing to the reliability exceeding all conceivable standards without foreseen replacement.

Therefore, the main purpose of this chapter is to remind the scientific community, engaged in superconductivity, to look at the materials science of low-temperature superconductors with the new positions, using modern possibilities for their research.

2. About Nb_3Al and $Nb_3(AlGe)$ compounds

Superconducting intermetallic Nb_3Al compound belongs to the class of compounds with a structure of Cr_3Si type (A15 lattice). In initial studies, its critical temperature T_c depending on the composition was changed in the range from 15.2 to 17.6 K, but in recent studies [1] was brought to the 18.4–18.9 K. Interest in Nb_3Al was designated because its upper critical magnetic field H_{c2} has a higher value than Nb_3Sn and V_3Ga compounds. Values of H_{c2} at 4.2 K, measured by pulse method, reached 29.5–32.0 *T* [2, 3], at 12 K $H_{c2} = 13$ *T* [1].

The temperature of the superconducting transition T_c of compounds of A15 type is maximal at their stoichiometric compositions, and during the preparation, there are important factors to consider like the stoichiometric composition of compound whether it is located in its region of existence at room temperature or not. For example, the compound Nb_3Sn has maintained its stoichiometric composition from the temperature of its formation, which is about 2100°C, to the room temperature. Therefore, to get it with 25 at. % of tin, it is not necessary to use a temperature close to 2000°C (**Figure 1*a***). The compound Nb_3Sn with maximum critical temperature is readily formed at temperatures of about 800°C.

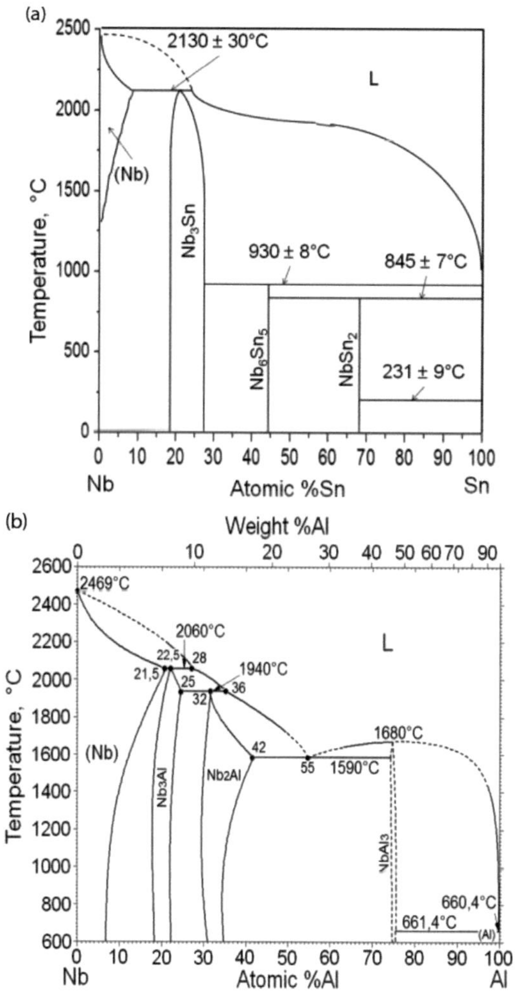

Figure 1. (*a*) State diagrams of Nb–Sn and (*b*) Nb–Al systems [4].

Compound Nb_3Al is another case. According to the state diagram, Nb–Al [5] (**Figure 1*b***) Nb_3Al with 25 at. % aluminum exists at temperatures above 1870–1940°C. Consequently, to obtain Nb_3Al with composition close to the stoichiometric composition and with high T_c ~ 18.5 K, it needs to use ~1850°C and higher temperatures. Therefore, with such value, the critical temperature has been obtained for him not immediately, and in later works.

In early 70s, when we began our researches, there were only a few works on the study of mutual diffusion of niobium and aluminum. Moreover, published works were devoted mainly to questions of formation of compounds $NbAl_3$ and Nb_2Al [6, 7]. Only in [7] at 1400°C with a 16-hour annealing in the diffusion layer the presence of interlayers of compound Nb_3Al was observed with 2 microns thick. The most complete data on Nb_3Al formation was given in [8], the authors who carried out a study of the interaction of niobium and aluminum to a temperature of 1550°C. The result of this work had a different slope of the linear dependencies of coefficients of mutual diffusion D dated the inverse value of the absolute temperature 1/T during the formation of compounds $NbAl_3$, Nb_2Al and Nb_3Al. In this case, the slope of the direct line D(1/T) was the greatest for Nb_3Al, however, with up to 1500°C, it was lower than the corresponding direct lines for $NbAl_3$ and Nb_2Al. If all the direct to extrapolate D(1/T) in the region are of higher temperatures, they cross in the range 1850–1900°C, i.e. the formation of compound Nb_3Al became predominant at higher temperatures (**Figure 2**). Subsequently, this was confirmed by our researches.

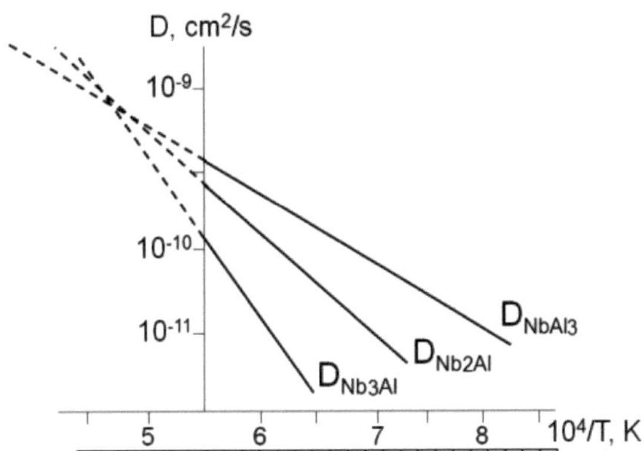

Figure 2. The coefficients of mutual diffusion for compounds $NbAl_3$, Nb_2Al and Nb_3Al depending on the inverse of the absolute temperature.

Consideration of compound Nb_3Al often occurs together with another compound in which a portion of the aluminum atoms is substituted with germanium – $Nb_3Al_{0.8}Ge_{0.2}$. It was discussed in 1967 by B.T. Matthias with colleagues. At that time, it had a record of critical temperature – 20.05 K and the upper critical magnetic field H_{c2} – 42 *T* at 4.2 K and 20 *T* at 14 K. The practical application of this material in the case of the development of manufacturing technology of wire or ribbon was to prove that it is more effective than the use of Nb_3Sn or V_3Ga especially with regard to the possibility to use it for cooling liquid hydrogen instead of helium.

The main disadvantage of intermetallic compounds is their natural brittleness. This is the reason why, in the case of application of the compound for the solenoid coils and magnets, it is prepared by methods that result in the intermetallic compound formed as a thin metal layer on a flexible substrate or inside the conductor. In this state, it retains its ability to deform while bending without destruction.

3. The joint rolling how laboratory technology for production of superconducting tapes on the basis of A15 connections

All the superconducting materials out of the A15 type compounds obtained by heat treatment in a tape or wire of thin section, one way or another can be attributed to the conductors of the composite type. Methods for the preparation of the composite, as we have seen, is very different: deformation of tube with the core out of components of the compound in the form of powder or of the compound (method Kunzler or "powder-in-tube"); coating of tape or wire out of niobium or vanadium with fusible component by passing them through a bath of molten tin or gallium (diffusion method from a liquid phase); niobium tape coated fusible component by spraying it in a vacuum (diffusion method from the gaseous phase) or by using gas-transport reactions (recovery with hydrogen of tin chloride); bronze technology and, of course, the method of deformation of dissimilar materials.

Among the laboratory technology of superconducting materials, we developed in the present research through the latter method. As mentioned, the lack of data on the superconducting characteristics in the obtained conductors of this type in the early 70s and the relatively small number currently suggest the specific difficulties are associated with this seemingly simple at first glance technology.

The first – is a great heterogeneity of metals, which is required to deform together. The second and the main difficulty are as follows. Manufacturing technology of bimetallic strip of any compositions must ensure durable bonding of layers. In practice, bimetals, the initial preform for making a flaky composite material, serve in a certain way assembled package is subjected to joint cold rolling. In this stage, lasting cohesion of layers must be achieved over the entire area of contact of joined metals. Only at this condition the process of further rolling is possible by preparing sufficiently long tape of thin section to avoid bundles.

To overcome the abovementioned difficulties on the initial stage of deformation, we used a vacuum rolling of bimetallic packets. With the help of vacuum rolling, we were able to obtain a monolithic planar blank for subsequent rolling at room temperature to the tape of thin-section subjected to a thermal treatment for solid phase formation of superconducting compound. This sequence of operations included the possibility of manufacturing the superconducting tape from almost all A15 compounds with high superconducting characteristics.

To obtain compounds such as V_3Ga, V_3Si, Nb_3Sn and $Nb_3(Al,Ge)$, we used copper alloys with gallium, silicon, tin (bronze) and aluminum alloy with germanium, respectively. However, it is difficult to roll bronze at room temperature. Additionally, their ductility decreases with increasing of gallium content, silicon and tin; also, for solid phase preparation of compounds, it is desirable to have copper alloys enrich with these elements. For ternary intermetallic compound $Nb_3Al_{0.8}Ge_{0.2}$ must have aluminum alloys 40–50 wt. % Ge, which are not deformed by rolling at room temperature without cracking.

Therefore, the joint rolling of bimetallic packets in a vacuum at elevated temperatures in the initial stage was what we needed. Firstly, the temperature was promoted firm adhesion of dissimilar metals. In most cases, when adhesion, or welding, cannot be achieved at room temperature, only the deformation in vacuum makes this possible. Secondly, bronzes and alloy Al–Ge are difficult deformable at room temperature, but when after vacuum rolling they are firmly cohesive in the bimetallic monolithic preforms with plastic metals – niobium and vanadium, we were able to deform this alloys at room temperature without intermediate annealing. The thickness of the bimetallic tape $Nb_3(Al,Ge)$ in the final stage equals 50–70 microns.

3.1. Questions about Joint Plastic Deformation of Metals

Practical achievements of metal welding by joint plastic deformation (rolling, pressing, explosion welding) so far were outstripped theoretical ideas about the nature and the mechanism of adhesion, although there are a number of hypotheses to explain this phenomenon. This question else in the 70s with insufficient detail was illuminated in a number of fundamental works. It should be noted that certain provisions of the proposed hypothesis well explain the phenomena in clutch area of metal strips when the joint deformation by rolling occurs. The essence of these provisions are as follows:

1. Clutch of similar and dissimilar metals in the process of plastic deformation is conditioned by "grasp" phenomenon, which occurs due to the formation of metallic bonds between atoms of adjacent physically clean (juvenile) surfaces when their rapprochement at a distance of action of interatomic forces.

2. The ability of metals to "grasp" is the physical property of juvenile surfaces and depends on their nature.

3. In ideal conditions, the grasping is a beneficial thermodynamic process and should occur spontaneously, as the energy of the system of two connected metal is reduced by the elimination of free surfaces. Under ideal conditions, it should be understood that rapprochement of atomically smooth surfaces free from oxide and adsorbed films, over a distance equal the sum of atoms radii of connected by metal.

4. The joint plastic deformation of metals is one of the ways of obtaining juvenile surfaces, their convergence and increase of the actual area of the clutch.

3.2. Preparation of Packages for a Joint Rolling of Dissimilar Metals in Vacuum

The initial workpiece for the experienced tape samples of superconducting compounds V_3Ga, V_3Si, Nb_3Sn, Nb_3Al and $Nb_3(Al,Ge)$ was served with package assembled from two plates of refractory and one plate of fusible component (**Figure 3*a***). In the case of Nb_3Sn compound, five-layer packages with copper plates from the outside were also collected (**Figure 3*b***).

http://dx.doi.org/10.5772/64500

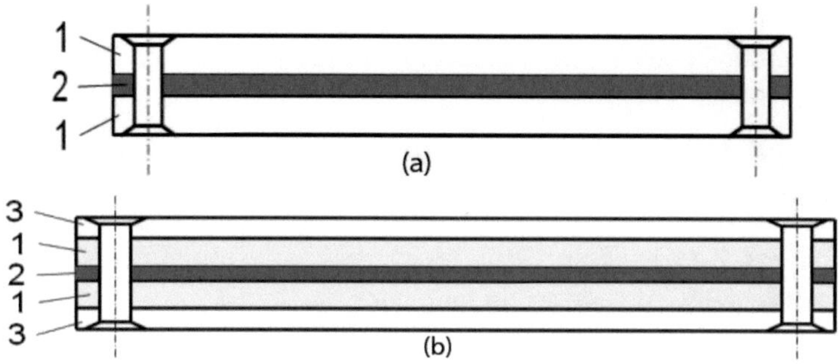

Figure 3. (*a*) Construction of collected packets without copper and a copper-clad; (*b*): 1 – niobium, vanadium, niobium alloys; 2 – aluminum, Al–Ge alloy, bronze; 3 – copper.

Bronze was smelted by the induction method in an argon atmosphere, Al–Ge alloy – by flash smelting. Massive flat copper molds were used for casting. Then, the ingots were rolled at room temperature with intermediate annealing in a vacuum, and, in some cases, with heating in air to 500–600°C.

To ensure grasping and obtaining of reliable clutch between connectable metals from the contact surfaces, oil and fat film were removed, and water vapor and oxides were adsorbed. This was achieved by immediately chemical etching of plates before assembling packages for rolling. The package was assembled from plates of the same width to prevent cracking of the tape edges. The ends of plates fastened conventional rivets (see **Figure 3**). Package size for laboratory tests: width 15–20, length ~70 *mm*.

3.3. Vacuum Rolling of Packages.

The rolling of packages in vacuum efficiently protects the surfaces of the connected metals against oxidation, which was one of the conditions for lasting clutch of dissimilar metals.

In our works, vacuum rolling of packages was made on the rolling mill DUO–170, basic design of which was developed in the Physical-Technical Institute of the National Academy of Sciences of Ukraine. The diameter of the work rolls – 170, length of barrel – 144 *mm*. Clearance between the rolls can be adjusted from 0 to 20 *mm*. The mill – two reverse, has 5 speeds: 8, 16, 22, 34 and 54 *rpm*. Permissible design pressure on the rolls is 30 tons. The pressure of residual gas in the chamber is $\sim 5 \cdot 10^{-5}$ *mm Hg*.

The most problematic object for joint rolling proved to be a pair of niobium/aluminum. This interests the study of the properties of relatively thin layers Nb_3Al since all of its connections by nature are very brittle at room temperature, where in such state, it retained the ability to bend without breaking. Assuming the possible technical use of the Nb_3Al in the future, it is necessary to obtain it in the form of a thin layer within a conductor.

Studies of optimal modes of rolling showed that the volumetric content of aluminum in the package:

- decreased from 9.8 to 0% with an increase of the heating temperature of package before rolling in vacuum of 400 to 700°C (reduction ratio per pass from 22 to 24.4%, the initial thickness of Al-foil – 1 *mm*);
- decreased from 18 to 9% with increasing of degree of reduction per pass from 10 to 22.5% (heating – 500°C; the initial thickness of Al-foil – 1 *mm*);
- increased from 4.7–5.6 to about 13% at an increase of the initial Al-foil thickness of 0.1 to 1 *mm* (heating – 470°C, the reduction ratio per pass 17–25%).

Optimal modes of vacuum rolling: the heating temperature of the package 450–500°C, the degree of reduction per pass of 15–25% and the initial thickness of the Al-foil of 0.5–0.8 *mm* at a thickness of Nb-plate equal 1.5 *mm*.

Selecting of reduction per pass at rolling of packages for compound Nb_3Al became looser after the replacement of pure aluminum on a more durable duralumin brand D16.

Vacuum rolling of packages Nb/(Al–Ge)/Nb to obtain intermetallic compound Nb_3(Al,Ge) was carried out at the temperature range of 390–410°C, which was limited by the melting temperature alloys Al–Ge. Unlike the packages of Nb/Al/Nb, they were allowed rolling with a reduction of 35 to 55% per pass. At this alloy content in packages after the rolling (or in composite tape of final thickness) varied from 4.8 to 10.0 vol. %.

3.4. Cold Welding of Packages Nb/Al/Nb

Apart from welding in vacuum at a temperature above the room temperature experiments by the cold grasping (cold welding) of packages Nb/Al/Nb using a twin-roll mill with 300 *mm* diameter rolls were performed. Reliable welding of packages, after further rolling into a ribbon the exfoliation was not observed, was achieved at a reduction per pass of more than 60%.

It should be noted that the grasping is dependent on the low-melting component. For example, the cold joint rolling of packages, which consisted of plates niobium and SAP-1 (sintered aluminum powder, which is much stronger than pure aluminum) did not give positive results. However, packages Nb/Sn/Nb attempt to obtain the intermetallic compound Nb_3Sn. (Packages of Nb/SAP/Nb grasp was implemented by joint rolling in a vacuum at a temperature of 400°C).

Eventually, we preferred to obtain a three-layer tape Nb/Al/Nb out of packets welded by rolling in vacuum.

3.5. Rolling of Thin Tapes at Room Temperature

The monolithic workpiece after welding with a vacuum rolling mill (or cold welding) was rolled at room temperature to the tape of a thin section. At present stage, it did not attempt to get long pieces of tape. The length of the piece was determined by the original thickness of package and final thickness of the rolled tape (**Table 1**). For example, length of tape Nb/Al/Nb could be from ~4 *m* if the initial thickness of the package and its final thickness were 4 *mm* and 65 *μm*, respectively, until ~11 *m* – at a thickness of the package of 8 *mm* and the final thickness of tape 50 microns.

For further studies, the obtained tape of width 15–20 *mm* was cut by the disk scissor for a few tapes 2 *mm* wide.

Initial thickness of package Nb/Al/Nb, *mm*	Thickness of tape, *μm*	Length of tape piece, *m*
4	50	5.5
	65	4.0
8	50	11.0
	65	8.5

Table 1. Geometric parameters of packages and pieces of the three-layer tape Nb/Al/Nb. The length of welded packages is 70 *mm*.

The rolling of packages at room temperature was simultaneously done by estimation of the grasp niobium with aluminum. With rare exceptions, an exfoliation of tape was not observed. This indicated that the joint plastic deformation of the packages in the selected parameter range of rolling provided a strong clutch between metals.

However, the quality of the outer surfaces of the 3-layer composite tape with aluminum depended on its volume content. It was found that up to 8–8.5 vol. % aluminum was not adversely affected. But, at higher content during the process of cold rolling on Nb-surfaces of tape, emerged local discontinuities through which was looked through aluminum. The tape preserved integrity, but after the heat treatment in these places, it broke down because of brittleness. Based on our observations, this was due to high ductility of internal component of the composite. For a more durable alloy Al–45 wt. % Ge such a phenomenon was not observed at 10 vol. % alloy in tape.

3.6. Heat treatment of tapes Nb/Al/Nb and Nb/(Al–Ge)/Nb

To obtain superconducting compounds Nb_3Al and $Nb_3(Al,Ge)$, heat treatment was carried out which resulted in the formation of inside tapes in a thin diffusion layers. It included high-temperature heating (h.-t. h.) and a low-temperature annealing (l.-t. an.).

Strictly speaking, intermetallic compounds were formed in the first stage. During this stage, the optimum temperature of heating was exceeded at 1700–1750°C, and the time was measured by a few seconds. This regime was carried out in a vacuum chamber (~10^{-5} *mm Hg*) by direct passing of electric current through the tape sample attached between two the massive copper contacts. Temperature t was monitored by magnitude of an electric current i through the sample.

Dependencies $t(i)$ for the tapes with thickness of 50, 60, 65 and 70 microns were constructed in individual test experiments in which the temperature of the sample was determined by (W–Re)-thermocouple WR 5/20 and was controlled by indications of the pyrometer. Thermocouple junction was joined by spot welding directly to tape. The diameter of thermocouple wire was 0.1 *mm*. In actual experiments, 10 tape samples of the same thickness connected in parallel were placed at once into the camera. The distance between contacts equaled to 100 *mm*. After heating the edge of the tape ~15 *mm* was cut, mid of 60–70 *mm* was used for research.

The low temperature (600–900°C) annealings of samples undertaken for the ordering of crystal structure of the compounds was carried out in vacuum furnace of resistance with a W-heater.

Figure 4 shows the results of the measurement of the critical temperature of Nb/Al-tapes depending on the volume content of aluminum after heating at 1750°C for 5 seconds, and 2-hour annealing at 800°C. It is seen that T_c is not dependent on the aluminum content and is ~17.5 K.

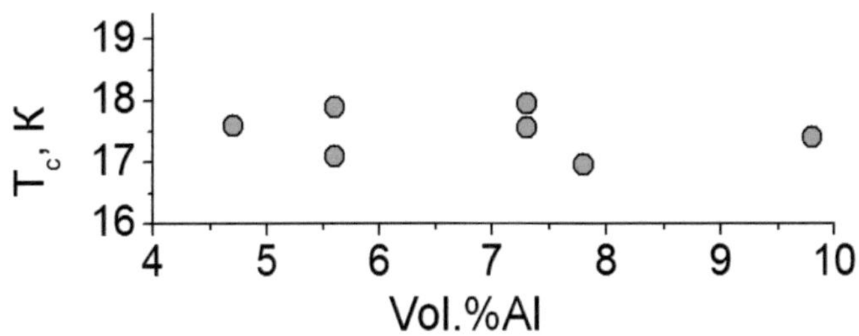

Figure 4. The dependence of the critical temperature T_c of tape Nb/Al/Nb from aluminum volume content Heat treatment: 1750°C, 5 *s* + 800°C, 2 *h*.

4. Microstructure and superconducting properties of the tapes Nb_3Al and $Nb_3(AlGe)$

4.1. The Cross-section Structure of Tapes

The phase structure of the diffusion layers was formed in the process of high-temperature heating and depended on the heating time and the volume content of aluminum in the initial tapes. Heating for 1 second at 1750°C led to the formation of the diffusion layer out of $\sigma(Nb_2Al)$- and $\gamma(NbAl_3)$-phase (**Figure 5*a***). Compound Nb_3Al (β-phase) is hardly viewed as separate inclusions on the boundary of σ-phase with Nb. With the increase of time and temperature of heating, layer structure is changed by increasing the thickness of the interlayer β-phase (see **Figure 5*b*** and **5*c***). Furthermore, ß-phase became too occupied with the entire volume of the diffusion layer (see **Figure 5*d***). A further increase of the heating time caused the diffusion of aluminum over the entire cross-section of the sample and the formation α-solid solution based on niobium (see **Figure 5*e***).

Figure 5*e* shows the cross-sectional microstructure of the composite tape out of niobium and alloy Al-30 wt. % Ge, which occupies 7.4% of its volume. After a 5-second heating at 1850°C in the tape, the diffusion layer with thickness of 8–5 microns was formed consisting of an intermetallic compound $\beta[Nb_3(Al,Ge)]$-phase with σ-phase splashes of compounds $Nb_2(Al,Ge)$.

After subsequent low-temperature annealing, the ratio of phases in the resulting diffusion layer did not change. However, in the beginning, an annealing at 900°C then formed γ-phase, that is, already before h.-t. h. It contained 35–43.5 wt. %Al and had a tetragonal lattice with the ratio $c/a = 1.63$. The X-ray patterns of specimens with β-phase set of lines corresponded to cubic lattice of type Cr_3Si with parameter a equaled to 5.178–5.207 Å. For chemical compound Nb_3Al in the cast state $a = 5.187$ Å.

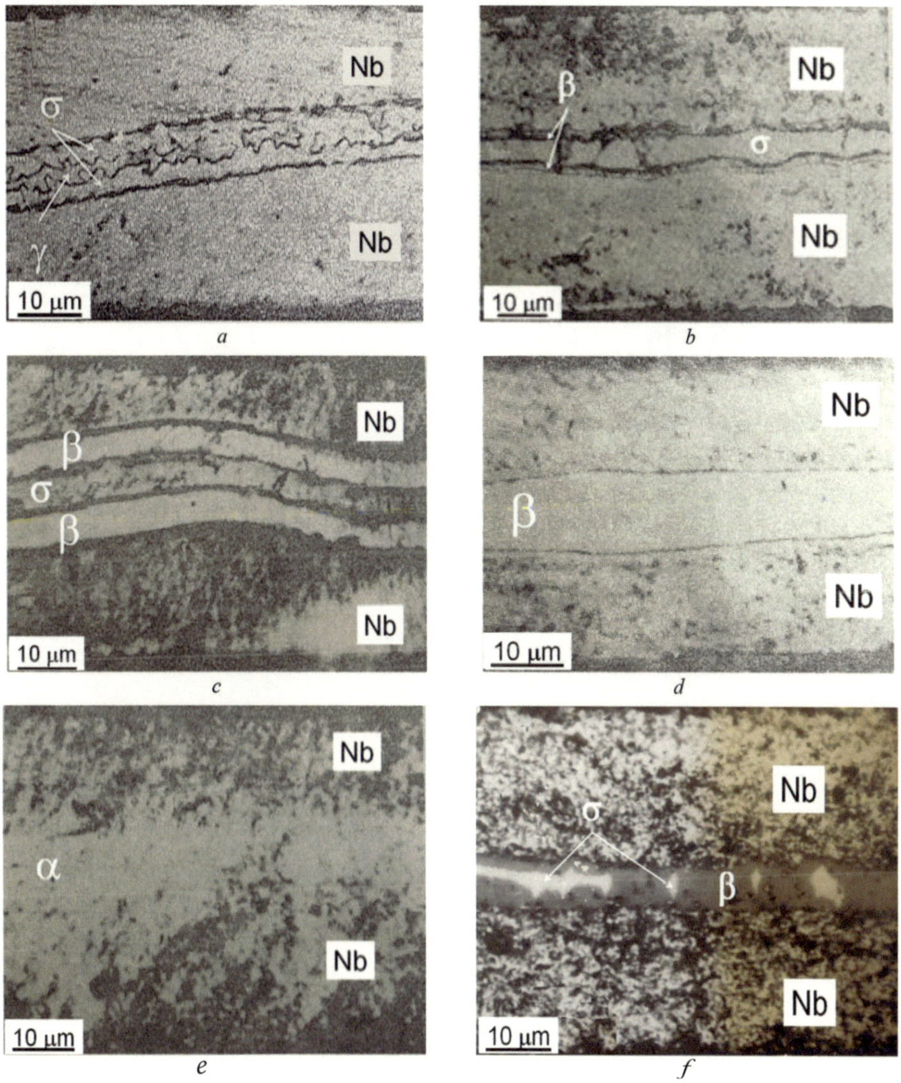

Figure 5. The structure of the cross-section of tape samples out of Nb/Al /Nb- (*a-e*) and Nb/(Al-30Ge)/Nb-composites (*f*) with thickness of 50 microns. H.-t. h, °C: *a* – 1750, 1 *s*; 7.4 vol. %Al; *b* – 1750, 5 *s*; 7.4 vol. %Al; *c* – 1850, 3 *s*; 8.1 vol. %Al; *d* – 1850, 5 *s*; 4.7 vol. %Al; *e* – 1850, 8 *s*; 4.7 vol. %Al and *f* – 1850, 5 *s*; 7.4 vol. %(Al–30Ge).

A layer of β-phase quickly arises, grows rapidly, and reaches a maximum value, ceased to exist. At temperatures of 1750°C and higher a heating, an excerpt and a cooling are continuing during several seconds (see **Figure 5**). However, even against the backdrop of such fleeting process, one may note that in the tapes, which contained about 7.4 vol. % Al, β-phase was maintained for a longer time than in tapes with 4.7 vol. % Al. In **Figure 6** (curve 2), in the samples with less volume content of aluminum equal to 4.7 vol. % aluminum layer, β-phase existed only for 5 seconds. During this time, it has grown to 5.5 microns, and then formed α-solid solution of aluminum in niobium. A similar case can be observed by comparing microstructures of the cross-sections of two identical samples *d* and *e* in **Figure 5**, which differed only by the duration of the heating at 1850°C: after 5 *s* – layer of β-phase, after 8 *s* – α-solid solution. In the tape samples with 7.4 vol. % Al (see **Figure 6**, curve 1), β-layer with thickness of 4.5 micron was observed after heating for 10 seconds.

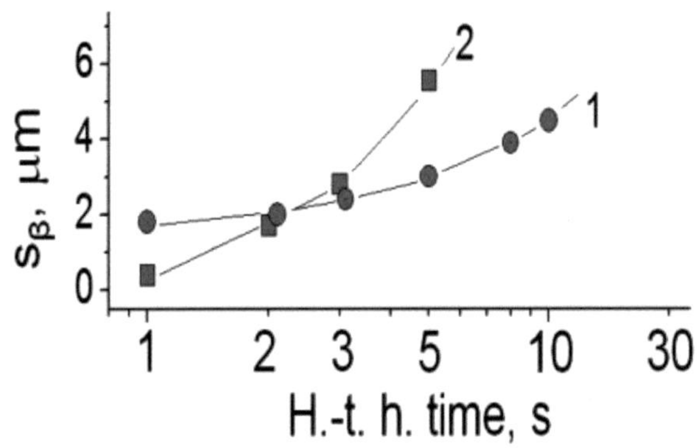

Figure 6. Layer thickness of β-phase versus time of h.-t. h. for samples of tape Nb/Al/Nb with 7,4 (1) and 4.7 vol. % Al (2) after heating at 1750°C. L.-t. an.: about 1 – 900°C, 2 *h*; 2 – 800°C, 2 *h*.

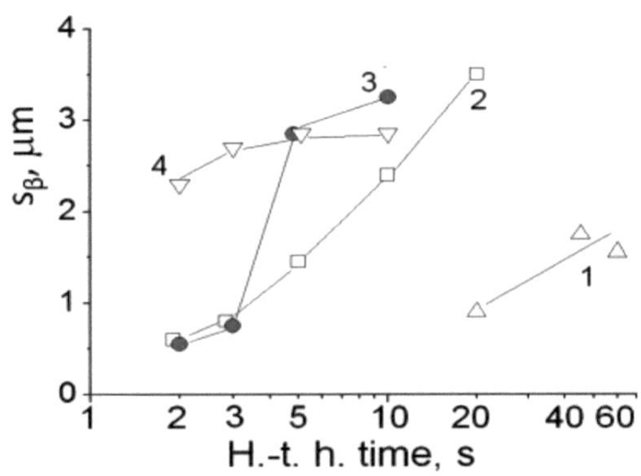

Figure 7. Layer thickness of β-phase versus time of h.-t. h. for samples of tape Nb/(AlGe)/Nb with 7,4 vol. % of alloy Al–30 wt. %Ge. Temperature of h.-t. h., °C: 1 – 1650, 2 – 1800, 3 – 1850, 4 – 1900.

Graphical dependence of β-layer thickness from the holding time at high-temperature heating of samples out of tape Nb/(AlGe)/Nb was shown in **Figure 7**. The layer thickness was calculated as the average of 10–15 measurements. At identical volume content of AlGe-alloy, β-layers after heating to a temperature in the small interval 1800–1900°C reached approximately the same thickness, ~ 3 μm at times of h.-t. h. equals 10 seconds. At 1650°C, to obtain the same thickness of β-layer, significantly longer times of the heating were required.

4.2. Tape Nb/Al/Nb. The dependence of the critical temperature from temperature and time the high-temperature heating

Before moving to an exposition of research results, dwell upon the interpretation the experimental data derived directly at measurements of the critical temperature T_c.

Figure 8 displays an experimental curve of the transition to the superconducting state, resulting to an inductive method of measurement T_c, which we appreciated. Two transitions were observed – the low-temperature transition at ~ 9.3 K, relating to the layers of niobium and the transition that begins and ends in the present experiment, at ~13.5 and ~18 K, respectively (this refers to the diffusion layer). In our studies of the transition temperature to the superconducting state, we took the beginning of the transition T_{CB}. This is caused by the fact that the received critical temperature in the literature as the middle of the transition does not entirely satisfactorily reflect its essence in the case of diffusion layers. As can be seen from the figure, the transition into district of high temperatures is greatly smeared to its left side, and a large portion of the signal occurs in the region near the beginning of the transition to the superconducting state. Therefore, the middle of the transition T_c, which is equal to 15.75 K, will not have any significance since it depends on the measurements of the critical temperature of the massive superconducting samples. In this study, to characterize samples of such, magnitudes were taken: T_{CB}, T_c and $\Delta T_c/2$ (see **Figure 8a**) Parameter $\Delta T_c/2$, correlating with a half-width of the superconducting transition, was calculated as the difference of T_{CB} – T_c,

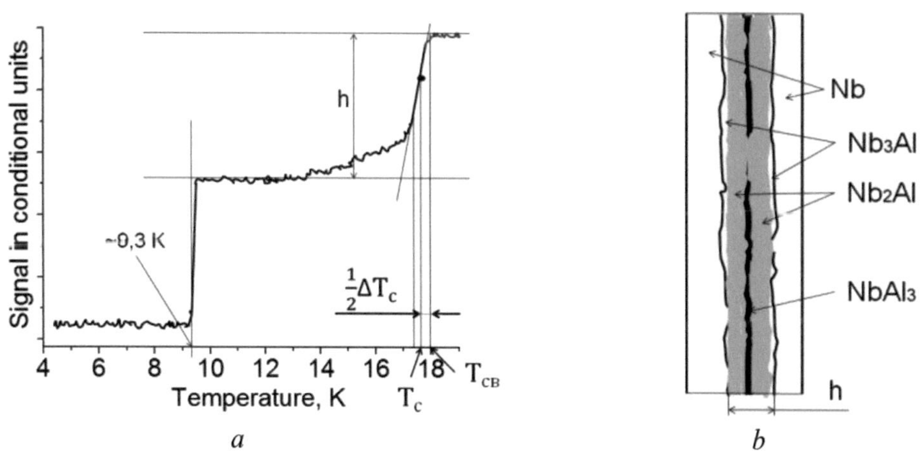

Figure 8. The experimental curve of the superconducting transition for the tape sample Nb/Al/Nb: 7.4 vol. %Al; 1750°C, with 1 *s* + 800°C, 2 *h* (*a*) and a schematic representation of the phase structure of the diffusion layer (*b*).

where T_c corresponded to the temperature of point of a flexion for the high-temperature transition on the experimental curve.

The height of the superconducting transition (designated as *h*) is determined by the volume of the superconducting phase in the sample the higher it is, the higher the jump of the signal. This also happens in bulk samples. Here, Nb_3Al is the superconducting layer with the highest critical temperature which is adjacent to the niobium and has bordered the whole diffusion layer (**Figure 8*b***). In the transition to the superconducting state, it screens the entire volume of the diffusion region h, where there are other compound Nb_2Al, non-superconducting at 18 K, and the intermetallic compound $NbAl_3$, not possessing superconductivity. Moreover, shielding occurs also if the layer Nb_3Al is not continuous.

Dependencies of the critical temperature T_{CB} and the parameter of $\Delta T_c/2$ from the time of high temperature heating at different temperatures were conducted in the range from 1400 to 1950°C. There are such dependencies for temperatures h.-t. h., equivalent to 1750, 1850 and 1900°C (**Figure 9**). Almost all dependencies characterized with a flat maximum for T_{CB} and minimum for $\Delta T_c/2$. Annealings at 800 and 900°C for 2 hours were followed after the short heating increased in T_c on 0.5–1.5 K (see **Figure 9**, top row, curves 1 and 2). There were excepted samples that were initially heated at 1500 and 1600°C for 1–5 seconds. Low-temperature annealing was decreased T_{CB} in these samples. The best samples had values of T_{CB} in the range of 18–18.5 K and minimum values of $\Delta T_c/2$ = 0.2–0.9 K. Such values of the superconducting parameters corresponded to tapes with time h.-t. h. from 3 to 5 seconds.

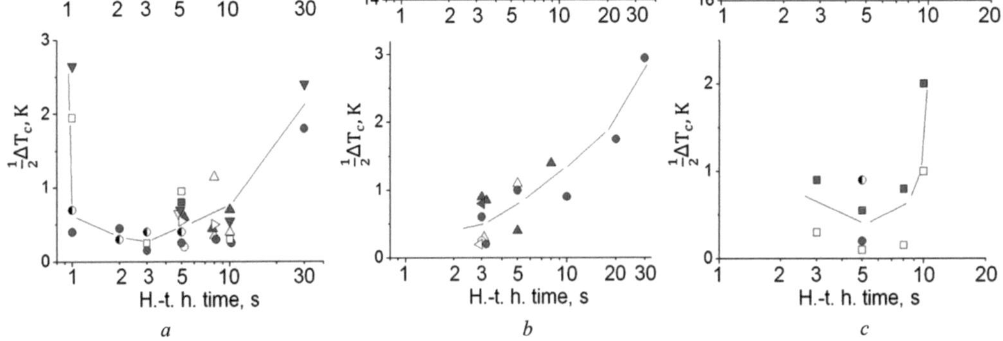

Figure 9. Dependence of the critical temperature T_{CB} (top row) and half-width $\Delta T_c/2$ of superconducting transition (bottom row) for tape samples Nb/Al/Nb from time h.-t. h. at 1750 (*a*) 1850 (*b*), and 1900°C (*c*): 1 – no l.-t. an., 2 – annealing at 800°C for 2 hours. Vol.% Al: △, ▲ – 4.7; ▽, ▼ – 5.6; ▷ – 6.3; ○, ● – 7.4; □, ■ – 7.8; ◁, ◀ – 8.1; ◐ – 7.4; 900°C, 2 h.

A dependence of T_{CB} from the temperature of h.-t. h. at a constant exposure during 5 seconds was presented in **Figure 10**. It shows that the maximum T_{CB} is equivalent to 18 K, observed at temperatures of the heating 1850–1950°C.

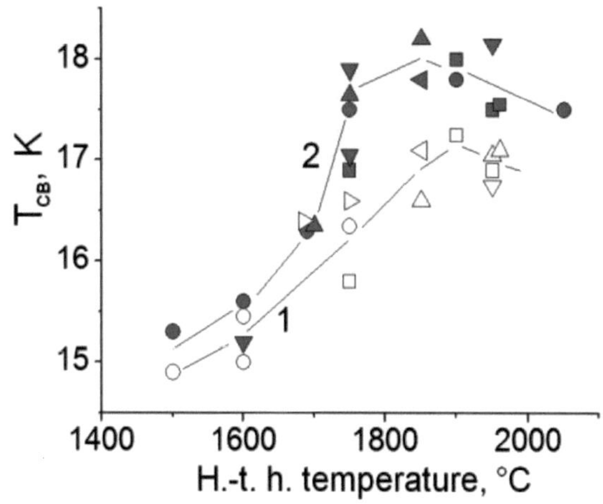

Figure 10. Dependence of the beginning temperature of the superconducting transition T_{CB} from temperature of the initial heating at the time of heating during 5 *s*: 1 – without l.-t. an.; 2 – annealing at 800°C for 2 *h*; designation similar to **Figure 9**.

Figure 11 shows that the maximum attainable (at a given temperature of h.-t. heating) temperature of the beginning of transition to the superconducting state T_{CB} (*a*) and the time of heating, at which maximum is reached (*b*), from the temperature of the h.-t. heating. On 1st curve is still more pronounced than in the previous figure, the maximum at 1850–1900°C, and on the 2nd – the bend in the range 1550–1700°C.

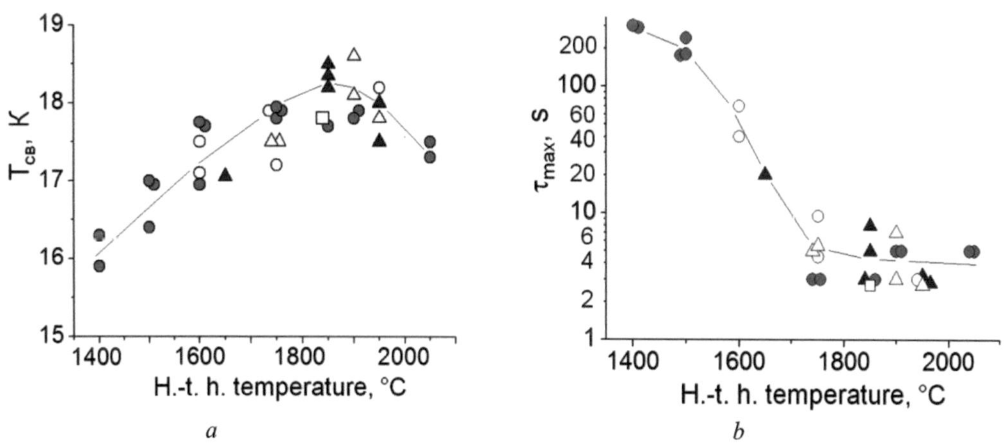

Figure 11. Dependence of maximum attainable critical temperature T_{CB} (*a*) and the time at which it is achieved τ_{max} (*b*) from temperature of the heating. L.-t. an. – 800°C, 2 hours. Vol. % Al: ▲ – 4.7; ○ – 5.6; • – 7.4; △ –7.8; □ – 8.1.

Monotonic increase of T_{CB} with increasing of heating temperature from 1400 to 1950°C is due to the approximation of the composition of β-phase to the stoichiometry. As already mentioned, this is due to a shift of existence region in the phase diagram Nb–Al (**Figure 12**) toward the increasing aluminum content. The maximum of $T_{CB} = 18.5$ K. Some drop T_{CB} beyond 2050°C connected with a transition in the region above the line of peritectic reaction of formation Nb_3Al. In this case, β-phase is formed by the decay of a supersaturated solid solution of Al in Nb. Composition of such β-phase is always <25 at. % Al.

The temperature interval of an inflection on the curve τ_{max} depending on the temperature of h.-t. h. was in the range of 1500–1700°C and coincides with a melting point of $\gamma(NbAl_3)$-phase. According to different authors, it ranges from 1550 to 1660°C. This affects the formation of Nb_3Al so much that the diffusion from liquid phase is faster than in the solid phase. At temperatures of heating 1400–1500°C τ_{max} = 3–5 minutes and at 1750°C and above – 3–5 s.

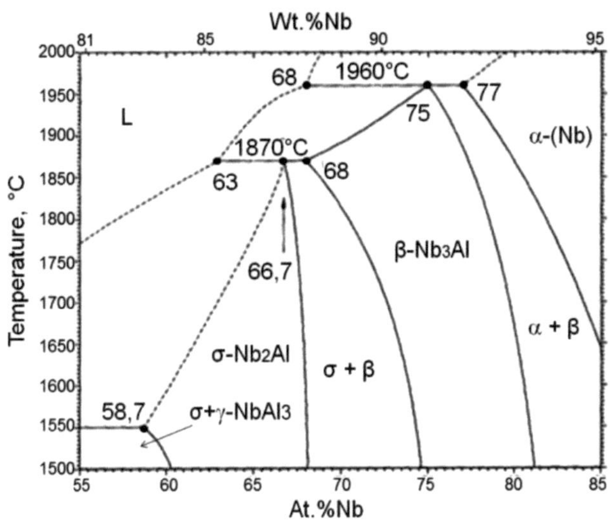

Figure 12. The state diagram Nb–Al system.

In the works of V. I. Arkharov, the appearance of new phases during reactive diffusion is described as follows. At first, in the regions adjacent to the phase boundary (in our case – to the boundary Nb-σ), the solid solutions are formed. When reaching of the saturation the solid solution lattice is rearranged to form a new phase. In a recent paper, the theory of "activated complex" was developed, which states that in the first stage of reactive diffusion, a border zone with a composition close to a new phase is formed, but the order of the arrangement of atoms differs from the order in the final structure of the product. This is the "activated complex", which is a highly distorted part of crystal lattice of product reactive diffusion with a concentration of point defects in excess of the equilibrium concentration. The accumulation of atoms within the reaction zone increases distortions and the achievement of their critical value is happening in the transmission of distortion into the reaction product. That is "complex" as a whole begins to move from reaction zone in depth of the layer of reaction product and to leave behind myself already an undistorted lattice. In order to establish this balance is needed time the less the, the higher the temperature at which there is mutual diffusion.

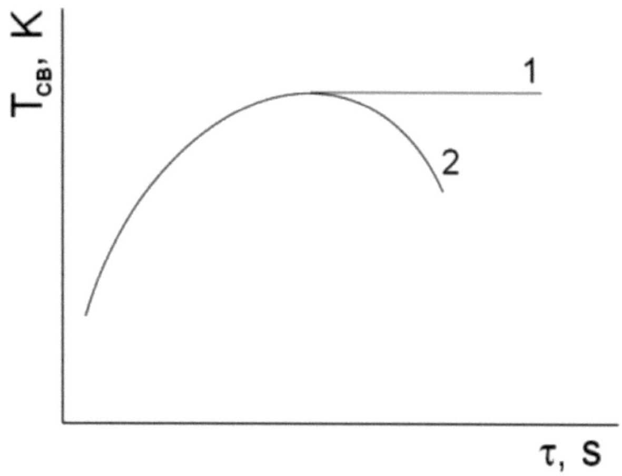

Figure 13. The dependence of the critical temperature T_{CB} from exposure time at a constant temperature of mutual diffusion in a semi-infinite diffusion pair (1) and in diffusion pair with a limited source of aluminum (2).

For diffusion couple Nb–Al, final reaction product is β-phase, the critical temperature depends on the concentration of point defects. At a constant temperature over time, T_{CB} should increase since the amount of defects in its crystalline lattice will become smaller. For the case of a semi-infinite diffusion couple (**Figure 13**, curve 1) T_c during the heating time τ should reach saturation. (Semi-infinite body – this body is extending from one side to a distance much greater than the diffusion length of on the other side is limited by plane x = 0). In investigated tapes, the aluminum is noticeably smaller than niobium and $T_c(\tau)$ has a maximum as a result the diffusion of aluminum on the entire volume of the tape. This was clearly observed mostly in the experimental T_{CB} depending on the heating time (see **Figure 9*a***).

Composition of the β-phase, corresponding to certain the temperature of heating is determined by the phase diagram. The higher the heating temperature, the faster equilibrium is established for the mutual diffusion. At a heating to 1850°C and above the time of stationary regime is so small that the maximum T_{CB} achieved almost immediately after switching on the electric current to heat the sample – for 1–2 seconds.

According to the local X-ray spectral analysis of σ- and β-phases layers were built on the dependencies T_{CB} of Nb_3Al tapes on the content of aluminum in σ- and β-phases (**Figure 14*a*** and ***14b***, respectively). Growth of T_{CB} with decreasing aluminum content in the σ-phase is evidenced on one hand, the directionality of flow the aluminum atoms from σ-layer, that is from the center of sample. On the other hand, due to the limited source of Al-atoms, the interface between σ- and β-phase is moved toward the central axis of the tape section. In line with this, it created a mistaken impression that niobium diffuses into the aluminum layer. In fact, the diffusion process is derived, at least in consequence of contrary streams of aluminum and niobium atoms and is not only owing to niobium as it may seem.

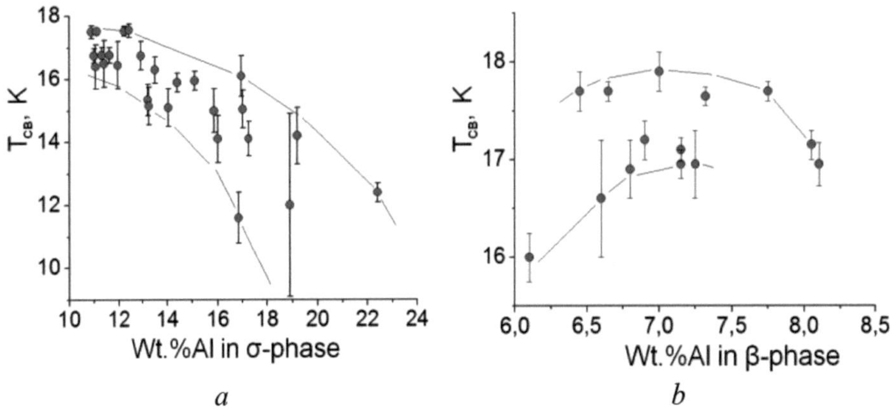

Figure 14. The dependencies of the critical temperature T_{CB} of tape samples Nb/Al/Nb from content of aluminum in the σ-phase (*a*) and in the β-phase (*b*). Vertical line segments are showed the doubled half-width $\Delta T_c/2$.

As the aluminum content in the σ-phase decreased, the content in β-phase increased, approaching the stoichiometric composition, which led to higher T_{CB}. Maximum values T_{CB} were observed at the content of aluminum in the β-phase, equal to 7.0–7.5 wt. % (see **Figure 14*b***). By stoichiometric compositions, Nb_2Al- and Nb_3Al-phases correspond to 11.0 and 7.6 wt. % Al, respectively.

In **Figure 14*a***, it is seen that the aluminum content in the σ-phase layer in a large number of measurements exceeds the content allotted by state diagram Nb–Al (see **Figure 12**). This is because real in σ-layer contained γ-phase (compound $NbAl_3$), mostly aluminum-rich. The data in **Figure 14*a*** were compiled for the samples in the diffusion layers which prevailed σ-phase. Afterwards, as the sample cooled down, it is very possible to observe an "imaginary σ-layer". In fact, it was a layer of σ + eutectic, which is a mechanical mixture of the two phases – σ and $NbAl_3$. Moreover, by using a micro analyzer, average composition of layer was determined.

The fact that in the investigated Nb/Al/Nb-tapes at temperatures of interdiffusion 1700–1900°C for formation a micron-layers of superconducting compound Nb_3Al enough 2–5 seconds time, there is nothing exceptional. This follows the Fick's 2nd law equation

$$\frac{\partial C}{\partial \tau} = D\frac{\partial^2 C}{\partial x^2}$$

for the finite body, where: C – the concentration, τ – time, x – the distance, D – the diffusion coefficient. When considering the diffusion from a layer in which the distribution of matter at T = 0 satisfies such boundary conditions:

http://dx.doi.org/10.5772/64500

$$f(x) = C_0 \text{ at } 0 < x < h$$

$$f(x) = 0 \text{ at } h < x < l \, (l > h),$$

a solution of Fick's looks

$$C = C_o\left(\frac{h}{l} + \frac{2}{\pi}\sum_{1}^{\infty}\frac{1}{h}\exp\left[-\left(\frac{n\pi}{l}\right)^2 D\tau\right]\cos\frac{n\pi x}{l}\sin\frac{n\pi x}{l}\right).$$

Here, $2h$ – Al-layer thickness in the initial tape Nb/Al/Nb, $2l$ – tape thickness, x – β-thickness layer. The time is defined in which a layer of β-phase with thickness of 1 micron grows. For our case: $h = 1 \cdot 10^{-4}$ *cm*, $l = 25 \cdot 10^{-4}$ *cm*, $D \approx 5 \cdot 10^{-9}$ cm^2/s at 1800°C [7], $C_o = 100$ at. %Al, C = 25 at. %Al. We restrict ourselves to n = 2. Having a numerical substitutions and calculations, we find that β-layer of thickness 1 micron is grown for 30 seconds. At 1300°C $\tau = 9$ *h* ($D \approx 5 \cdot 10^{-12}$ cm^2/s [7] and with the same values *h, l* and *x*). In our experiments, in samples of niobium-aluminum tape at 1300°C β-phase was formed during 5 *h*. However, T_{CB} of these samples was equal to 15.7 K. At temperatures of 900–1000°C, a layer of β-phase is formed only during 900 hours (with $h \sim 1$ micron).

In the present studies, layers of β-phase were formed after heating for 2–5 (at 1850°C) and not after 30 seconds as what we have been calculated using the Fick equation. This discrepancy is possible since values of mutual diffusion coefficients are taken from the literature. But our tapes should expect higher values of D because the interacting metals were contacted physically clean – juvenile – surfaces which were formed by their joint rolling. Based on our data, that is, $x \sim 2 \cdot 10^{-4}$ *cm*, $h \sim 1 \cdot 10^{-4}$ *cm*, $l \sim 25 \cdot 10^{-4}$ *cm* and $\tau \sim 3$ *s*, the diffusion coefficient $D \approx 6 \cdot 10^{-8}$ cm^2/s at 1850°C. This is an order of magnitude higher than in [7].

4.3. Tapes Nb/AlGe/Nb. Dependence of T_{CB} on the content of germanium in the alloy Al–Ge

For obtaining of tape samples with intermetallic compound $Nb_3(Al_{0.8}Ge_{0.2})$, instead of aluminum, AlGe-alloys c 30, 40, 45 and 50 wt.% of germanium were used. As a trial alloy, alloy Al–7.5 wt. % Ge was used. The alloys were prepared by melting in a suspended state with casting in a massive copper mold in the form of flat bars of thickness about ~2.5 *mm*, which prior to assembly of three-layer packages were rolled with heating to 400–410°C on air: the temperature close to the melting temperature of eutectic. To form a layer of the intermetallic compound Nb_3(Al,Ge) and to plot of the dependencies critical temperature on the germanium content in the alloy Al–Ge, the tape samples were heated to a high temperature heating at 1650–1850°C. Then, there was low-temperature annealing in two regimes: 750°C, 50 hours and 800°C for 2 hours. Part of the results obtained on measuring critical temperature is shown in **Figure 15**. Almost for all the samples T_{CB} values after low temperature annealing lie above the 19 K. The greatest value of critical temperature was equated to 20.2 K, obtained from the tape with the alloy Al–45 wt. % Ge after treatment on the regime: 1850°C, 2 *s* + 750°C, 50 *h*.

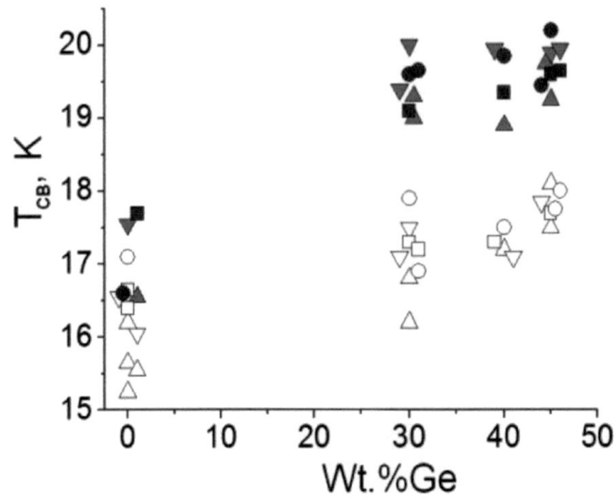

Figure 15. Critical temperature T_{CB} of tapes Nb/AlGe/Nb depending on the alloy composition of Al–Ge. Heat treatment, °C: △ – 1650, 20 s; ▽ – 1650, 30 s; □ – 1750, 3 s и ○ – 1850, 2 s; ▲, ▼, ■ и ● – the same + 750, 50 *h*.

With the addition of germanium in an amount of 7.5 wt. % T_{CB} is decreased a little compared with samples of Nb/Al/Nb-tape. This was observed on several samples with different modes of h.-t. h.

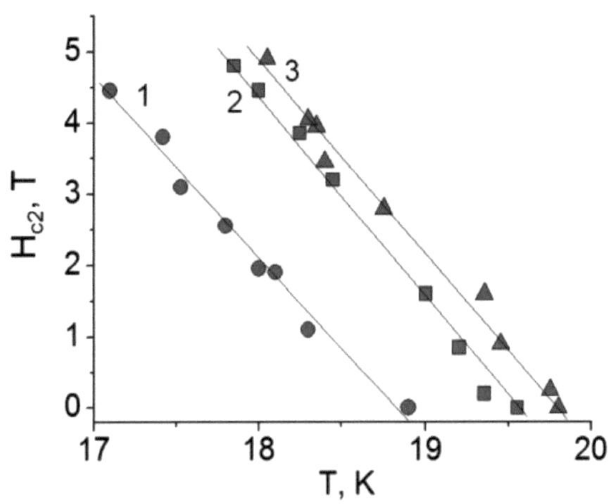

Figure 16. Experimental dependence H_{c2} on temperature near T_c for Nb/Al30Ge/Al tape after h.-t. h. at 1900°C. Heating time, *s*: 1 – 2, 2 – 3 and 3 – 10. L.-t. an.: 800°C, 2 *h*.

4.4. The critical magnetic fields of tapes Nb/Al/Nb and Nb/AlGe/Nb

The upper critical magnetic fields H_{c2}(4.2 K) at liquid helium temperature were determined by the method proposed in [9], measuring its near the critical temperature, and calculating by the following formulas:

http://dx.doi.org/10.5772/64500

$$H_{c2}(0) = -0,69 \cdot T_c \cdot (dH_{c2}/dT)_{T=T_c}$$

and

$$H_{c2}(\mathrm{T}) = \mathrm{H}_{c2}(0) \cdot [1 - (T/T_c)^2],$$

where: T = 4.2 K.

The numerical value of the slope $(-dH_{c2}/dT)_{T=T_c}$ is calculated from the experimental direct $H_{c2}(T)$. Such dependencies are shown in **Figure 16** for niobium-aluminum-germanium the tape after heat treatment at 1900°C.

The calculated values of H_{c2} at 0 and 4.2 K for Nb/Al/Nb tape and Nb/AlGe/Nb tapes with 50-hour annealing at 750°C are summarized in **Tables 2** and **3**, respectively.

Heat treatment, °C	T_{CB}, **K**	$(-dH_{c2}/dT)_{T=T_c}$, *T/K*	H_{c2}, *T*	
			0 K	**4.2 K**
1650, 10 s	15.4	2.1	23	21.3
1650, 20 s	15.9	2.0	22	20.5
1650, 10 s + 800, 2 *h*	16.0	2.2	24	22.3
1650, 20 s + 800, 2 *h*	17.4	2.4	29	27.3
1750, 8 s	16.0	2.6	28	26.1
1750, 10 s	16.4	2.3	23	21.5
1750, 5 s + 800, 2 *h*	17.0	2.5	32	30.0
1750, 8 s + 800, 2 *h*	16.5	2.4	28	26.2
1750, 10 s + 800, 2 *h*	17.0	2.4	29	27.2
1850, 3 s	16.3	2.4	27	25.2
1850, 5 s	16.3	2.3	26	24.3
1850, 8 s	17.0	2.2	25	23.5
1850, 3 s + 800, 2 *h*	17.5	2.2	26	24.5
1850, 5 s + 800, 2 *h*	17.7	2.0	25	23.6
1850, 8 s + 800, 2 *h*	17.4	2.0	23	21.7

Table 2. Results of measurements of the upper critical magnetic field H_{c2} near the critical temperature for the tape samples Nb/Al/Nb c 4.7 vol. % Al.

A tilt of straight lines $H_{c2}(T)$ for Nb/Al/Nb tape varies in the range of 2.0–2.6 *T/K* regardless of heat treatment. Low-temperature annealing, raising T_{CB}, is increased H_{c2}. Its best values at 4.2

were accounted for tapes with heating at 1750°C. After annealing at 800°C, H_{c2}(4.2 K) is equal 30 *T* by heating for 2 seconds and with the increase of its duration to 8–10 *s* it was reduced to 26–27 *T*.

The niobium-aluminum-germanium tape with alloy of Al–45 wt. % Ge having the maximally out of achieved critical temperature of 20.2 K (see **Table 3**), possesses and the highest H_{c2}(4.2 K), equal to 50.7 tesla. As a consequence of annealing at 750°C for 50 *h*, the value of the slope $(-dH_{c2}/dT)_{T=T_c}$ increases to 3.8 *T/K* in comparison with samples without the low-temperature annealing.

Wt.%Ge	L.-t. annealing, °C	T_{CB}, K	$(-dH_{c2}/dT)_{T=T_c}$, *T/K*	H_{c2}, *T*	
				0 K	4.2 K
30	Without annealing	17.8	3.4	42	39.5
30	750, 50 *h*	19.8	3.8	51	48.7
40	Without annealing	17.5	3.5	42	39.6
40	750, 50 *h*	19.7	3.8	51	48.7
45	Without annealing	18.1	3.3	42	39.7
45	750, 50 *h*	20.2	3.8	53	50.7

Table 3. Results of measurements of the upper critical magnetic field H_{c2} near the critical temperature for the tape samples Nb/AlGe /Nb after h.-t. h. at 1850°C for 2 seconds.

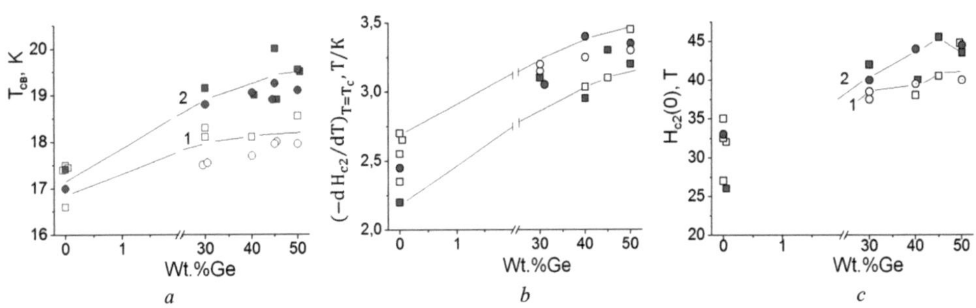

Figure 17. The dependence of the transition temperature to the superconducting state T_{CB} (*a*), the slope of the straight $H_{c2}(T)|_{T=Tc}$ (*b*) and the upper critical magnetic field at 0 K $H_{c2}(0)$ (*c*) on the germanium content with the alloy Al–Ge for tapes Nb/AlGe/Nb after h.-t. h. and after h.-t. h. + l.-t. an. Heat treatment, °C: ○ – 1750, 5 *s* and □ – 1850, 3 *s*; ● and ■ – the same + 800, 2 hours.

Towards the values of H_{c2}(4.2 K) ~ 50 *T* should be treated critically, since direct measurements of H_{c2} with use the pulse technique were not carried out.

Dependencies of tilt $(-dH_{c2}/dT)_{T=T_C}$ and H_{C2} as well as T_{CB}, measured in experiments to determine the H_{C2}, depending on the germanium content in the alloy Al–Ge for the second series of Nb/AlGe/Nb tapes are shown in **Figure 17**. Concentration diapason of germanium in the alloy was extended up to 50 wt. % Ge.

The maximum superconducting characteristics were for the samples with 40–50 wt. % Ge. The critical temperature after heating and annealing was 19–19.5 K, $H_{c2}(0)$ = 42–45 *T*. At 4.2 K H_{c2} = 40–43 *T*. The two-hour annealing at 800°C significantly increases the T_{CB} and H_{c2} compared with samples without annealing, but any effect on the slope of $H_{c2}(T)$. The latter circumstance is somewhat contrary to the results obtained for samples previous series (see **Table 3**). In those tapes 50-hour annealing increased the slope of dependences $H_{c2}(T)$ from 3.3–3.5 to 3.8 *T/K*.

Increasing of the upper critical magnetic field of the diffusion layers Nb_3Al and Nb_3AlGe was explained by the collapsing of the superconducting phase with allocation σ(Nb_2Al)- and σ(Nb_2AlGe)-phases accordingly. The upper critical field of superconductors II-kind equal $H_{c2} = \sqrt{2}\cdot\varkappa\cdot H_{cm}$, where $\varkappa$ is the parameter of the Ginzburg-Landau and H_{cm} is the thermodynamic critical field. The parameter $\varkappa$, defined by the ratio of the penetration depth of the magnetic field λ to the coherence length ξ ($\varkappa = \lambda/\xi$), in alloys and compounds can be great. In dirty superconductors II-kind with a high concentration of impurities, coherence length will decrease with decreasing the mean length of free path of electrons in a normal state l_e, what and should occur during the decay β-phase in the process of low-temperature annealing. According to the theory of Gorkova: $\varkappa \sim l_e^{-1}$.

Wt.%*Ge*	Regime of temperature treatment, °C	T_{CD}, K	$H_{c2}(0)$, *T*	Content of element in β-layer, *wt.%*			Chemical formula of β-phase
				Nb	Al	Ge	
30	1750, 5 s	17.6–17.7	36–37	90.4	5.2	1.8	$Nb_{3.2}Al_{0.7}Ge_{0.1}$
45	1750, 5 s	17.8	-	91.5	5.9	2.3	$Nb_{3.1}Al_{0.8}Ge_{0.1}$
45	1850, 3 s	18.1	41	90.8	5.1	3.7	$Nb_{3.2}Al_{0.6}Ge_{0.2}$
50	The same + 800, 2 *h*	19.6	48	90.9	5.7	4.6	$Nb_{3.1}Al_{0.7}Ge_{0.2}$

Table 4. Data local of X-ray spectral analysis for Nb/Al/Nb-tape samples with the alloys Al–Ge of different composition and their values of T_{CB} and H_{c2}.

On the decay of the superconducting β-phase, we will get back when considering critical currents, but now, let us compare the change of superconducting parameters T_{CD} and H_{c2} of compound $Nb_3Al_{1-x}Ge_x$ with a change in its composition. Increasing of T_{CB} compound of niobium with aluminum and germanium is due, firstly, to the approximation of composition β-phase to stoichiometric and, secondly, to the approximation of composition β-phase to a range of concentrations of aluminum and germanium, which is considered optimal in terms of the critical temperature and critical field (x = 0.2). If a greater or smaller difference in the composition of the superconducting phase from the stoichiometry is rigidly dependent on the temperature of the high-temperature heating, then the correlation of germanium and alumina into the β-phase, we can adjust by arbitrary change the germanium content in the alloy Al–Ge in the range from 30 to 50 wt.%. To determine the composition of the β-phase, a local X-ray spectral analysis was used. These data and values T_{CB} and H_{c2} for the two tested tape samples are summarized in **Table 4**.

With the increase of germanium concentration in the Ge–Al alloy with 30 to 45–50 wt. % Ge its content in the β-phase was increased from 1.8 to 3.7–4.6 wt. %, and the ratio of Al and Ge in the phase varied from 0.7/0.1 to 0.7/0.2. With this, T_{CB} has risen by 17.6–17.7 to 19.6 K and $H_{c2}(0)$ – from 36–37 to 48 Tesla. Note that niobium in the β-phase containing germanium was always more than its stoichiometric composition. It is very likely that such a deviation explained by the fact that the compound Nb_3Ge at 1750–1850°C is stable in the range of compositions sufficiently far from the stoichiometric. Then, it is possible to conclude that high temperature of transition to the superconducting state and the critical magnetic field of compound $Nb_3(AlGe)$ are caused advantageously by how close the relation of aluminum and germanium to the ratio of 0.8/0.2.

4.5. Current-carrying capacity of superconducting tapes Nb_3Al and Nb_3AlGe. Influence of temperature and time h.-t. h. (on example tapes of Nb/Al/Nb)

Measurements of the critical current I_c of tape samples was performed in a superconducting solenoid at a temperature of 4.2 K. The sample was placed in such a manner (**Figure 18**), that the direction of electric current I, flowing through the sample was perpendicular to the magnetic field H, which was produced by a solenoid (I ⊥ H), and plane (of rolling) of sample (*ab*) was parallel H – (*ab*) | | H. In this position, the Lorentz force F, acting on the superconducting vortices in the layer of superconductor is directed perpendicular to the (*ab*). Width of the tape is 2 *mm*.

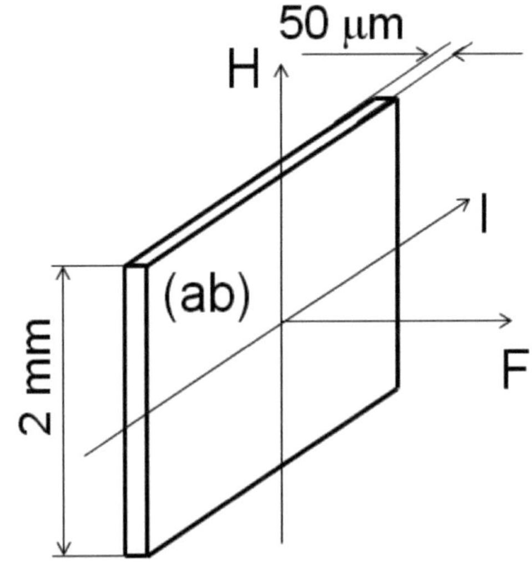

Figure 18. Disposition of tape specimen in the working area of the superconducting solenoid: H – magnetic field, I – transport current through the sample, (*ab*) – plane of the surface of the tape, F – Lorentz force.

http://dx.doi.org/10.5772/64500

Numerous measurements of the critical current I_c showed (**Figure 19**) that their maximum values in the tapes Nb/Al/Nb fall on the a fairly narrow temperature range of the initial heating of 1700–1750°C and reach 15 *A* (or 7.5 *A/mm* – amperes per millimeter of the tape width).

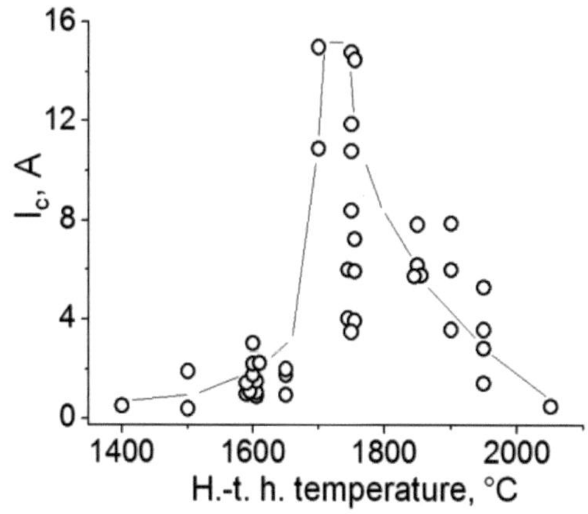

Figure 19. The dependence of the critical current I_c tapes Nb/Al/Nb on temperature of h.-t. h. Heating times were corresponding to the larger of the achieved values I_c at a given temperature. Regimes of l.-t. an.: about 800 and 900°C for 2 *h*.

From the viewpoint of the critical current I_c and the critical current density j_c most optimal are the shortest times of heating of 1–3 *s* at 1700–1800°C (**Figure 20**). The value I_c in the magnetic field of 5 *T* in the best samples were the same 15 *A*. This corresponded to $j_c = 2 \cdot 10^5$ A/cm^2 in calculation on layer Nb_3Al.

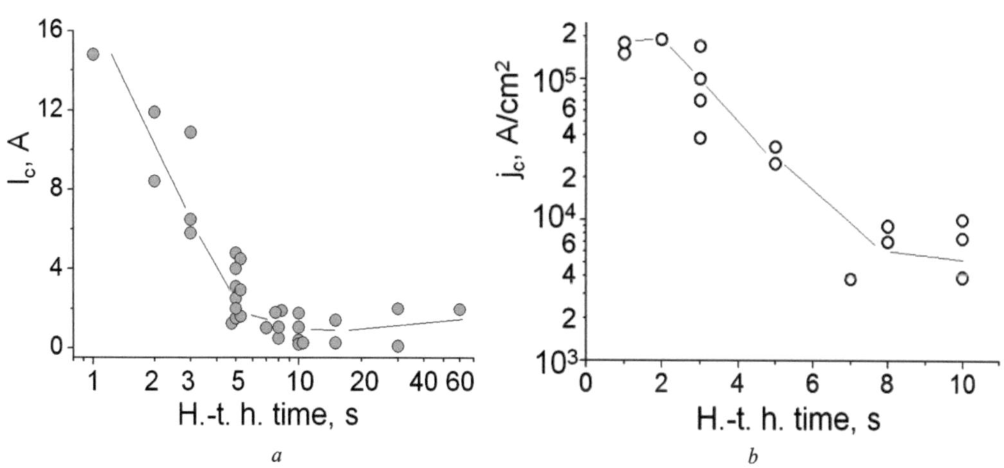

Figure 20. The dependence of the critical current I_c (*a*) and the critical current density j_c (*b*) tapes Nb/Al/Nb on the time of h.-t. h. in a magnetic field of 5 Tesla. Temperatures of the heating 1600, 1750, 1850 and 1900°C were corresponding to the larger of the achieved values I_c for a given duration of heating. Regimes of l.-t. an.: 800 and 900°C for 2 hours.

Increasing the heating time to 8–10 *s* leads to a decrease of the critical current to ~1 *A*, and the critical current density to (4–8)·10^3 A/cm^2 in a magnetic field of 5 Tesla.

4.6. Influence of the diffusion layer structure on the critical current and critical current density.

In considering the dependence of the critical current I_c on the structure of diffusion layer Nb–Al, namely the thickness of β-phase, the layer s_β might be expected to increase in a monotone. However, measurements of I_c showed that the current was rapidly increased with s_β only in a narrow, 0.5 to 2 microns, range of layer thickness, and then just as quickly fallen even below its initial level (**Figure 21*a***).

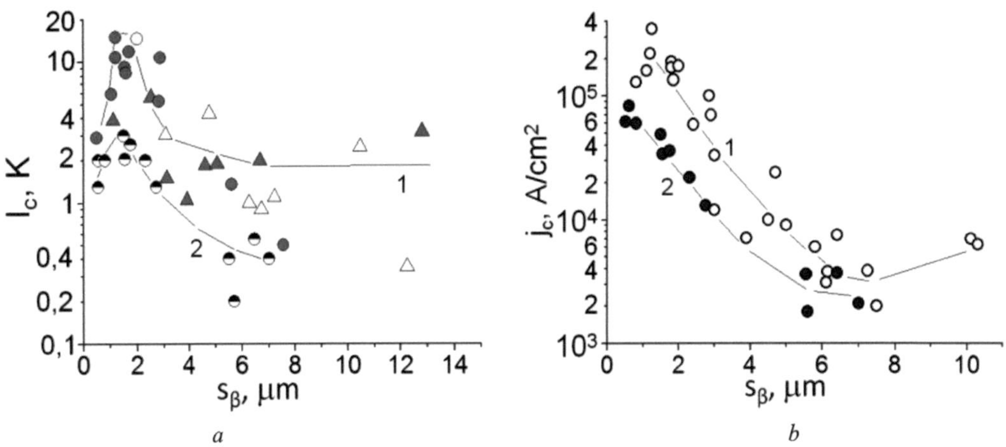

Figure 21. The dependence of the critical current I_c (*a*) and the critical current density j_c (*b*) on layer thickness of β-phase in tapes Nb/Al/Nb (1) and Nb/AlGe/Nb (2) in a magnetic field of 5 *T*.

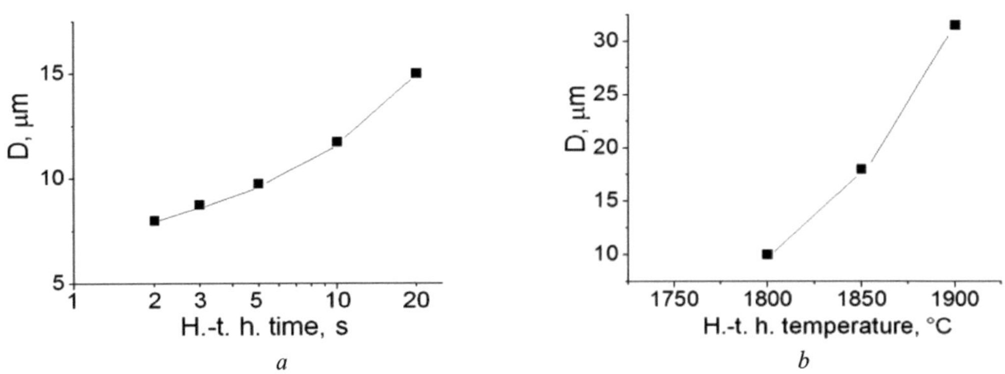

Figure 22. The dependence of the average grain size of β-phase in the tape Nb/Al30Ge/Nb on the temperature (*a*) and time (*b*) of the high temperature heating: *a* – 1800°C; *b* – 5 *s*.

That is, the critical current in tapes with diffusion layers containing, along with the β-phase, σ- or σ- + γ-phase, was significantly greater than in the tapes with diffusion layers only of β-phase. Maximum value of I_c observed in the tapes Nb_3Al with thickness of its layer equal 1–2 *μm*, and then dropped to 1–2 *A* in the layers of 10–14 microns thick. It was sharply decreased with the increase of β-layer thickness and critical current density (**Figure 21*b***). With s_β, equal to 1–2 microns, j_c reached ~ $2 \cdot 10^5$ *A/cm*^{2} but in a layer of thickness of greater than 5 microns current density dropped two orders of magnitude.

Similar behavior demonstrated tape out of compound Nb_3AlGe: the critical current in the tapes with $s_\beta = 2$ *μm* equal 3 *A* was decreased to 0.2–0.5 *A* at a thickness of β-layer 5 and more microns, the critical current density at this point ranged from ~$7 \cdot 10^4$ to about $3 \cdot 10^3$ *A/cm*^{2} in a field of 5 Tesla.

4.7. Effect of the grain size of β-phase on the critical current density Nb_3AlGe

Critical current in A15 compounds such as Nb_3Sn and V_3Ga, is dependent on the grain size, and the centers of fastening the superconducting vortices in them are mainly grain boundaries. Therefore, the smaller the grains of the compound, the higher the critical current density. This was shown in compound Nb_3AlGe, obtained in composite tapes out of niobium and alloy Al–30 wt. % Ge.

In studied tapes, which were subjected to heat treatment at relatively high temperatures, the grain size of β-phase in formed diffusion layers is very sensitive to temperature and time of the high temperature heating. Based on the data of metallographic analysis (**Figure 22**), the average grain size Nb_3AlGe D depending on the heating ranged from about 8 micrometers after h.-t. h. at 1800°C for 2 seconds (**Figure 23*a***) to 33 micron after 5-second heating at 1900°C (**Figure 23*b***). That is why, the increase of critical current was observed at the initial stage of formation of the compound and associated with an increase of thickness of its layer (see **Figure 21**) are not accompanied by the increase of the critical current density, since at an increase s_β very rapidly grew and grain of β-phase (**Figure 24**). It comes to the size of grains in a plane parallel to the rolling tape. It is understood that the grain size of the phase in the perpendicular direction cannot exceed the thickness of its layer.

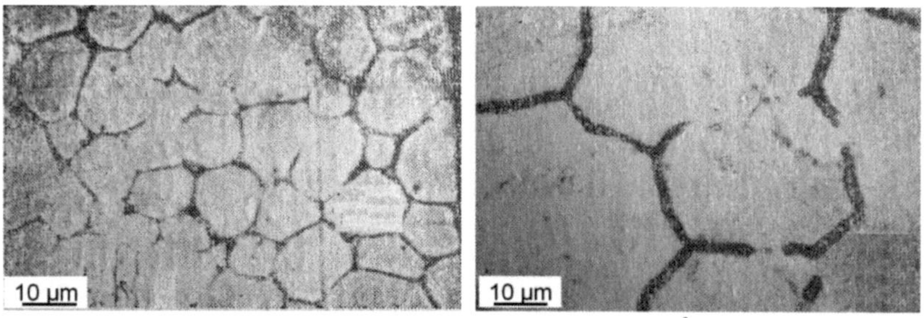

a *b*

Figure 23. The microstructure of β-phase layer in the tape Nb/Al30Ge/Al after h.-t. h. at 1800°C for 10 seconds (*a*) and 1900°C for 5 seconds (*b*). L.-t. an.: 800°C, 2 *h*.

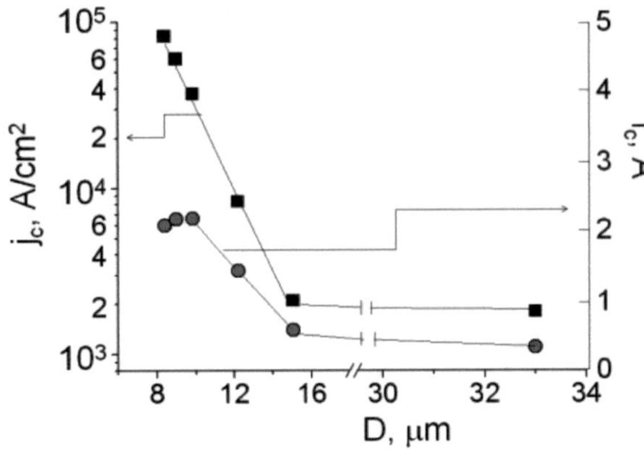

Figure 24. The dependence of critical current density j_c and a critical current I_c on the average grain size of β-phases in tapes Nb/AlGe/Nb with the alloy Al–30 wt. % Ge. L.-t. an.: 800°C, 2 *h*.

4.8. Effect of low-temperature annealing on critical temperature and critical current of tapes Nb/Al/Nb

Dependencies T_{CB} and I_c on temperature of l.-t. an. at constant time of annealing of 2 hours are shown in **Figure 25**. Compared to tapes without annealing, the critical temperature after low temperature annealing in the range of 500–700°C was on 1 K less, and the critical current was equal to 0.4 *A*. In samples without annealing $I_c = 1.5–3$ *A*. However, basing on the graphs for the starting point to take the room temperature, then both dependencies in that temperature interval will have a minimum. At the next increase of the annealing temperature T_{CB} and I_c are increased, can reach maximum and further fall. The maximum value of the critical current, relative to the maximum value of critical temperature is shifted to a much higher temperature, 900–1100°C, which depends on the high-temperature heating. The largest I_c equal to 7–10 *A*.

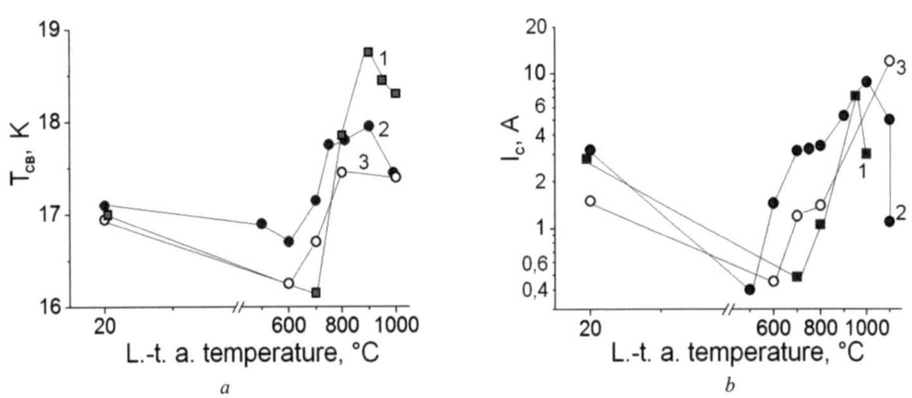

Figure 25. Critical temperature T_{CB} (*a*) and the critical current I_c in magnetic field of 5 *T* (*b*) for the tapes of Nb_3Al depending on temperature of l.-t. an. Time of the annealing – 2 *h*. Temperature of h.-t. h., °C: 1 – 1850, 2 *s*; 2 – 1850, 3 *s*; 3 – 1950, 3 *s*.

http://dx.doi.org/10.5772/64500

Dependencies T_{CB} on the annealing time were of two types (**Figure 26**).

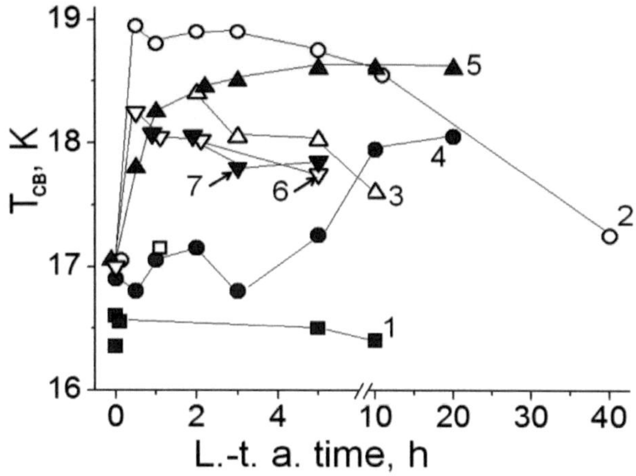

Figure 26. Dependencies of the critical temperatures T_{CB} on the time of l.-t. an. for tapes Nb/Al/Nb and Nb/Al30Ge/Nb (2). Temperature treatment, °C: 1 – 1750, 3 s + 1000; 2 – 1850, 2 s + 900; 3 – 1850, 2 s + 1000; 4 – 1850, 3 s + 700; 5 – 1850, 3 s + 800; 6 – 1850, 3 s + 900; 7 – 1950, 3 s + 900.

1. At annealing temperatures of 700 and 800°C (curves 4 and 5) the critical temperatures monotonically increase and, reaching a maximum, remained on constant level.

2. The tapes, annealed at 900°C (2, 6, and 7) have a gently sloping maximum. Pronounced maximum was of sample with temperature of annealing at 1000°C (3). Their T_{CB} came back after 5 hours annealing to a level of the critical temperatures for the samples not subjected to annealing. It is very likely that the rise, a maximum and drop of T_{CB} have occurred earlier than 5 hours (1). The time at which the maximum values of the critical temperatures were reached, with the increasing of annealing temperature was decreased. At 700 and 800°C maximum of T_{CB} advanced, after annealing for 10 and 3–5 hours. At 900°C was after a half-hour annealing.

To trace the dependencies of the critical currents on the annealing time for the samples with the same regimes of the high-temperature heating failed. From all the number of measurements, it is possible to confidently state that only the general level of I_c values increased after 10 hours of the annealing.

4.9. Low-temperature annealing – a necessary tool to enhance critical temperature and critical current of compounds Nb_3Al and Nb_3AlGe

As a result of the low-temperature annealing in the range of 700–1000°C in the layers of superconducting compounds Nb_3A land Nb_3AlGe should be observed in two processes:

- an increase of order degree in the arrangement of atoms in knots of the crystal lattice and
- decay with allocation σ-phase, due to the non-equilibrium of resulting β-phase.

If the initial and final crystal structures of the compounds are though somehow the same type, the ordering process consists of an exchange of atoms places and does not require diffusion over long distances. The rate of process depends on the temperature and by the parameter of the estimation of speed its change will the change T_{CB} on time and temperature of annealing. The rate of process depends on the temperature. Then the change T_{CB} on time and temperature of the annealing will be the parameter of the estimation of speed its change. Effect of ordering on T_{CB} of Nb_3Al and Nb_3AlGe is reflected by curve 1 in **Figure 27**. The maximum established Nb_3Sn and V_3Ga, is located at 700–750°C. For similar compounds of niobium with aluminum, it is achieved at somewhat higher temperature, about 900°C (see **Figure 25*a***). It is thought that the ordering processes are actively occurring in the initial stage of annealing. Since the diffusion mobility of atoms increases with increasing temperature, then and T_{CB} reach their maximum values at a given temperature in more short time – at 900°C for 30 minutes, at 700°C for 10–20 *h* (see **Figure 26**, curves 4 and 6).

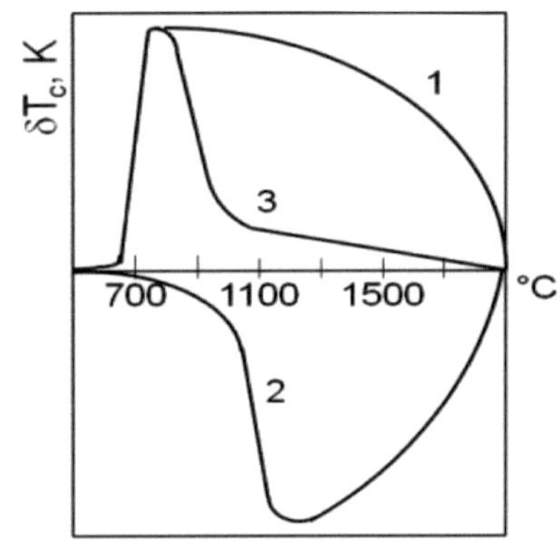

Figure 27. Change the critical temperature T_c depending on the annealing temperature under the influence by the degree of ordering (1) and the equilibrium composition (2): 3 - summary curve.

Precipitation of σ-phase, in contrast to the ordering process should reduce T_{CB} (see **Figure 27**, curve 2). It was already mentioned that in the systems of Nb–Al and Nb–Ge, the temperature boundaries of homogeneity areas of β-phases are changed and so the percentage composition of phases also changed. Characters of the change in composition such that:

- at 1700°C Nb_3Al stably in the ranges of 23–26 at. % Al, at 1000°C – in the ranges of 19–23 at. % Al;
- Nb_3Ge is stable with compositions that are far from the stoichiometric, namely, at 1700°C – in the range of 16–19 at. % Ge, at 1000°C – in the ranges of 15–18 at. % Ge.

Therefore, at low-temperature annealing Nb_3Al- and Nb_3AlGe-phases necessarily are undergoing the disintegration with a precipitation of σ-phase. A matrix consisting of β-phase is depleted with aluminum or aluminum and germanium, resulting in the removal of β-phase composition from stoichiometric composition and decrease of T_{CB}.

The present study shows that in the tapes of Nb_3Al at 2 hours of exposures at annealing up to 900°C, the ordering affects to a greater extent than the process of decay. The ordering completely is dominated at relatively long annealing in the range of 700–800°C. Falling of T_{CB} in this temperature range is not observed at exposure up to 20 hours (see **Figure 26**, curve 5). This accounts for the decrease of T_{CB} beginning from 900°C observed on its annealing temperature (see. **Figure 25*a***) and on depends on the annealing time at 900°C (see **Figure 26**, curve 2).

Unlike the critical temperature, the critical current is important in the decay process since it is accompanied by the precipitation within the grains of the superconducting phase particles of σ-phase which, together with the boundaries of the grains should be additional superconducting vortex pinning centers. In the scientific literature on superconducting materials the fastening of vortices accepted to call "pinning" and fastening centers – "pinning centers". Therefore, I will continue to use this term.

Stages of decay process with the precipitation of σ-phase particles, how they appear to us, will be considered later. Here the same we note the following. For vortex pinning more favorable is the stage when the final phase (in this case – it σ) becomes incoherent with the matrix (β-phase), that is, when there is the interphase boundary. By this explains the earlier marked shift of the critical current maximum on 50–200°C with respect to the maximum of critical temperature towards higher temperatures l.-t. an. (see **Figure 25*a*** and ***25b***). It can be assumed that after 2-hour annealing at 900°C, particles of σ-phase are the intermediate product coherently associated with β-matrix, although it already have a tetragonal structure and a composition near to composition of equilibrium precipitates. This stage of the disintegration process is critical for T_{CB}. Matrix is already enough depleted in aluminum in order for T_{CB} to drop, but still is such that the ordering remains predominant. At this stage, the pinning force on σ-precipitates has not yet reached the maximum value. Its maximum will come at a temperature slightly higher when the particles of σ-phase will separate and the interphase boundaries are formed. But, becoming isolated, σ-particles begin to grow and rapidly to coagulate. The particles size becomes more than the efficient size from the standpoint of the pinning. Upon reaching the maximum value, the critical current drops sharply.

The minimum of I_c at 500–700°C (see **Figure 25*b***) explained by a simple removal of thermal stresses in the diffusion layer of β-phase, typically occurs at low temperatures of the annealing.

Explanation of complex dependencies of the critical current and the critical temperature on regimes of low-temperature annealing, and even earlier, and increasing the upper critical magnetic field as a result of l. t. an. are based on classical concepts that the non-equilibrium alloys until the formation of the precipitates during annealing was preceded changes inside of a supersaturated solid solution (matrix). By Genie with employees on the basis of results of the analysis of a thin X-ray structure studies first it was concluded about education of the certain zones at a very early stage of decay process.

1. Initially are formed Genie-Preston zones 1 which were presenting itself areas of matrix with an increased content of the dissolved element but with the same arrangement of atoms as in the matrix. The difference in the size of components atoms initiates an occurrence of distortion of the crystal lattice around the area. Exactly by the existence of

significant distortions in matrix are explained diffraction effects, which were observed in the study areas. In our case the matrix was been by β-phase supersaturated with an aluminum or aluminum and germanium.

2. In the process of annealing in zones, a new crystalline structure corresponding to arrangement of atoms in the crystalline structure of precipitates (tetragonal – for the case of decay β-phase) is formed. The precipitates on the second stage of the low-temperature annealing or Genie-Preston zones 2 are similar to the coherent precipitates. While they are not in equilibrium and represent itself an intermediate product coherent with the β-matrix and having tetragonal structure and composition near to the composition of the equilibrium precipitates σ-phase . In the supersaturated solid solutions, such zones (clusters) are formed very quickly with the most significant changes in the physical properties taking place till the fixed structural changes.

   When implementing such mechanism disintegration of β-phase, a decrease in mean length of free path of electrons in a normal state must occur at these two stages of the structural changes in β-matrix. At relatively low 750–800°C, temperatures of the annealing the disintegration β-phase was not completely realized.

3. Increase annealing temperature to 900–1100°C leads to the formation of the final σ-phase. It is already non-coherent with the matrix. Phase boundaries were formed, leading to a marked increase in the critical current.

So our reasoning on the effect of low temperature annealing on superconducting characteristics of tapes compounds Nb_3Al and Nb_3AlGe can be summarized in the following.

1. The stage of disintegration with the formation of Genie-Preston zones 1 and 2:

   - increase of the upper critical magnetic field H_{c2}, due to a decrease in mean length of free path of electrons in a normal state l_e;
   - increase of the critical temperature T_{CB} due to the predominant influence of the ordering process in comparison with the process disintegration;
   - a small increase of the critical current I_c.

2. Stage of disintegration of with the formation of σ-phase:

   - maximum of I_c;
   - drop of T_{CB} in resulting of depletion β-phase toward aluminum or aluminum and germanium.

3. Coagulation stage of particles σ-phase: drop I_c because of that the size of σ-precipitates become larger its optimum size with point of view for the pinning of superconducting vortices.

Likely the results of the local X-ray spectral analysis, which inflicted on the fragment of the triple phase diagram Nb–Al–Ge (**Figure 28**) correspond to one of three stages beta-decay in the tapes of superconducting compound Nb_3AlGe.

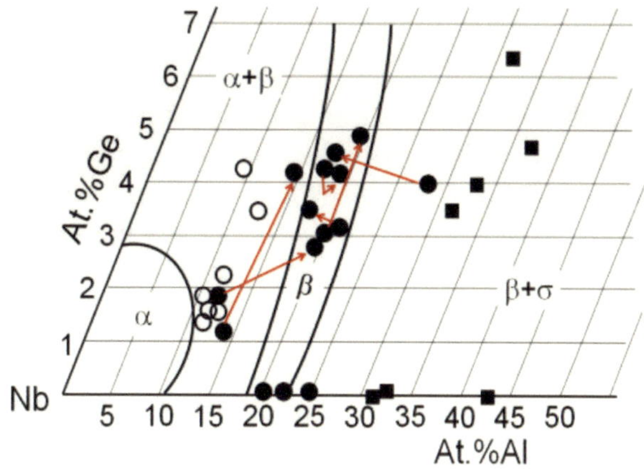

Figure 28. Isothermal section of the niobium angle of state diagram of the system Nb–Al–Ge at 1000°C: ○ – α, ● – β, ■ – σ. Heat treatment, °C: 1800–1900, 3–20 *s* + 800, 2 *h*. The changing of the phase contents of elements in β-phase after Low-temperature annealing is shown by arrows.

It is unclear why points that concerns the β-phase, after high-temperature heatings found themselves on either side of its region of existence. It is important to note that the "β-points" are going to the existence region of β-phase. But the points of β-phase, which already before the annealing were in its homogeneity region, were moving within it. According to the selected time and temperature of the annealing, and a change in the content of elements in the β-phase can be assumed that the ordering here was the dominant process.

4.10. Conclusions

1. The diffusion layers of compounds Nb_3Al and Nb_3AlGe in three-layer composite tapes Nb/Al/Nb and Nb/AlGe/Nb with high superconducting characteristics with thickness of 1–2 microns are formed at temperatures of 1750–1850°C for 2–3 seconds. In these cases, the diffusion layers, in addition to β(Nb_3Al)- or β(Nb_3AlGe)-phase, are also σ(Nb_2Al)- or σ(Nb_2AlGe)-phase and also σ- + γ($NbAl_3$)- or σ- + γ[$Nb_5(AlGe)_3$]-phases. With the increase of heating time, the thickness of the β-layer was increased. The diffusion layer only consisted of one β-phase after the heating during 5 seconds. After the heat treatment, tapes were flexible and able to bend with a certain radius without destroying the superconducting layer.

2. For Nb/AlGe/Nb tapes dependencies of the critical temperature and the upper critical magnetic field from the germanium content of the alloy Al–Ge were studied. The best parameters of superconductivity were observed in tapes with an alloy of 40 to 50 wt. % Ge.

3. An influence of regimes the low-temperature annealing was investigated for tapes Nb/Al/Nb. It is shown that there is a principle possibility by choosing regime of the annealing to increase the critical current of tapes by 2–3 times, compared to non-annealed tapes. Dependencies T_{CB} and I_c from time and temperature of annealing are explained by the

laws of reactive diffusion and by features of the diagrams state of systems Nb–Al and Nb–Al–Ge.

4. It is found that the critical current and critical current density j_c decrease with increasing layer thickness of β-phase. Their highest values are reached in tapes with layers of β-phase, with thickness of 1–2 microns. With the help of metallographic analysis for samples of Nb_3AlGe it is found that with the increase in thickness of β-layer, the average size grain of the compound is increased. This was the reason for the fall of j_c at the growth of thickness of the layer of β-phase.

5. It is established that at low-temperature annealing of the samples Nb_3AlGe, 750°C for 50 h, the slope of the straight $H_{c2}(T)$ near T_c is increased with 3.3–3.5 to 3.8 *T/K*. Then, the calculated values of H_{c2}(4.2 K) are reaching ~50 *T*. The increasing H_{c2} as compared with samples without annealing was the consequence of the decrease in mean length of free path of electrons in a normal state l_e. The decrease l_e is explained in the light of the Genie-Preston theory about the decay of supersaturated solid solutions.

6. Tapes Nb_3Al and Nb_3AlGe, made according to the best regimes have values of the superconducting parameters are presented in **Table 5**.

The tape	T_{CB}, K	H_{c2}(4.2 K), *T*	At 4,2 K in field поле 5 *T*	
			I_c, *A/mm*	j_c in β-layer, *A/cm²*
Nb/Al/Nb	18.0–18.5	30	7.5	$2\cdot10^5$
Nb/AlGe/Nb	19.5–20.2	43	1.5	$9\cdot10^4$

Table 5. The values of the superconducting parameters of the tape samples from Nb_3Al and Nb_3AlGe compounds prepared according to optimum regimes of heat treatment.

5. Influence of copper on the structure and critical current of superconducting tapes from Nb_3Al

5.1. Samples and determination method of the grain size of β-phase

Studies were performed on composite tape that was produced by joint rolling of two plates made of niobium and one plate of an alloy of Al–Cu. Aluminum alloys with a sufficiently high content of copper are not deformed by rolling without breaking. Therefore, the concentrations range of copper in alloys Al–Cu was from 0.7 to 12 wt. %. The width of the tape is 2 *mm*, the thickness – 50 *μm*. To obtain layer Nb_3Al tape was heat treated by mode: 1650°C for 20 and 30 *s*, 1750°C for 3 and 5 *s*, and 1850°C for 3 *s*. Then annealing at 800°C for 2 hours was followed.

The grain size of the β-phase was determined on surfaces of the tapes parallel to the rolling plane. For this, surfaces was ground and polished until appearance of the intermetallic compound layer. Polishing and successive etching was performed several times. The layer β-phase is bordering on niobium. A layer of β-phase borders with niobium. But to distinguish β-phase from niobium was easy, since niobium is not etched at those times at which layer of β-phase is etched. Etching time was 5–10 seconds.

Furthermore, measurements of micro hardness showed that micro hardness of niobium considerably smaller micro hardness of portions β-phase. To reveal the grain boundaries of β-phase, chemical reagent of equal parts of hydrofluoric and sulfuric acids and water with a small amount of hydrogen peroxide was used. The grain size was calculated through "cutting".

5.2. Critical current

The best results were obtained in tapes with the heat treatment at 1850°C. A critical current I_c equaled 38 *A* is observed in samples with alloy Al–12 wt. %Cu in the magnetic field of 5 *T*. After heating at 1650 and 1750°C I_c are smaller.

Dependence of I_c on the alloy composition of Al–Cu must divided into two regions (**Figure 29**). The area 2 is the increase in critical current of tapes with increasing concentration of copper from 4–8 to 30–38 *A*. The area 1 is the same, but has the values of I_c are much lower – I_c is increased from ~1.5 to 3.7 *A*. The width of tapes is of 2 *mm*.

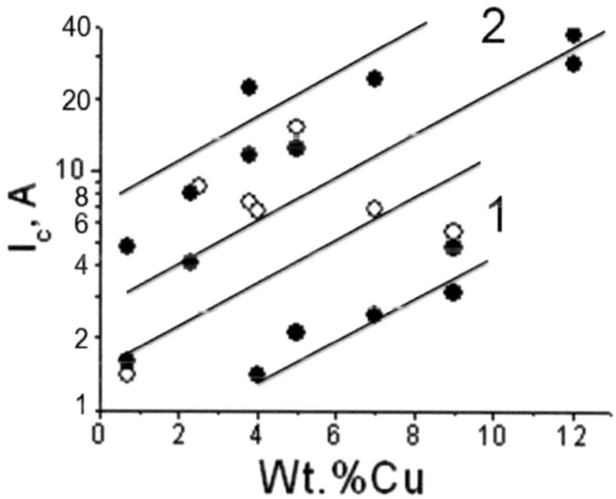

Figure 29. The critical current I_c of Nb/AlCu/Nb-tapes on the copper content in magnetic field of 5 *T* and at 4.2 K after the h.-t. h 1850°C for 3 *s* (○) and the same as + 800 for 2 *h* (•): 1 – samples with the diffusion zone of the β-phase, 2 – samples with zone of β + α. The width of tape – 2 *mm*.

5.3. Discussion

The existence of two ranges of critical current values is explained by the different microstructure of the diffusion zones in the samples. In samples of the first range the diffusion zone consists of only layer β-phase (**Figure 30*a***), samples in the region 2 have the diffusion zone out of the two layers β(Nb_3Al)-phase and one layer σ(Nb_2Al)-phase (**Figure 30*b***). Copper is present in the layer β-phase in quantity of 0.1–0.2 % if the diffusion zone is two-phase and 0.7 wt. %, if it is a single-phase.

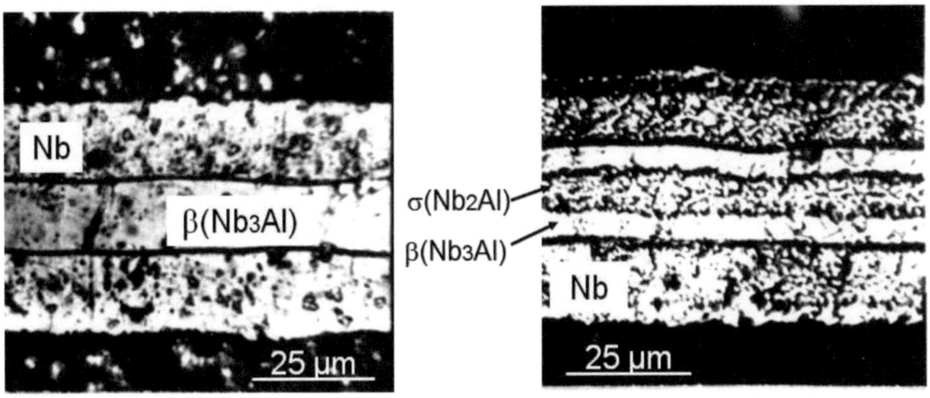

Figure 30. Microstructure of the cross-sectional Nb/AlCu/Nb-tape, wt. %Cu: *a* – 5, *b* – 12. Heat treatment: 1850°C for 3 *s* + 800°C for 2 *h*.

The diffusion region is one third the volume of the tape. It can be observed that after high-temperature heating for 3 *s*, the diffusion zone can consist of two layers of β-phase with a thickness of 5 *µm* and one σ-phase layer with a thickness 8.6 *µm* so and only one layer of β-phase with thickness 13–15 *µm*. Such cases are explained by deviations of ratio Nb-thickness to the thickness of the alloy Al–Cu along the length of the tape and the inaccuracy of compliance 3 seconds exposure at heating. At the second exposure, even a slight deviation could lead to significant deviations of the microstructure of the diffusion zone.

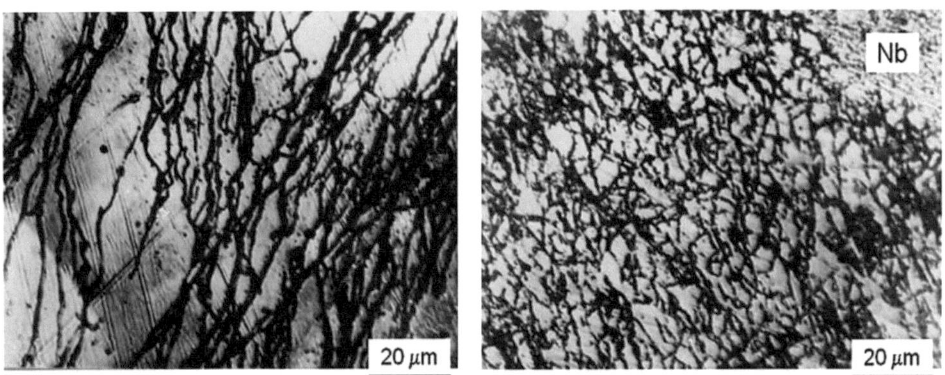

Figure 31. The microstructure of β-phase layer in a plane parallel to the rolling plane of the tape Nb/AlCu/Nb after h.-t. h.: 1850°C for 3 *s* and l.-t. an.: 800°C for 2 *h*. Visible small area of no etched Nb, wt. %Cu: *a* – 5, *b* – 3.8.

Metallographic studies showed that the average size of grains of intermetallic compound Nb_3Al D for the samples which belong to the one- and the two-phase areas are significantly differed. The grains of β-phase in samples out of zone 1 are greatly differed in size and many of them are elongated along the rolling direction (**Figure 31*a***). In the direction of the minor axis, the average grain size of β-phase is equal to 4 *μm* and in the direction of the major axis it is equal 25–30 microns. The grains of β-phase in samples from second zone were significantly smaller (**Figure 31*b***).

It is known that the pinning of the superconducting vortex in compounds A15 occurs mainly at grain boundaries. Here, it can be seen that though the copper improves the critical current of tape, but the transition from two-phase structure of the diffusion zone (β + ∇) to a single phase dramatically reduces the critical current, since the size of grains has become much greater.

The critical current density j_c for the tape with the alloy of Al–12 wt. %Cu and with the maximum I_c, equaled to 39 *A*, is $1.35 \cdot 10^5$ *A/cm²* in the β-layer with thickness of 5.5 *μm* (**Table 6**). Note that in the tape with aluminum in β-layer of the same thickness j_c was equal to ~$6 \cdot 10^3$ *A/cm²*. The increase in critical current density of layers Nb_3Al doped with copper is due to fineness of compound grains. The method we used for determining the grain size D has allowed to receive β-phase accurate value D ~ 3 *μm* (see **Figure 31*b***). In samples with pure aluminum, D was 7–8 *μm* [10].

The tape	The thickness of β-layer, *μm*	5 *T*; 4.2 K; H ⊥ I	
		I_c, *A/mm*	j_c in β-layer, *A/cm²*
Nb/Al/Nb	1.5	5.0	$1.7 \cdot 10^5$
Nb/Al/Nb	5.5	0.7	$6.0 \cdot 10^3$
Nb/AlCu/Nb	1.5	17.0	$5.7 \cdot 10^5$
Nb/AlCu/Nb	5.5	14.5	$1.3 \cdot 10^5$

1. Notes: I – transport current in the sample.
2. In the calculation should take into account the two β-layers.

Table 6. The influence of the thickness of the β-layer on I_c and j_c.

The result of the alloying of β-phase by copper deserves special attention since the critical current and critical current density with increase of β-layer thickness is reduced much smaller than in samples which do not contain copper. This is shown in **Table 6** – an increase of the thickness of β-layer from 1.5 to 5.5 *μm* is causing a decrease of j_c in the tapes with aluminum at about ~30 times but in the samples with copper – only at 4–5 times.

5.4. Conclusions to section

1. The influence of copper on the current carrying capacity Nb/Al/Nb- apes was investigated. Copper was added to the aluminum in the concentration range from 0.7 to 12.0 wt. % and tape Nb/AlCu/Nb was examined. The value I_c is increased to 19 *A/mm*, the critical current density in a layer of β-phase with thickness ~1 *μm* – to $5.7 \cdot 10^5$ A/cm². In samples out of the tape without copper, with the conditions of being equal, I_c was equaled 7.5 *A/mm* and $j_c - 2 \cdot 10^5$ *A/cm²*.

2. The increase of I_c in the samples with copper due to the fineness of grain β-phase. In the samples with the alloy Al-12 wt. %Cu it was 3 *μm* after the high temperature heating by regime: 1850°C for 3 *s*. In the tapes without copper D was more than 8 microns.

3. In layers β-phase copper-doped the critical current density with the increasing of β-layer thickness is reduced to a much lesser extent than in β-layer does not contain copper.

6. Manufacture, structure and critical current of multilayer Nb_3Al

We would like to summarize the presented research in the following form:

I. The effect of copper on T_c of diffusion layers Nb_3Al is as follows. Firstly, T_c for samples with the addition of copper equalled to 18.6-18.8 K, is somewhat higher than in the samples without copper, and secondly, the high values of T_c were achieved after heat treatment at 1650°C, 20–30 *s* + 800°C, 2 *h*. The samples without copper after this treatment regime had T_c from 17.0 to 17.7 K. This suggests that copper makes the formation temperature of stoichiometric Nb_3Al less on 150–200°C.

The lowering of the formation temperature of the compound Nb_3Al is important to achieve a high current-carrying capacity in its layers because that the heating of a tape in the range of 1850–1950°C, even for a few seconds, is causing a rapid growth of β-phase grains. It was shown (see **Figure 24**) that the critical current density decreases with ~10^5 to $2\cdot10^3$ A/cm^2 in magnetic field 5 *T*.

II. The second conclusion is based on the first one. In terms of the formation of compound Nb_3Al by reactive diffusion for reducing the building time of its layers, it is advantageous to have small diffusion distances, which will be overcome by aluminum atoms. Taking into account that the aluminum atoms before meeting with the niobium surface, it must be overcome the distance equal to the thickness of σ(Nb_2Al)-phase layer formed earlier let us estimate the thickness of Al-interlayer in the composite Nb/Al-tape in order to the formation of β-phase at 1000°C for 10 *h*. Solving the second diffusion equation for final of the body in the assumption that the thickness of β-layer will be equal to 1 micron, we find that the thickness of Al-interlayer to be 0.6 *μm*. The diffusion coefficient is ~$7\cdot10^{-14}$ cm^2/s [8]. Layers of aluminum with such thickness can be obtained in multilayer tapes by rolling.

III. An advantage of the multilayer tape in comparison with the three-layer tapes consists of the possibility to create in tape the structure of alternating layers from superconducting compound and niobium (strictly speaking – solid solution of aluminum in niobium). In the multilayer tape with large number of layers Nb_3Al and due to their mutual shunting the critical current will be not sensitive to change in the thickness of β-layers or even to their local discontinuities. However, the mechanical properties of aluminum and niobium are very different. Therefore, to achieve laminar (right) structure along the rolling direction is challenging enough. Therefore, instead of aluminum for the manufacture of multilayer tapes Nb_3Al, it was decided to try an industrial aluminum alloy with 4.4 wt. %Cu of class D16. It is more durable than pure aluminum at room temperature.

6.1. The initial preform

Thus, to improve the process of joint deformation of metal components of the composite for the obtaining of the multi-layered tapes was used duralumin D16 instead of pure aluminum.

The starting preform for extrusion was assembled as follows. Niobium strip 0.25-0.30 *mm* thick and duralumin ribbon 0.06-0.09 *mm* thick were folded together, and by means of special clamping device tightly wound on Nb-rod with diameter 10 *mm* until an outer diameter of the received twist equal 35 *mm*. The twist was placed in thick-walled Nb-tube and the whole assembly was laid in the Cu-glass with an outer diameter of 75 *mm*. The glass was closed densely with copper lid. The fully assembled preform had height of 100 *mm*. Copper at the extrusion is fulfilled a useful role of lubricant, as niobium have an ability to stick on the walls of a tool.

An extrusion of copper glass with embedded in it a twist happens so, that niobium and duralumin tapes are drawn out across the direction of their rolling. Mechanical tests on stretching for specimens of the deformed cold Nb-tape cut transversely and along its rolling (**Table 7**) shown that an elongation $\delta_\perp$ in the direction perpendicular to the rolling direction was equal to only 0.4%. Therefore, to improve ductility of the metal foils Nb-foil prior to winding in twist was annealed at 1100°C and the duralumin foil was annealed at 400°C for 1 *h* and quenched. Then $\delta_\perp$ of Nb-tape was increased to 9–10%.

Heat treatment, °C	Along the rolling direction			Across the rolling direction		
	σ_b, *MPa*	σ_t, *MPa*	$\delta_{\|\|}$, %	σ_b, *MPa*	σ_t, *MPa*	$\delta_\perp$, %
deformed at 99%	520	340	3.0	390	260	0.4
1100, 1 *h*	255	180	22.0	265	140	9.0
1100, 5 *h*	250	190	23.0	260	150	10.0

Table 7. The mechanical properties of niobium foil: σ_b – ultimate strength, σ_t – yield strength, $\delta_{||}$ and $\delta_\perp$ – elongation of Nb-foil in a direction parallel and perpendicular to the rolling direction respectively.

The extrusion of preforms were conducted on a hydraulic press with a maximum force of 630 tons in a container of diameter 76 *mm*. Before the extrusion preforms were heated to 200-300°C in air. As a working tool the matrix with channel of 30 *mm* diameter was used. The obtained rod is cut into chunks, which in the first experiments extruded again into rods of 10 *mm* diameter and in the subsequent experiments – into billets (sutunkas) ~6×20 mm^2. A macro-structure of the cross-section of one of them is shown in **Figure 32**. Rods and billets were rolled at the room temperature to 50–70 μm thick tape without intermediate annealing. On one of the initial stages of rolling both edges of the band were cut off and the outer Cu-covering quite easily was removed by mechanical means. To facilitate the last operation, the inner surface of the cupper glass before the assembly of billet for extruding was lubricated with fine powder alumina. The edges were cut off using a disc scissors.

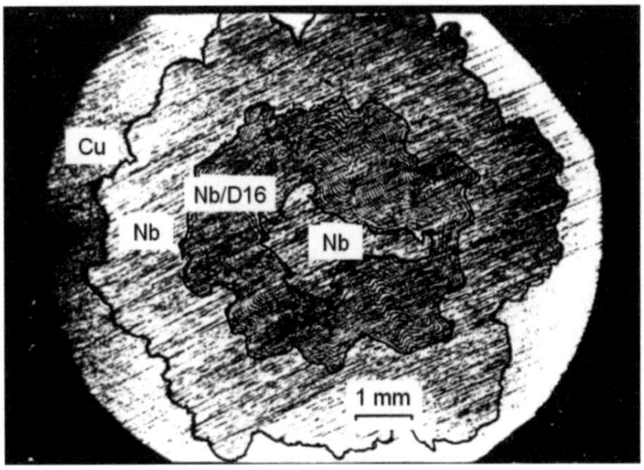

Figure 32. Macrostructure of the Nb/Al-rod cross-section in Cu-shell diameter of 10 *mm*.

For the formation of Nb_3Al-layers samples of the tape were subjected to heat treatment in a vacuum: high-temperature heating + low-temperature annealing. For research purposes, high temperature heating of the multilayer tape was performed in analogy with the three-layered tapes – by direct passing of an electric current.

6.2. The microstructure of the multi-layered tapes

After rolling and cutting of edges tape had a width of 10–15 *mm*. To research a wide ribbon was cut into strips 2 *mm* wide (**Figure 33**).

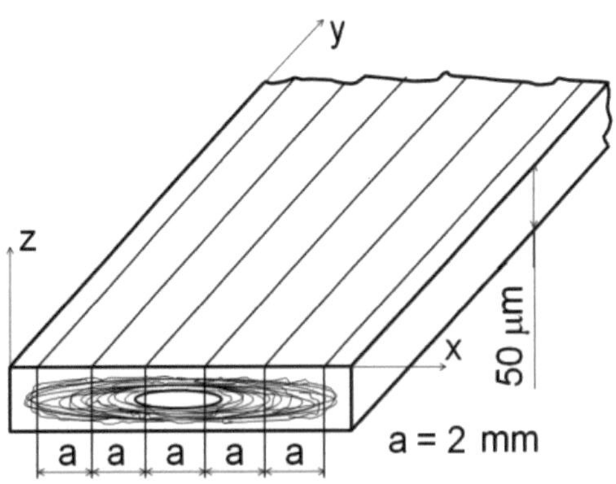

Figure 33. Schematic representation of the cross-sectional structure of a wide tape, being cut into tapes of 2 *mm* wide.

In the multilayer structure of the tape cross-section may be a three- and five-layer. If narrow ribbon was cut from the middle of a wide tape, in the cross section were two multilayer layers

http://dx.doi.org/10.5772/64500

Nb/D16 derived from the twisting, two outer wall layers of Nb-pipe and the middle layer of Nb-rod (**Figure 34*a***). Narrow tape was cut at a certain distance from the middle, have only three layers: the average Nb/D16-layer formed from twisting and two extreme Nb-layers formed from the walls of Nb-tube (**Figure 34*b***).

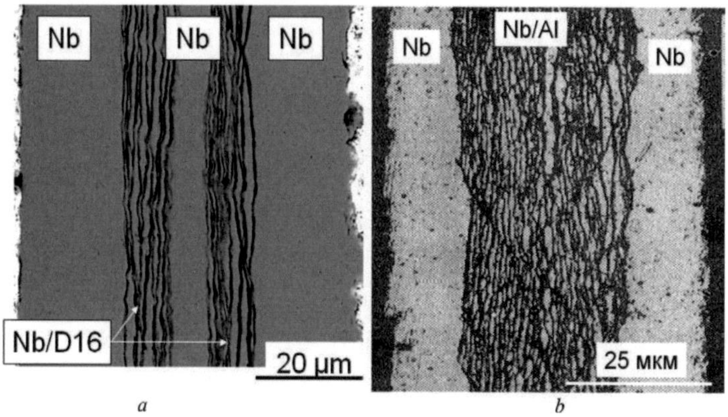

Figure 34. The microstructure of the cross section of the multilayer tape after rolling width of 2 *mm* cut from a wide tape in its middle (*a*) and at a distance from the middle of the (*b*).

After heat treatment at 1200°C for 30 minutes, the layers Nb/D16 transformed into a layered structure of two phases differing slightly light contrast (**Figure 35**). According to the local X-ray analysis in dark elongated areas was contained ~29.5 at. % Al, which corresponded to the σ-phase, in lighter areas was contained 18.7–19.9 at. % Al, which were been β-phase.

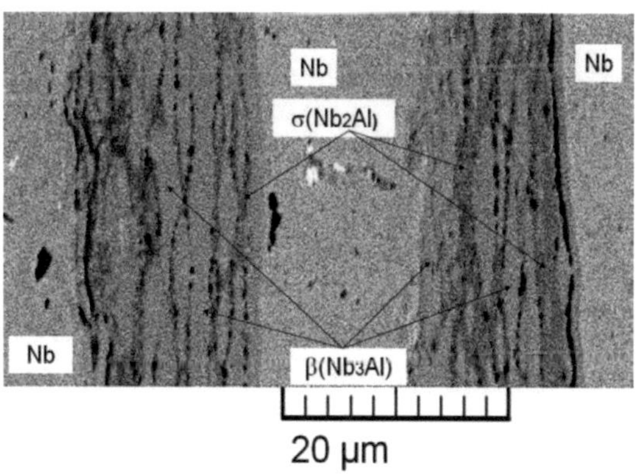

Figure 35. The microstructure of the cross section of the multilayered tape after heat treatment at 1200°C for 30 minutes.

After high temperature heating at 1850°C for 2 *s* in diffusion zone was only β-phase. Layers of diffusion zone were differed so little contrast from niobium layers. Therefore, for this it will show the concentration dependencies of niobium and aluminum over the cross section of tape

(**Figure 36**). In a tape of 2 *mm* width, cut from the middle of the wide band, clearly looked two layers of β-phase. In the tape cut at some distance from the center of a wide band, observed one β-layer with thickness approximately equal to the previous two layers.

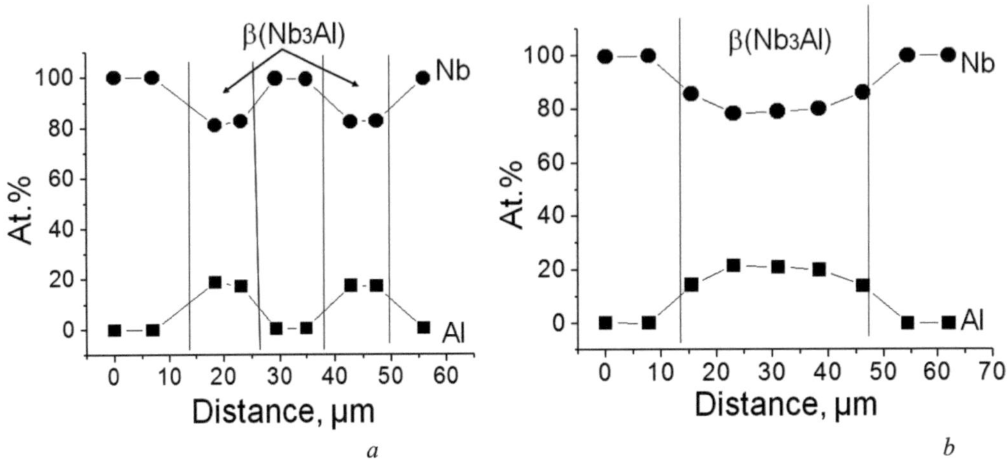

Figure 36. Concentration Nb- and Al-sectional profiles of multilayer Nb/D16-tape with a width of 2 *mm* after the heating at high temperature for 1850°C for 2 *s* cut from the middle of (*a*) and at some distance from it (*b*) of wide tape.

Concentration profile shows that the aluminum content of the β-layers is substantially less stoichiometric composition. It has changed from 17.5 to 21.5 at. %. But the values of the superconducting transition temperature after the heat treatment at such a regime were ~18 K. This indicated that the composition formed β-phase is close enough to 25 at. %. From this it can be assumed that in β-layer there are interlayers of niobium unreacted with aluminum, which is risen content of niobium in diffusion zone.

6.3. Superconducting properties of multilayered Nb/D16 tape after heat treatment at 1700°C

High temperature heating was conducted with times ranging from 1 to 30 seconds. Dependence of T_c and the width of the transition from ΔT_c from heating time are shown in **Figure 37**. Each value T_c and ΔT_c was calculated as the average of 6 samples. It is seen that the dependence of T_c on the heating time has slightly pronounced maximum after heating for 7 *s* equal 17.2–17.4 K when $\Delta T_c = 3.0$ K. However, after exposure exceeding 10 *s* the transition width abruptly is increased. This is indicative of "blurring" of the concentration of elements in the superconducting β-phase and the degradation of its superconducting parameters. On the curve of transition to the superconducting state, conforming to the sample with a 30-second heating, observe a broadening of Nb-transition toward lower temperatures. This is evidenced by the formation of a significant amount of solid solution of aluminum in niobium in a concentration sufficient to reduce the T_c of pure niobium.

The critical current I_c were measured for a 2 *mm* wide strips cut from the middle of the wide tape. I_c dependence on the heating time (**Figure 37c**) showed that his best its values were observed in the samples after heating for 1 second and equal to 9–10 *A* in a magnetic field of 5 *T*. Increasing time to 3 and 5 *s* to reduced $I_c = 7$ *A*.

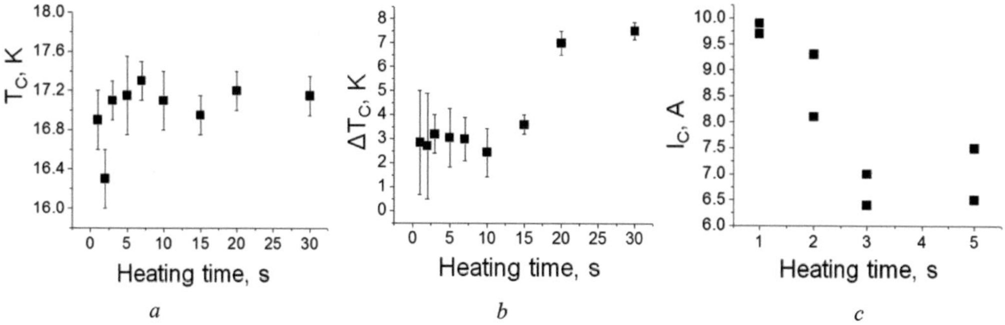

Figure 37. The dependencies of the critical temperature T_c (*a*), the width of the transition from ΔT_c (*b*) and critical current I_c at 4.2 K in magnetic field 5 *T* (*c*) of multilayered Nb/D16-tape on the heating time at 1700°C.

The upper critical magnetic field $H_{c2}(0)$ and the slope $(-dH_{c2}/dT)_{T=Tc}$, depending on the heating time are changed respectively between 24 and 30 *T* and within 2.1–2.5 *T/K*. An additional 5-hour annealing at 800°C is increased $H_{c2}(0)$ to 32–35 *T*.

7. Lengthy tapes Nb_3Al

The laboratory technology developed for the preparation of superconducting tape Nb_3Al can be successful if used for manufacturing and other functional materials, for example, alloys with shape memory effect or heat-resistant composite materials with a layered structure. It is somehow impossible to obtain monolithic preforms with the use of the rolling in vacuum – process much better than the cold welding using rolling at room temperature. But at the same time, this technology cannot provide the making of long superconducting tape at least 100 meters.

Furthermore, it was clear that the heating up to ~1850°C by direct passing of an electric current through the tape specimen, clamped in the fixed contacts, unacceptable for heat treatment of a lengthy tapes. From here, there were two big problems: to develop the technology of long pieces of the composite Nb/Al-tape and to create equipment for the high-temperature heating of ribbons in a continuous mode by manner "to passage".

More or less fully technology of manufacture of a multilayer tape of large length was described in [11]. But in that work multi-layered Nb/Al-tape was investigated with the aim of testing and heated as before – directly by passing an electric current through the sample. Here we denote the stages of technology and present the value of current-carrying capacity of the multilayer Nb_3Al-tape. The manufacture of tape was included: collection of the preform, its extrusion, and rolling at room temperature. First let us dwell on the hydro-extrusion for manufacture of the long multilayer tape of Nb_3Al.

7.1. Hydro-extrusion in production process of multilayer tape Nb_3Al

The high hydrostatic pressure has been widely use in a metalworking. Hydro-extrusion method is ideal for the cold processing of metal materials in general and composite materials in particular. Implemented through the hydro extrusion a comprehensive schema of high pressure allows to process the materials with a high degree of deformation without destruction and to deform the materials which in normal conditions difficult or not at all deformed. For the last the hydrostatic extrusion is the only possible way their processing.

In the manufacturing of superconductors using hydro-extrusion are solved two problems with the original rod-preform for the rolling from which must to obtain a long tape. This is its length and relatively small diameter of ≤10 *mm*. And especially actually it for the compounds with the structure of the A-15 is prepared by co deformation of dissimilar metals. The tape length is conditioned by the size of the rod-preform.

Conventional pressure by the extrusion of multilayer preform Nb/D16 cannot be affected by a high degree of deformation, for example, from 75 to diameter of 10–15 *mm*. In the process of pressing the workpiece warms up to the 200–300°C. The resistance of deformation due to such a low temperature is too large, and pressing of workpiece cannot be executed. But if the pressing been success to carry out, then due to the high reduction ratio the workpiece in the deformation zone so was heated that led to the melting of duralumin.

For obtaining of long pieces of multilayer tape Nb/D16, a tentative technological cycle was conducted using hydro extrusion, which consisted of the following steps: manufacturing-assembly of an original workpiece with diameter of 75 *mm*, conventional pressing of workpiece in rod with diameter ~30 *mm*, hydrostatic extrusion of it into a rod of 10 *mm* in diameter and a rolling of last rod into a strip. At the assembly of the initial workpiece, the thickness of copper glass wall was calculated so that after pressing it in the rod diameter of 30 *mm* and processing of the rod on the lathe to a diameter of 20–22.5 *mm* on it still remained the copper layer with thickness of 0.5–1.5 *mm*. Thereafter, rods were cut into bars 200 *mm* in length, which were subjected to hydro extrusion on an installation of the Institute of High Pressure Physics, Russian Academy of Sciences [12]. As a lubricant of rods before the extrusion is used ktiol CT-15 or MoS_2. Working matrix has the angle $2\alpha = 30°$, degree of deformation – 76%, the working pressure of a fluid is 12.5–16.0 *kbar*, final diameter of rods – 9 *mm*.

Hydro extrusion positively characterized at any of its stage, the multilayer preform is not required to pre-heat despite the great degree of deformation. The length of rods after hydro extrusion was ~0.8 *m*. It should be noted that the length of the starting bars at the pressing with the use hydro extrusion are limited by the size of the reception chamber which, if desired, can be increased. A meltdown of duralumin, located inside bar, was not observed as not been and discontinuities of the material as the whole.

http://dx.doi.org/10.5772/64500

After hydro-extrusion rods of 9 *mm* in diameter were rolled at room temperature without intermediate annealing's in the multilayer tape with the breadth of 10.5 *mm* and the thickness of ~50 microns.

7.2. Facility of the high-temperature heating of long Nb/D16-tapes

Thus, the scientific results have brought us a purely engineering task to create an apparatus similarly designed for the thermal treatment at temperatures higher than 1500 (one) for a few seconds (two) long (three) thin multilayer tapes Nb/D16 by mean of "to pass ". Others simply do not give . A description of such installations in the foreign and domestic literature of course did not meet. Itself task is hidden the solution of many complex issues – from the choice of the type and the material of the heating device to the sagging of tape due to its elongation under the influence of high temperature heating. Approximately similar can be found in Müller's research [13] of Nb/Al- and Nb/AlGe-tapes after the heat treatment at 1700–1900°C. As can be understood from a brief description of the method of heating, the latter is similar to the heating realized in the proposed installation.

The main components of high-temperature heating installation we developed are vacuum chamber, vacuum pumping system and the control unit. The vacuum chamber is made of pipe with an inner diameter of 300 *mm* horizontally laid on a welded frame, with chamber length is 800 *mm* (**Figure 38**). The ends of the chamber is equipped with flanges that can be opened. The end covers and the pipe itself have four viewing windows made of quartz glass for observing of movement and heating of tape. The pipe in its middle part has a water jacket for protection against thermal radiation emitted by the heating device. There were electrical insulation required from the hull, a pyrophyllit was used. Its details before the assembly were subjected to calcination at 1200°C for 2 hours.

Figure 38. Installation for the high-temperature heating of the long multilayer Nb/D16-tapes.

A pressure in the chamber ~10^{-5} *mm Hg* was achieved within 40 mines with the moment of pumping the inclusion. In the process of high-temperature heating it was fallen, but not lower than the ~$5 \cdot 10^{-5}$ *mm Hg*.

7.3. Heating device

The heater is tantalum tube with a diameter of 10 and a length of 150 mm, welded out of the foil with thickness 0.3 *mm*. Copper bars are served at its ends and an electrical current of industrial frequency is discharged. The installed length of a working zone is 50 *mm*. To achieve temperature of 1800°C current flowing through the heater is ~200 *A*. To control the temperature of the heater, W-Re-thermocouple has been used, tucked away in the isolation of beryllium oxide. The signal from the thermocouple was fed to a regulator of temperature maintenance. The heater is surrounded by a system of screens collected from the W-foils.

The heater with screens for the isolation against walls of the chamber has been mounted on the pyrophyllite plate with thickness of 15 mm, which is fixed inside the chamber.

7.4. Kinematic scheme of tape passage

The movement of the tape was carried out by means of two pulleys (for supply and receiving coils). Their rotation was set by DC servomotors arranged on the same axis with the coils. On the way from the coil filled with tape (supply) to the receiving coil the tape, except heater, passes still through pinch, tension and guide rollers. By means of a reducer the speed of tape movement can vary in a fairly wide range. The friction device allows for the receiving coil reduce the angular velocity at an increase of the winding radius of the tape. An initial radius of winding is 30 *mm*. It exceeds the critical radius at which the tape after the heat treatment can be bent without affecting on the current carrying capacity of the tape.

7.5. The electrical circuit of installation

Passage of the tape through the heater (or rather the speed of passing) is adjusted automatically. Electric circuit of the speed regulator is built on the principle of the tracking system. It includes the following: a sensor, a control for the circuit of speed of servo motors, a high frequency

Figure 39. Apparatus for a speed control of tape passage.

oscillator, an amplifier, the subtracting device, power supply and a safety unit. Exterior view an apparatus which contains the entire electric scheme is shown in **Figure 39**.

Mismatch voltage of the two circuits depends on the bias of tape ± Δy relative to the average situation.

The sensor depends on the accuracy and stability of the whole tracking system. Specifically for the occasion, a proximity (noncontact) sensor is designed. It works on the principle of change of the coupling coefficient. The sensor represents two resonant contours, interacting on of differentiated communications. Contours are spaced apart at some distance. For protection against extraneous electromagnetic fields contours are enclosed in screens with cutouts facing each other. The tape moves freely between the contours and on it through the brush is transmitted a current from the high-frequency generator. The tape is the third circuit, generating its electromagnetic field of a certain frequency. And together with the two previous differentially coupled contours the ring circuit is obtained. As soon as the tape is shifted to ± Δy from the middle position between the first two contours a mismatch of their voltages arises.

The amplifier receives signals from the sensor, amplifies, filters, and transmits them to the inputs of the subtracted device and the safety unit.

A signal proportional to the displacement Δy of the walking tape is transmitted to the input of the subtracted device, is recorded and reproduced the micro ammeter, located on the front panel of the apparatus (see **Figure 39**).

If the heated tape is cut off, the safety device disables the amplifier and the subtracted device. Then the servomotors stops and a light signaling includes.

7.6. Test results of tape after heat treatment in continuous mode

Installation for heating high-temperature long pieces Nb/D16 layered strip in a continuous mode examined for three modes (**Table 8**). The output data was the critical current of tapes (**Table 9**). Alloy D16 is an industrial aluminum alloy with 4.4 wt. %Cu. The residence time of the tape portion in a working zone (zone length – 50 *mm*) was calculated from the rate of movement of the tape given by the receiving coil. Initial angular velocity ω of the coils was set depending on the linear velocity υ which is remained constant during the entire heating of tape for the selected mode.

Temperature of high-temperature heating, °C	ω, *turn./min*	υ, *mm/s*	Heating time, *s*	Performance, *m/h*
1700	0.2	0.8	62.5	2.9
1800	5.4	25.8	2.0	92.9
1800	8.6	39.0	1.3	140.4

Table 8. Modes of tape passage.

The first trial pieces of tapes 2 *mm* wide have a length of ~5.5 *m*. After the high temperature heating, from their middles cut out piece of ~3 *m* long, which was cut in half. The obtained pieces of tape 1.5 *m* long in its turn cut into six lengths of 250 *mm*. In threes such lengths of 250 *mm* from each of two 1.5 *m*-pieces of tape were selected for measurement of the critical current immediately after h.-t. h. in a continuous mode and for measurement I_c after h.-t. h. and subsequent low-temperature annealing at 800C for 3 hours. The l.-t. an. was performed in furnace of resistance at vacuum.

Mode of high-temperature heating, °C	The average value of I_c, *A*		The range of changes of I_c, *A*	
	Without the l.-t. an.	800°C for 3 *h*	Without the l.-t. an.	800°C for 3 *h*
1700, 62.5 *s*	1.1 ± 0.2	1.8 ± 0.3	0.4 ÷ 1.8	1.2 ÷ 2.6
1800, 2.0 *s*	4.1 ± 1.5	71.0 ± 12.0	1.8 ÷ 11.0	36.0 ÷ 96.0
1800, 1.3 *s*	10.8 ± 4.0	106.0 ± 8.2*	4.1 ÷ 27.5	92.0 ÷ >120

*For samples which became non-superconducting

Table 9. Results of measurements of I_c of tape Nb_3Al (5.5 *T*, 4.2 K, tape width – 2 *mm*).

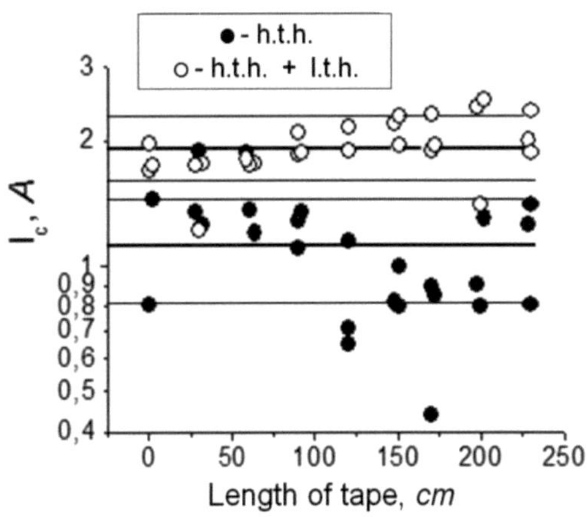

Figure 40. The values of I_c of the tape after h.-t. h. at 1700°C for 62.5 *s* (υ = 0.8 *mm/s*): • – without the l.-t. an. and ○ – after l.-t. an. at 800°C for 3 *h* (5.5 *T*; 4.2 K).

From each 250 *mm*-part of the tape, 9–10 samples of the length 15 *mm* for I_c were selected. Three parts of tapes intended for the measurement of critical current only after heating and three after heating and annealing. Therefore, the average value I_c, in **Table 9** is calculated from samples in an amount of from 54 to 60 taken from a piece of tape 1.5 *m* long.

The values of I_c along the length of the three 250 *mm*-parts of tape, heat treated for one of three modes are summarized in a single graph and presented in **Figure 40**–**Figure 42** respectively. Heating at 1700°C for 62.5 seconds gave a small critical current value I_c, equal to about ~1 *A* in a magnetic field of 5.5 *T*. Low-temperature annealing increased the critical current twice (see

Figure 40). Judging by our studies [14, 15], heating for 1 minute at 1700°C is far from optimal. It may be assumed that layer of compound Nb_3Al is not continuous. Most probably the β-phase preserved only as islands and strongly depleted by aluminum. Most of the tapes volume has already turned into a solid solution of Al in niobium.

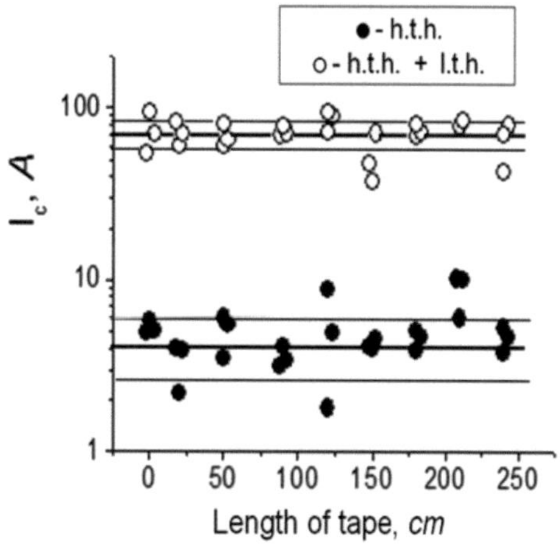

Figure 41. The values of I_c of the tape after h.-t. h. at 1800°C for 2 s (υ = 25.8 *mm/s*): • – without the l.-t. an.) and ○ – after the l.-t. an. at 800°C for 3 *h* (5.5 *T*; 4.2 K).

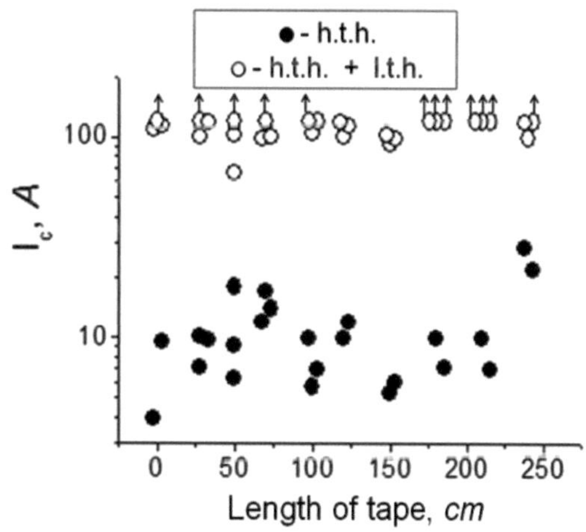

Figure 42. The values of I_c of the tape after h.-t. h. at 1800°C for 1.3 s (υ = 39.0 *mm/s*): • – without the l.-t. an. and ○ – after the l.-t. an. at 800°C for 3 *h* (5.5 *T*; 4.2 K). An arrow pointing upwards indicates that I_c is greater than the current shown by the arrow, that is, an electric current of this magnitude does not put the sample in the normal state.

A completely different pattern was observed in the tape after the h.-t. h. at 1800°C for 2 seconds (see **Figure 41**). However, the level of critical current values immediately after heating was not high total 4–5 *A*. But the subsequent annealing at 800°C was increased critical currents of more than an order of magnitude – up to 70–80 *A*. Higher I_c were obtained in tapes after 1800°C for 1.3 seconds (see **Figure 42**). Without annealing is 9–10, with annealing is 100–120 *A*. A significant number of samples are not transferred to the normal state by electric current 115–120 *A*. This is the maximum current of source, which we possessed at that time.

Analyzing the results, we can say that the critical current is strong enough and varied along the length of tapes. For this, we need to turn to the 4th and 5th columns of **Table 9**. In samples without the low-temperature annealing, minimum value of critical current is different from the maximum in 4.5, 6.1 and 6.7 times. But after a low-temperature annealing, this difference is reduced to 2.2, 2.7 and ≥1.3 times, respectively. The higher the current carrying capacity of tapes, the more times the monitored parameters are improved.

The average arithmetic deviation from the mean values I_c for tapes without low temperature annealing which has undergone the heating in a continuous mode according to the second and third embodiment, is 37%. The lowest value of the critical current of the sample are different from the maximum, as we have already established, an average of 6.5 times (see **Table 9**). But be noted that out of the total number of samples that heated at 1800°C for 2 *s*, the samples with critical current equal 2 *A* or less there were only two. The number of samples after heating for 1.3 seconds, having I_c about 6 *A* or less there are only three.

The most common reason for the change of the critical current over the length of the tape became intelligible from **Figure 45** and is explained by change of the cross-sectional structure of the tape in the direction of axis *y*. This is a consequence of the method of the tape manufacturing. Because of the fact that the tape is narrow – only 2 *mm* width – under excision out of wider tape its structure of cross-sectional owing to random circumstances may have as 3-layer so and 5-layer. The case is very unfavorable when it is found on the outer regions of a wide tape. They account for the smallest number of layers D16. The use of a wider tape will be smooth out and level the structural changes along the length of tapes and thereby decrease the distinction of the critical current values.

The common property of the analyzed data is a significant decrease of the arithmetic average after the low-temperature annealing. That is, the scatter of the critical current value after annealing becomes smaller. For samples with heating at 1800°C for 2 *s* the scatter of I_c values after annealing at 800°C was decreased from 37 to 17%. At that the average critical current value has increased from 4.1 to 71.0 *A*, or in 17 times.

In these studies, we have used 2 *mm* in width ribbons in order to have to deal with wittingly lower values of electrical current. This was done to facilitate the measurement of the critical current at liquid helium temperature. The narrower the tape, the smaller its critical current is needed for the transfer of sample in the normal state. From here, wires will be less heated, lead-in electric current to the sample, which is in liquid helium, and will be less losses of helium due to its evaporation.

7.7. Conclusions to part

One. Quite a promising factor is that already in the first experiments have been received very high values of the superconducting characteristics for the tapes of compound Nb_3Al. Of particular importance is the fact that the high values has structurally dependent characteristic, which is the critical current. The critical current of tapes Nb_3Al was equal 50–60 *A* in 1 *mm* of width tape in a magnetic field of 5.5 *T*, which corresponded to a current density in the β-phase (1.0–1.2)·10^5 *A/cm*^{2}. The critical temperature (at the beginning of the transition to the superconducting state) is equal to 18.5 K. The upper critical magnetic field H_{c2} (4.2 K) = 30–32 *T*. Considering that the alloy of D16, is commercially available, and is much cheaper than tin, could have been expected that the multilayer Nb_3Al, would be very competitive with Nb_3Sn. Further are seen the following paths of improving of the critical current Nb_3Al:

- modes optimization of high temperature heating and low-temperature annealing;
- optimization of the construction and structure of the cross section of composite tapes Nb/D16;
- alloying of niobium [16, 17].

Two. The optimal modes of high-temperature heating for the obtaining of maximum possible the critical current in the three-layer tapes Nb/Al/Nb and multi-layer composite tape Nb/D16 (1800–1850°C, 1–2 *s*) by means the passing of an electric current through a short immovable samples [16] are matched with the third mode of high-temperature heating of the long multilayer tape Nb/D16 (1800°C, 1.2 *s*) by means "to pass" through permanently installed Ta-heater.

Three. At fast heating of multilayer tapes by passing an electric current not infrequently was observed the so-called "swelling" of tapes the surface. To avoid this, before the high-tempera ture heating is conducted 1 min-calcination of the tape at 1000°C. At the heating by "to pass" this procedure is not required. Evidently in order to "swelling" is not occurred, for tape was sufficiently smooth entry into the heating zone. Evidently, the time there was enough for formation out of aluminum and niobium the intermetallic compound $NbAl_3$.

Four. The presented results demonstrated that the developed high-temperature heating technology at some improvements of individual units of installation can be used for manufacture of the long (~100 *m* or more) pieces of multilayer tape Nb_3Al. Implementation of short-term (1–2 *s*) and high temperature (1800–1850°C) heating of a moving tape was quite solvable problem. Short times allow the leading of the heating process with high performance around 100–150 meters of tape per hour.

8. The dislocation structure of the superconducting layers in multilayer tape Nb_3Al

Electron microscopic studies [18] were conducted on three-layer tapes Nb/(AlGe–Cu)/Nb, in which Al–Ge alloy contain copper. In layers Nb_3(AlGe) was discovered precipitation of fine particles of the second phase as inside so and on grains boundaries of the compound. The samples were subjected to heat treatment: 1650–1800°C, short time + 800°C, 2 *h*. By this was explained a significant (more than an order of magnitude) increase in the critical current after the low-temperature annealing. In the tapes without annealing of the second phase was not observed, and I_c was remained small.

The difficulty of preparation of the foils was "hit" on the layers of Nb_3Al, concluded between the Nb-layers (**Figure 43**).

Here, were studied the layered tapes Nb/D16 of 50 microns thickness and a width of 5 *mm*, after heating at the high temperature at 1850°C for 1.2 to 6.4 seconds. The structure of β(Nb_3Al)-phase layer was investigated using transmission electron microscopy. Foils for studies were prepared by thinning at room temperature in an acidic electrolyte of composition H_2SO_4 : HNO_3 : HF = 3 : 2 : 5. The cathode was stainless steel. The difficulty in preparing of foils was in that to appear in process of the thinning in layer of the β-phase, which is between the layers of Nb-solid solution (**Figure 43**).

Figure 43. Microstructure of the cross-sectional of multi-layered strips of Nb_3Al after the h.-t. h. at 1850°C for 1.3 s. Actually designations of Nb_3Al and Nb are: Nb – two outer layers and one central layer of the solid solution of Al in Nb, Nb_3Al – two multilayer layers out of alternating considerably more thin layers of Nb_3Al and Nb-solid solution.

Heating at 1850 C for 1.3 s (**Figure 44*a***): beta-layer has structure of equiaxed grains. Its average size is 0.2 *mm*. Period of crystal lattice is 5.185 *Å*. Increasing the heating time to 4.6 seconds leads to the development of polygonization with the manifestation of its symptoms from education "honeycomb" structure, dislocation walls and dislocation grids to small-angle boundaries. The result is the formation of elongated sub-grains perpendicular to the former boundaries (**Figure 44*b***). A size of blocks is of 0.2–0.3 *µm*. A further increase of heating time to 6.4 *s* leads to formation of a three-dimensional network Frank. This is accompanied by an increase in the thickness of the layer of β-phase and increasing of its grains size to 2 or more microns (**Figure 44*c***). The contrast of the dislocation image made it possible to judge about atmosphere of impurities accompanying dislocation.

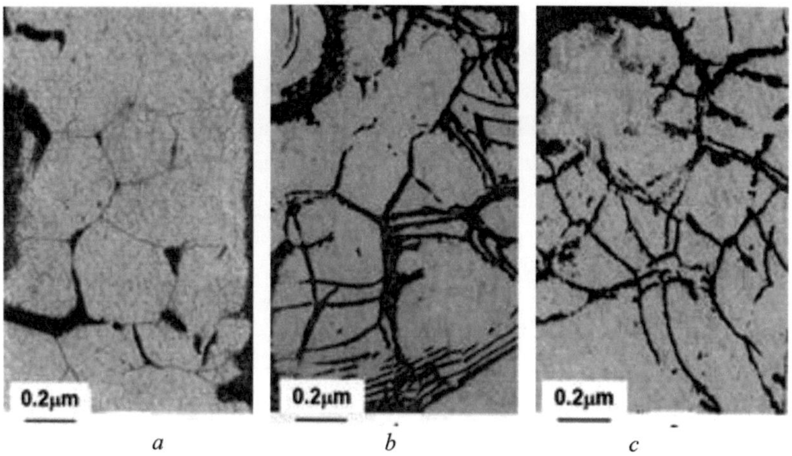

Figure 44. The microstructure of β-layer in a multi-layered tape Nb/Al after the h.-t. h. at 1850°C for 1.3 (*a*), 4.6 (*b*) and 6.4 *s* (*c*).

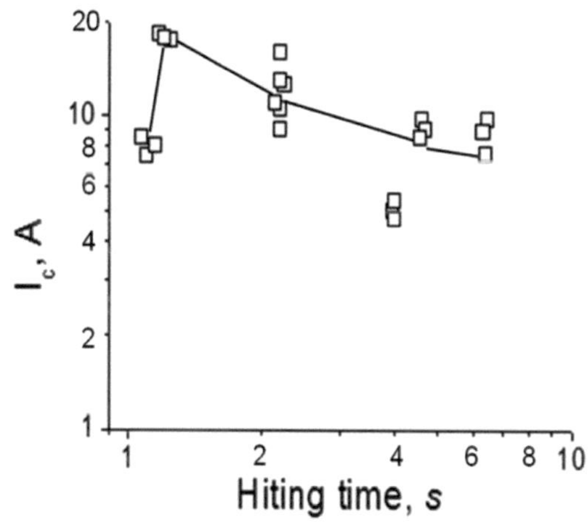

Figure 45. Critical current I_c of Nb_3Al multi-tape depending on the time of h.-t. h.

The maximum current carrying capacity corresponded to the heating time for 1.3 *s* (**Figure 45**). At the following grain growth and evolution of the dislocation structure I_c was decreased. It was another confirmation of concept that in compounds with A-15 structure grain boundaries are the most effective pinning centers of superconducting vortices.

9. Conclusion

1. Research works which are described in the presented chapter began in the 70s of the last century. The aim was to develop and test the semi-industrial technology to produce a

superconducting material on the basis Nb_3Al connection. At that time industrial production of superconducting materials were presented alloys Nb–Ti with T_c little more than 9 K. Technological developments of for superconducting material out of the more advanced at the time of the compound A-15 Nb_3Sn already were, but none of them has been implemented in the industrial of variant as the superconducting material. Compound Nb_3Al with a critical temperature equal to T_c of Nb_3Sn, has a much greater than Nb_3Sn the upper critical magnetic field of 32 *T* and, therefore, seems more promising for use in magnetic fields exceeding 10–15 *T*.

2. The samples of superconducting tape Nb_3Al were tested in magnetic fields up to 22 *T*. The so-called "constructional" or "engineering" critical current density calculated on the entire cross-section of the superconducting tape at 4.2 K in an external magnetic field of 14 *T* (**Figure 46**), and 18–22 *T* equal to (7–8) 10^4 [19] and (3.4–4.5) 10^4 A/cm^2 [20], respectively. Measurements in magnetic fields up to 14 *T* of Bitter magnets were conducted in the Laboratory of strong magnetic fields and low temperatures (Wroclaw, Poland), measurements in magnetic fields up to 22 *T* of the combined superconducting solenoid – in the Institute of Atomic Energy of name I.V. Kurchatov (Moskow, Russia).

3. The multilayer tape is capable of withstanding very small critical bend radius without degradation of critical current equal to 6–7 *mm* (**Figure 47**).

4. Semi-industrial of technology for long pieces production of multi-layer composite tape Nb/D16 developed and tested on one of a plant. The technological cycle consists of the following major steps:

   - the hydro extrusion of composite rod Cu/Nb/D16 on the press of horizontal type,
   - a profiled cold rolling of the resulting rod of diameter 10 *mm* on 8-stand rolling mill to the profile of the flat section,
   - a removing of the copper sheath,
   - an intermediate rolling of the profile to tape with thickness of 0.3 *mm* and
   - the rolling at room temperature using a four-roll mill of thin rolling equipped by coilers with regulated tension (**Figure 48**).

   One piece of the tape Nb/D16 of final thickness is shown in **Figure 49**.

5. Superconductivity, which was before the discovery of high-temperature superconducting ceramics and which is dedicated to this chapter, now called low-temperature superconductivity. Was a short time when on the wave of HTS-euphoria an opinion existed, that the previous superconductivity would lose prospect for its practical application. But this did not happen and could not happen. Low-temperature superconductivity was used up to HTS, even wider uses at present and will peacefully co-exist with high-temperature superconductivity in the foreseeable future.

http://dx.doi.org/10.5772/64500

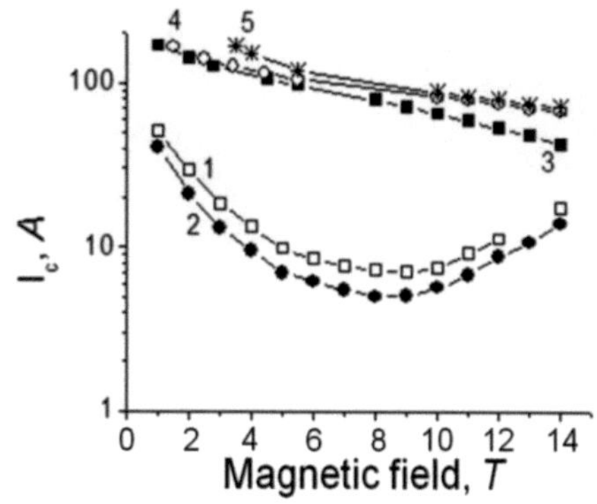

Figure 46. Critical current I_c of multi-layered Nb_3Al tape depending on the external magnetic field. Width of tape is 2 *mm*. Heat treatment, C: 1 and 2 – 1000, 1 *min* + 1850, 2 *s*; 3, 4 and 5 – the same + 900, 0.5 *h*, + 800, 3 *h* and + 900, 1 *h*.

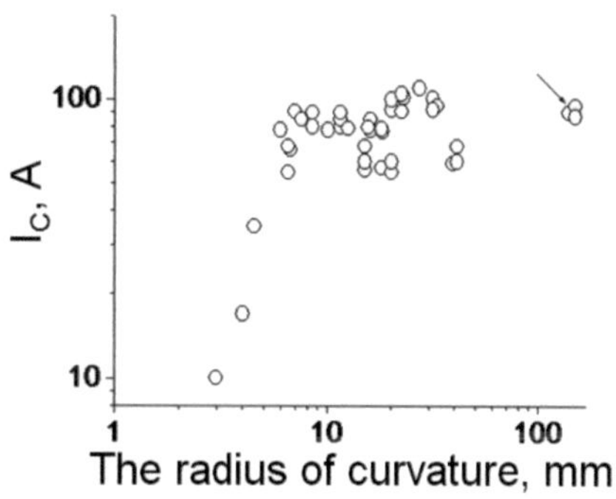

Figure 47. I_c on the radius of curvature (4.2 K, 5.5 *T*). Heat treatment, °C: 1750, 2 *s* + 800, 2 *h*. The tape thickness of 50 *μm* and width of 2 *mm*.

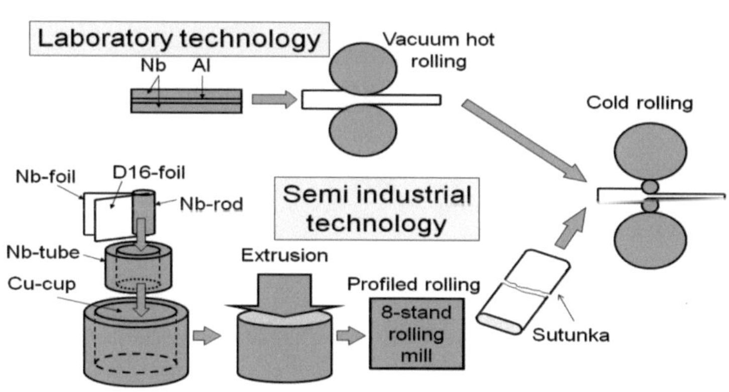

Figure 48. Simplified schemes of operations of the laboratory and semi-industrial technology for obtaining composite tape of Nb_3Al.

Figure 49. Segment of multilayer Nb/D16 tape.

Acknowledgements

When performing this work at various stages of its implementation my colleagues and my teachers-mentors attended and got involved on research activities and on the life. Some of them has changed the circle of their scientific and vital interests. Many are no longer among us. To all of them I express the deep gratitude and retain the fond memories of the time when I was with them. They are F. S. Goncharov, Ju. F. Bychkov, V. V. Schmidt, Ch. V. Kopezky, A. K. Gapeev, G. M. Korzhova, A. M. Markov, M. I. Karpov, V. A. Marchenko, N. A. Tulina, V. S. Kruglov, I. S. Dukchovnii, A. V. Nikulov.

Author details

V.P. Korzhov

Address all correspondence to: korzhov@issp.ac.ru

Institute of Solid State Physics, Russian Academy of Sciences, Moscow region, Chernogolovka, Russia

References

[1] Kohr J.G., Strauss B.P. and Rose R.M. 1971. "Development of a new practical high-T_c superconducting materials", IEEE Trans. Nucl. Sci., 18:716-717.

[2] Foner S., Mc Niff E.J. Jr. and Matthias B.T. 1970. "Upper critical fields of high-temperature superconducting $Nb_{1-y}(Al_{1-x}Ge_x)_y$ and Nb_3Al: measurements of H_{c2} more than 400 kG at 4,2 K", Phys. Lett., A31:349-350.

[3] Foner S., Mc Niff E.J. Jr., Matthias B.T., Geballe T.N., Willens R.H. and Corenzwit E. 1971. "H_{c2}(4,2 K) of high-temperature superconducting alloys", J. Appl. Phys., 42:58.

[4] Dergounova E.A., Sudev S.V. and Aliyev R.T. 2009. "Fundamentals of superconductors Materials Science based on compound A-15." Textbook, Moscow, Moscow Engineering Physics Institute, – 48 p.

[5] "Diagrams of state of binary metallic systems." Handbook. In the 3 vol.: vol. 1. Under commonly the editorship N.P. Lyakisheva. 1996, Moscow: Machine building, 177-180.

[6] Vignes A.G. 1971. "Coating of niobium and niobium alloys with aluminium. Part I: Pack-cementation coatings", J. Less-Common Metals, 23:375-393.

[7] Slama G. and Vignes A. 1971. "Coating of niobium and niobium alloys with aluminium. Part II: Hot-dipped coatings", J. Less-Common Metals, 24:1-21.

[8] Slama G. and Vignes A. 1972. "Diffusion dans les aluminiures de niobium", J. Less-Common Metals, 29:189-202.

[9] Hake R.R. 1967. "Upper-critical-field limits for bulk type II superconductors", Appl. Phys. Lett., 10:189-192.

[10] Korzhov V.P., Tulina N.A. and Schmidt V.V. 1976. "Superconductivity of compounds Nb_3Al and Nb_3(Al,Ge), obtained by means of diffusion," Issues of Atomic Science and Technics. Series: fundamental and applied superconductivity. Kharkiv, KhPTI of UkrSSR Academy of Sciences, 1(4): 8-22.

[11] Kopezky Ch.V., Korzhov V.P., Korzhova G.M. and Schmidt V.V. 1979. "Superconductivity of multilayer composite Nb_3Al," Physico-chemical analysis of superconducting alloys. M. "Science", 119-123.

[12] Kurdyumov G.V., Vereshchagin L.V., Entin R.I., Gurevich Ya.B., Konyaev Y.S., Dmitriev V.N. and Polyakov E.V. 1970. "Deformation aging of martensite by using hydro extrusion," Physics of Metals and Metallovedenie, 29:869-873.

[13] Müller P. 1975. "Superconductivity of doped layers of A-15-phase niobium–aluminium and niobium–aluminium–germanium", J. Less-Common Metals, 42:29-41.

[14] Korzhov V.P. 2011. "Superconducting properties Nb_3Al, resulting in a wide temperature range in a multilayer composite Nb/Al-tape," Materialovedenie, 11: 28-34.

[15] Gapeev A.K., Kopezky Ch.V., Korzhov V.P., Tulina N.A. and Schmidt V.V. 1977. "Formation and the superconducting properties of the diffusion layers of compound Nb_3Al," Superconductivity. Proceedings of the Conference on the technical use of superconductivity. T. IV. Superconducting materials. M. Atomizdat, 61-72.

[16] Korzhov V.P. 1979. "On the strength of the pinning in layer of compound Nb_3Al alloyed by copper," Physico-chemical analysis of superconducting alloys. M. "Science", 127-130.

[17] Korzhov V.P. 1977. "On the critical current in the layers of superconducting compound Nb_3Al alloyed with copper and zirconium," Issues of Atomic Science and

Technics. Series: fundamental and applied superconductivity. Kharkiv, KhPTI of UkrSSR Academy of Sciences, 1(5):29-32.

[18] Dukchovnii I.S., Korzhov V.P., Fridman A.A. and Schmidt V.V. 1981. "The critical current and the microstructure of the diffusion layers $Nb_3(Al,Ge)$," Metallophysica, 3:43-51.

[19] Kopezky Ch.V., Korzhov V.P., Korzhova G.M., Marchenko V.A., Bazan Ch. and Grenier B. 1981. "The critical current of Nb_3Al tapes in magnetic fields up to $1.11 \cdot 10^7$ A/m," Physics of Metals and Metallovedenie, 52:994-997.

[20] Dukchovnii I.S., Korzhov V.P., Krasnoperov E.P. and Cheremnych P.A. 1982. "The critical current of Nb_3Al- and $Nb_3(AlGe)$-tapes in magnetic fields up to 22 Tesla," Issues of Atomic Science and Technics. Series: General and Nuclear Physics. Kharkov: KhPTI of UkrSSR Academy of Sciences, 1(1):49-51.

Chapter 7

Niobium in Cast Irons

A. Bedolla-Jacuinde

Additional information is available at the end of the chapter

http://dx.doi.org/10.5772/64498

Abstract

This work reviews the effect of niobium on different cast irons. This alloying element has a high affinity by carbon forming carbides of the type MC; therefore, the main effect is to strengthen the matrix of the alloys due to a hard-phase dispersion mechanism. Even at low levels of niobium (<0.2%), MC carbides are found. The presence of such carbides increases hardness, yield and tensile strength in ductile irons, but the most important role is to increase wear resistance. Therefore, the use of niobium has been focussed mainly in alloyed white irons for wear resistance applications and in mottled cast iron used for rolls in hot rolling mills.

Keywords: cast iron, niobium, carbides, wear

1. Introduction

1.1. Cast irons

Cast irons are ferrous alloys containing carbon contents generally between 2 and 4 wt.%. Ferrous alloys with carbon contents below 2 wt.% are called steels. The high amount of carbon in cast irons exceeds the solubility of this element in ferrite and/or austenite; the excess of carbon promotes the precipitation of a carbon rich phase during solidification, which remains during cooling to room temperature. This phase can be a carbide or graphite. Therefore, the final structure of cast irons is composed by a metallic matrix and carbide or graphite, depending on the chemical composition and on the solidification rate. According to the microstructure, cast irons are classified as white, grey, ductile and mottled irons.

1.1.1. White iron

Without alloying elements, a Fe-C alloy solidifies mostly as a eutectic austenite–cementite (γ-Fe_3C). During cooling, austenite transforms to pearlite and the carbide remains. Thus, the typical structure is a continuous network of the brittle cementite containing islands of pearlite as shown from **Figure 1**. This structure makes this material too brittle and practically not useful for engineering applications.

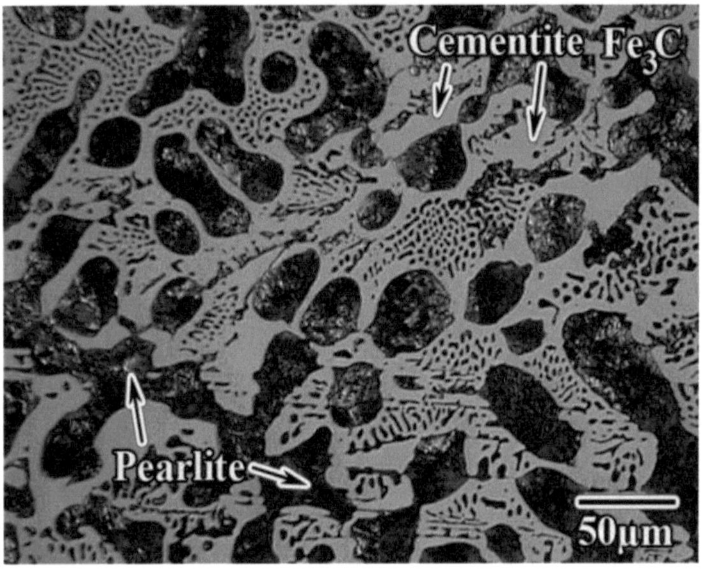

Figure 1. Typical structure of unalloyed white iron consisting of a network of cementite and pearlite.

1.1.2. Alloyed white iron

Since the continuous network of cementite promotes brittleness in white iron, it is common to alloy such materials to get better mechanical properties. A very successful element for this purpose is chromium. In amounts between 12 and 30wt.%, chromium changes the stoichiometry of the carbide phase from the orthorhombic M_3C to the hexagonal M_7C_3, which is less continuous. In addition, the presence of other elements such as nickel and molybdenum, which increase hardenability, promotes the presence of the austenite in the final as-cast structure. These irons are called high-chromium cast irons, and the as-cast structure consists of eutectic carbides of the type M_7C_3 in a matrix of austenite (**Figure 2**). The presence of austenite and the less interconnected carbides gives some toughness to the alloy, and the harder M_7C_3 carbide gives a better wear resistance to these materials.

If the amount of chromium is less than 12wt.%, the stoichiometry of the carbide does not change and the structure consists of eutectic M_3C carbides in a matrix of austenite plus martensite. These irons are called low-chromium cast irons. **Figure 3** shows the typical as-cast structure of this latter iron.

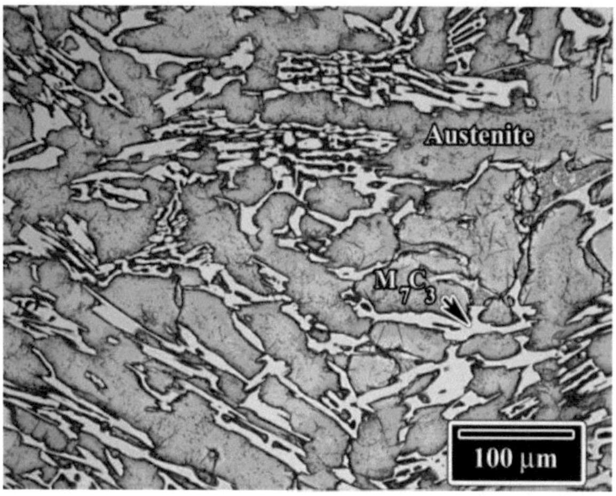

Figure 2. Typical as-cast microstructure of a high-chromium white iron consisting of eutectic M_7C_3 carbides in a matrix of austenite.

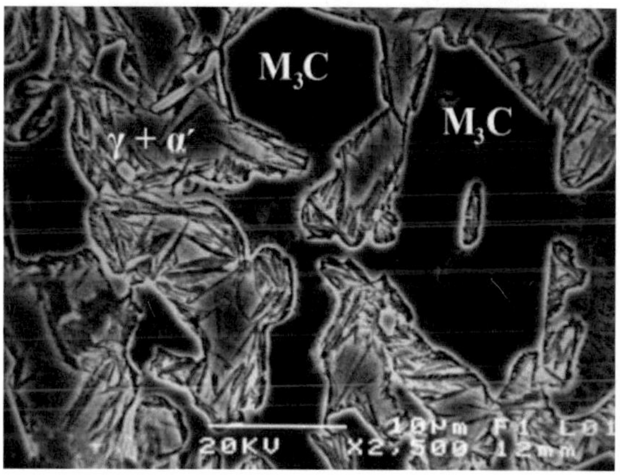

Figure 3. Typical as-cast microstructure of a low-chromium white iron consisting of eutectic M_3C carbides in a matrix of austenite (γ) plus martensite (α').

1.1.3. Grey iron

Without carbide-forming elements and the presence of silicon in cast irons, the carbon excess precipitates as graphite instead of carbide during solidification. Under these circumstances, the as-cast structure is composed by graphite "flakes" in a mainly pearlitic matrix. **Figure 4** shows the as-polished (a) and as-etched (b) structure of a 2%Si 3%C grey iron. **Figure 5** shows the tridimensional "flake-like" morphology of graphite at the fractured surface of the iron shown in **Figure 4**. One of the main characteristics of grey iron is the size, type and distribution of graphite according to the ASTM A247 standard [1]. Such characteristics depend on the solidification rate and on the effectiveness of the inoculation process before pouring.

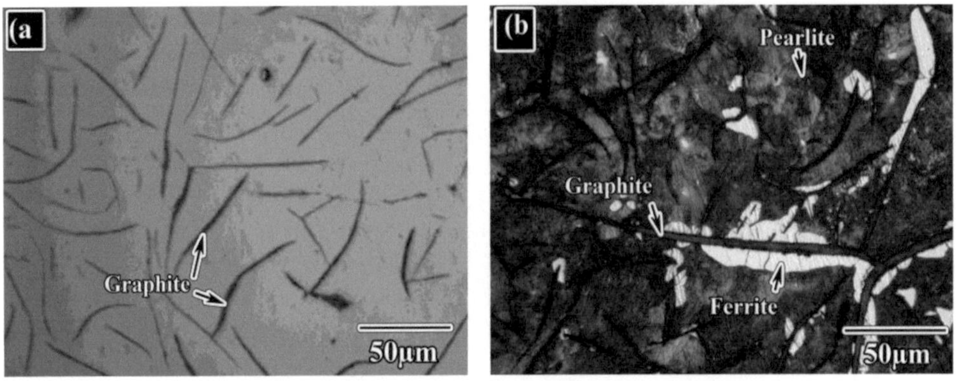

Figure 4. Microstructure of grey iron (a) as-polished, showing the flake-like graphite, and (b) 2% nital etched, showing the matrix to be composed by pearlite and small amounts of ferrite.

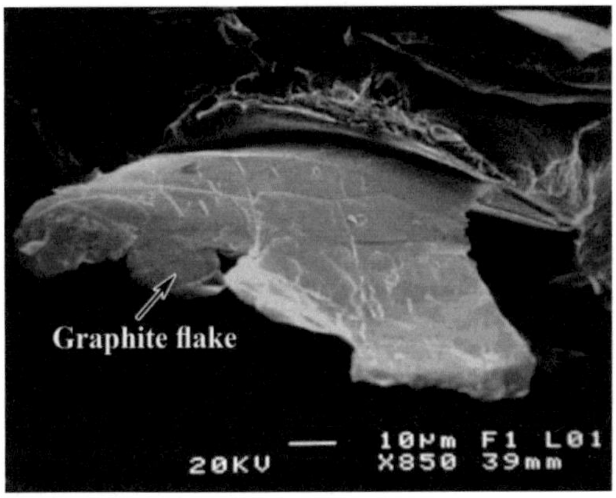

Figure 5. Tridimensional morphology of a graphite flake in a grey cast iron.

1.1.4. Ductile iron

Ductile iron has the same chemical composition than grey iron; the difference is a particular treatment with small amounts of magnesium or cerium which is applied to the liquid alloy. Such a treatment promotes the excess of carbon to precipitate as spheres during the eutectic solidification. Then, the microstructure is composed by a continuous matrix of pearlite plus ferrite and isolated spheres of graphite. The amount of ferrite and pearlite depends on the solidification rate, the alloying elements and on the effectiveness of the inoculants. The continuous matrix increases mechanical properties and makes these irons comparable with some steels. **Figure 6** shows the microstructure of two ductile cast irons evidencing the effect of copper on the amount of pearlite. The base chemical composition for both irons is 3.14%C, 2.5%Si and 0.5%Mn. It is common to find the ferrite phase surrounding the graphite spheres in ferritic/pearlitic ductile cast irons due to the solidification nature of the alloy.

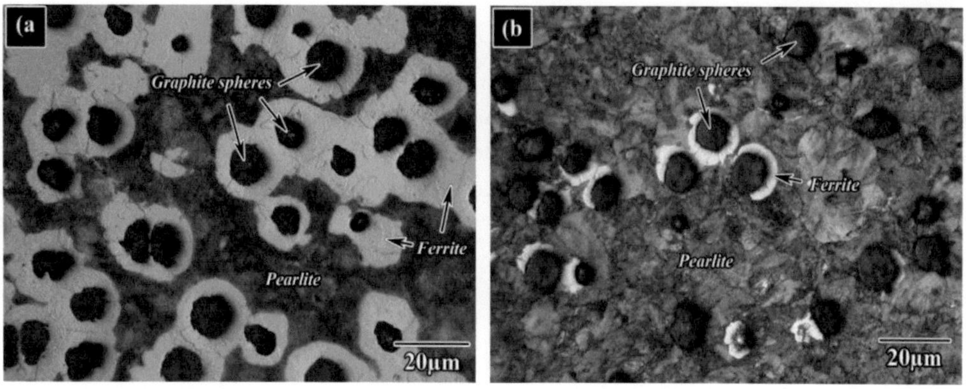

Figure 6. As-cast microstructure of ductile iron, (a) 3.14%C, 2.5%Si and 0.5%Mn showing a matrix composed of 50% ferrite-50% pearlite; (b) the same base composition of (a) but with the addition of 0.2%Cu, showing a matrix composed of 90% pearlite and 10% ferrite.

1.1.5. Mottled iron

Mottled iron is a mixture of white and grey/ductile iron; the excess of carbon precipitates both as graphite and as carbide. The presence of carbide-forming elements (chromium, manganese, molybdenum, etc) is common in these irons in order to form the M_3C carbide; in addition, the presence of silicon and nickel is also necessary to promote the precipitation of graphite. If treated with magnesium, the graphite phase can be obtained as "spheres" as in ductile iron; other way, the graphite will be present as "flakes" as in grey iron.

Figure 7 shows two micrographs showing the typical structure of a mottled iron 3.2%C-0.52%Mn-2.24%Si-0.53%Cr-3.49%Ni-0.56%Mo-0.2%Cu. The iron was magnesium-treated to obtain nodules of graphite, and the main application of this alloy is for mill rolls [2].

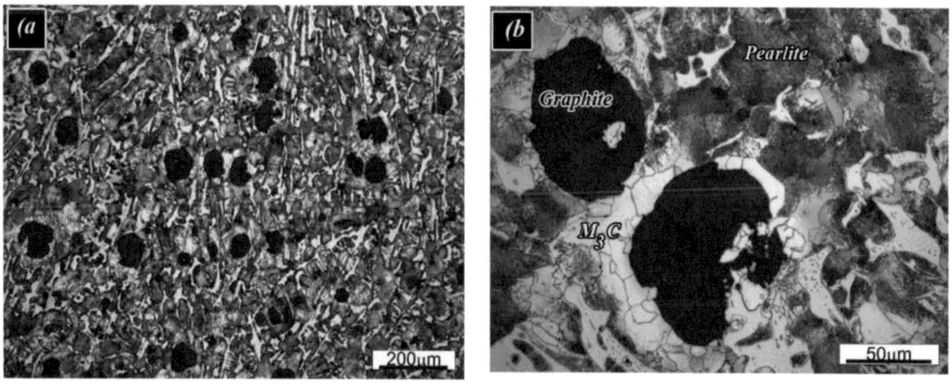

Figure 7. Microstructure of a mottled iron showing the microstructure to be composed by graphite nodules and M_3C carbides in a matrix of pearlite (micrographs kindly provided by Prof. Janus Krawczyk, from the AGH University of Science and Technology, Krakow, Poland).

1.2. Niobium properties and applications

According to Takai [3], the discovery of niobium can be attributed to two people; first of all, to an English chemist named C. Hatchett, the first name that was given to this new element was "*Columbium*" since Columbia was the place where it was discovered. Secondly, by the year 1844, H. Rose, a German chemist, managed to separate a new element from tantalum and named it "*Niobium*". Two years later, it was proven and worldwide recognised that both elements were the same. The International Union of Chemistry Congress adopted the name niobium for the new element, which is the principal name used around the world; the name Columbium is used just in the USA. Niobium always coexists with tantalum in tantalite, the ore which is the main raw material for extracting tantalum.

Niobium is a refractory metal, such as tantalum, tungsten and molybdenum, and has a high melting point (2468±10°C), being chemically stable against acid and alkali. Its physical properties, such as superconduction, low thermal neutron capture cross section, low corrosion under some environments, make this element an attractive candidate for so many applications in the fields of nuclear industry [3, 4], superconducting [5–7], acoustics and electro-optics [8], in biomedical devices [9], and other extreme applications as in corrosive environments [10] and cemented carbides [11].

Due to its soft magnetic properties and its ability to form glassy alloys [12], niobium has been also used in several alloys for magnets either cast [13, 14] of bond [15]. In addition, the chemistry of niobium is appropriate to form intermetallic compounds with iron such as FeNb, Fe_2Nb_3 depending on the content of other alloying elements [16, 17].

However, the main use of niobium as alloying element has been in iron alloys such as steels and cast irons. In the case of steels, the development of high-strength low-alloy steels more than 40 years ago, the use of niobium as a microalloying element was widely exploited. The role of niobium (in amounts <0.1%) in refining the grain size of steels during the controlled thermomechanical processing has been deeply studied since then. It is well known that when dissolved in austenite retards recovery in hot deformed steels and avoids the deformation-induced ferrite transformation [18, 19] and also retards the static and dynamic recrystallisation [20]. One of the most important phenomena about niobium in steels is the strain-induced precipitation of NbC or NbCN during the thermomechanical processing [21–25]. The precipitation of such nanometric particles delays the grain wrought during hot deformation, and more refined structures are obtained as a result of niobium microalloying and this in turn represents an increase in strength [26].

The ability of niobium compounds to precipitate during heat treatments has also been used as a tool for increasing the strength of some iron alloys, for example the precipitation of niobium compounds in HP heat-resistant cast stainless steels for creep applications [27] and the dispersion hardening in aged steels [28]. Furthermore, it has been observed that niobium additions modify the cast structure of niobium alloyed hot work tool steels and reduces the size and volume of eutectic cells and increases the maximum hardness of the steel [29].

1.3. Cast iron production

Cast iron is usually melted in a cupola furnace, an electric arc furnace, an induction electric furnace or an air (reverberatory) furnace. When melting in a cupola, the charge is composed of coke, steel scrap and pig iron in alternate layers of metal and coke. Sufficient limestone is added to flux the ash from the coke and from the slag. The amount of coke depends on the amount of carbon of the metallic charge (or the melting point); during melting, the coke burns as air is introduced through the furnace tuyeres and the metallic charge absorbs carbon from the coke. The amount of some other elements in the metallic charge are modified due to the oxidation process, for example, some manganese and silicon are oxidised and also some manganese combines with sulphur forming MnS, and the product of these reactions is incorporated to the slag [30]. A clear disadvantage of melting in a cupola is the oxidation reaction of carbon whose CO_2 emissions are not environmentally acceptable.

When melting in electric furnaces, the process is cleaner since no coke is added to the charge. When the metallic charge is melted, the chemical composition is adjusted by adding the proper amounts of ferroalloys according to the desired composition. The metallic charge may be high-carbon steel scrap or pig iron (see **Figure 8**).

Figure 8. Induction furnace charged to produce cast iron, (a) when using pig iron as raw material and (b) when using steel scrap.

Once melted the metallic charge, the slag is mechanically removed to clean the surface in order to add the alloying elements. Under these conditions, the incorporating efficiency of such elements into the melt is higher. It is recommended to add the alloying elements as ferroalloys (ferro-manganese, ferro-silicon, ferro-niobium, etc), since their melting points are relatively low and this facilitates their dissolution in the liquid metal. Working temperatures of about 1500°C are common in many foundries since this allows to transport and treat the liquid metal before pouring it. Most of the alloying elements are added into the furnace; however, some others highly oxidised are added late in the ladle or even in the furnace but just before pouring into the ladle. In the particular case of niobium, it has a high affinity for carbon forming carbides even at very low levels of niobium. Such carbides of high melting point may be the first to solidify (depending on the carbon content of the alloy) and segregate in the liquid or even incorporate to the slag; thus, the recovery of this element may be too low. Due to this, ferro-niobium injection techniques have been developed to increase the recovery of niobium in the solidified castings [31].

Once the chemical composition has been adjusted, the liquid metal is commonly poured into a ladle (**Figure 9**), where the inoculation process takes place (in the case of grey and ductile iron)or it is nodulised (magnesium treated in the case of ductile iron; **Figure 10**). Finally, when the melt has been already treated, it is commonly poured into sand moulds (**Figure 11**) for a slow solidification rate. Once solidified, the castings are extracted from the moulds by destroying them. Details of different iron production practices are found elsewhere [32].

Figure 9. Pouring the liquid iron into the preheated ladle in the ductile iron making.

Figure 10. Nodulising ductile iron with magnesium, note the intensity of the reaction when magnesium is added into the liquid metal.

http://dx.doi.org/10.5772/64498

Figure 11. Pouring the liquid iron into sand moulds.

2. Niobium effect on the structure and properties of cast irons

2.1. Niobium in grey iron

The use of niobium in steels has a long tradition; in most of the cases, it is restringed as a microalloying element (<0.1%). Its use in cast irons is not as wide as in steels, and one of the main applications is in castings for making grey iron castings for the automotive industry. Grey cast iron has a very low ability to bend and low ductility, due to the presence of graphite flakes which act as discontinuities, but it has a low pouring temperature, high fluidity, low liquid to solid shrinkage, etc., that make it suitable for casting.

With the correct chemical composition, grey iron may be used for engine blocks and cylinder heads due to its excellent properties of thermal fatigue, while other chemical composition may improve its thermal conductivity and specific heat and may be used for disk brake rotors.

According to James et al. [33], the new performance requirements and more stringent automotive standards, engines are required to run hotter and at higher pressures. These requirements demand increase in strength and enhanced thermal fatigue resistance of grey iron castings, particularly in engine cylinder heads due to their proximity to the combustion chamber. During combustion, temperature in the chamber may be increased to 700°C and pressure as high as 160 MPa. This heat may be conducted to the cylinder head which may be cooled by circulating a coolant through channels therein. This cycling heating and cooling combined with the mechanical stresses due to the pressure on the cylinder wall make the cylinder head highly susceptible to thermal fatigue and creep. These properties may be enhanced by alloying elements in the grey cast iron.

Molybdenum and vanadium have been traditionally used in grey cast iron to increase thermal fatigue resistance. However, the use of niobium has been also investigated with some success [33]. This element has been added as a partial substitute of molybdenum (a more expensive element) and may be present in the alloy from 0.05 to 0.3% by weight. Very small amounts of carbides may be present in the structure and the main matrix is pearlite (**Figure 12**) although bainite and martensite may be present in small amounts. The strength of this as-cast alloy may be as high as 360 MPa. Other patented work [34] reports the use of niobium in amounts of 0.38 to 0.45% and the formation of fine carbides well dispersed and embedded in a pearlitic matrix which reduce wear during braking operations.

The replacement of molybdenum with niobium has also been studied in similar grey irons for brake disks but focussed on the effect of this alloying element in the low cycle fatigue [35]. According to the results, the fatigue behaviour of the alloy is improved when increasing the amount of niobium and decreasing molybdenum. They conclude that niobium can be replaced by molybdenum for this particular application and suggest to replace 0.32%Mo with 0.1%Nb. This benefit is attributed to the ability of niobium to refine the eutectic cells and graphite type, to the pearlite refinement (**Figure 13**) and to the precipitation of small niobium carbides that increase wear behaviour. In addition, the cost of the alloy is reduced.

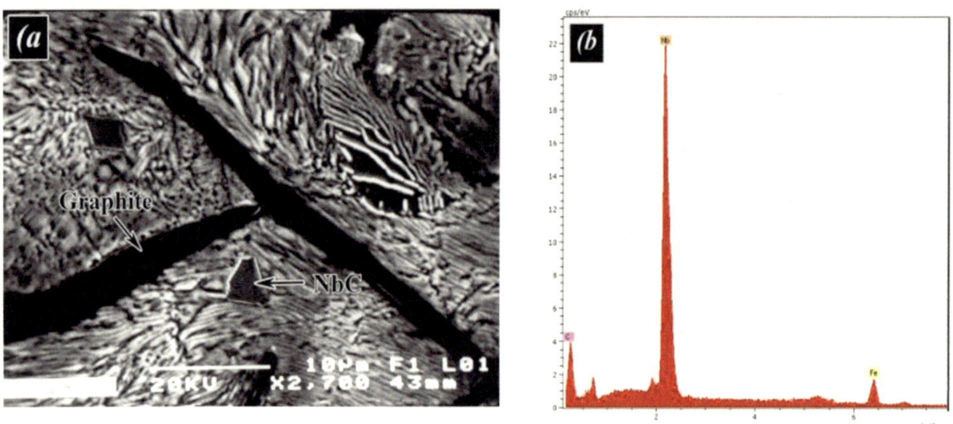

Figure 12. (a) Microstructure of 0.48%Nb grey cast iron showing a mainly pearlitic matrix and (b) EDS of the carbide particle arrowed in (a).

General aspects on the use of niobium in grey irons indicate amounts lower than 0.5% due to its high affinity for carbon. With this amounts, Nylen [36] highlights that the effect of this alloying element is a higher stability of austenite, an increase in microhardness, a small refinement of the graphite structure and the precipitation of small carbides. These microstructural features contribute to increase the tensile strength in the as-cast conditions from 205 MPa for an iron without niobium, to 270 MPa for an iron with 0.8%Nb.

Similar results have been also observed when adding up to 1.48%Nb in high-carbon equivalent cast irons for brake disks [37]. After a clear refinement of the graphite phase, the authors argue that some niobium carbide particles may act as heterogeneous nucleation for graphite in the eutectic reaction. As a result, the increase in the nucleation rate produced a refined graphite morphology. Also, the pearlite lamellar spacing is reduced and it is attributed to the decrease in the eutectic temperature with the niobium additions. These microstructural features contribute in turn to increase hardness and wear resistance of the alloy after the addition of niobium.

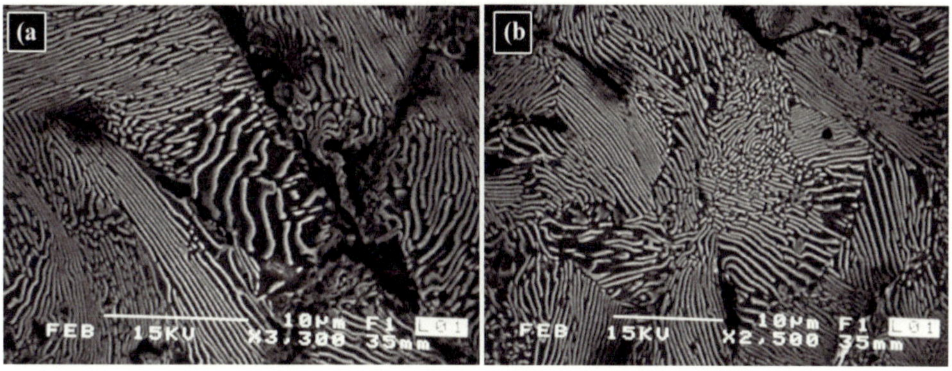

Figure 13. Refinement of the pearlite structure by the niobium addition. (a) Grey cast iron without niobium and (b) 0.32%Nb grey cast iron.

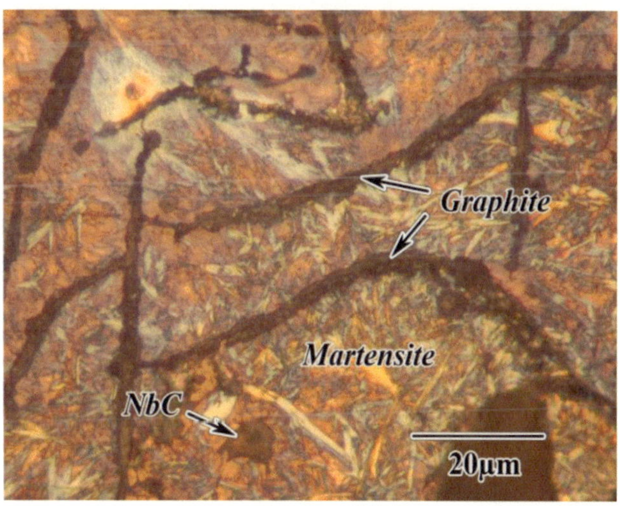

Figure 14. Martensitic matrix in a quenched and tempered grey iron containing 0.32%Nb.

The effect of niobium on strength and wear resistance of grey irons is noted only in the as-cast conditions, since when the alloy is quenched and tempered or austempered, the effect of niobium is minimised by the transformed matrix. This has been observed for high-carbon equivalent [38] and for hypereutectic grey irons [39]. The martensitic matrix (**Figure 14**) in the quenched and tempered conditions increases strength and wear resistance. In the case of the austempered alloy, the ausferritic structure with 20% of retained austenite enhances the wear resistance through a strain-induced martensite transformation effect when the alloy is under frictional contact. The presence of niobium carbides prevents the loss of material during sliding by reinforcing the matrix and graphite serves as a lubricant lowering friction. However, when a carbide particle is released from the matrix, three-body abrasion may take place [40]. Similar results have been observed if the alloying element is titanium [41].

Grey iron has been also used as a base material for making NbC particulate reinforced iron matrix composites produced in situ [42]. The technique is to mixture grey iron powders with niobium wires of 0.7 mm diameter. After about 25 min at 1172°C, the niobium wires dissolved completely and combine with carbon of the alloy producing small well-distributed niobium carbides in an iron matrix.

2.2. Niobium in ductile iron

As mentioned above, the addition of niobium to liquid cast iron whose carbon content is high, which promotes the primary precipitation of solid niobium carbides before the rest of the alloy starts solidifying. Such precipitation may produce segregation and agglomeration of such particles, and it is also common to eliminate them if they get incorporated to the slag. Then, niobium additions must be done at later stages of the melting process [36].

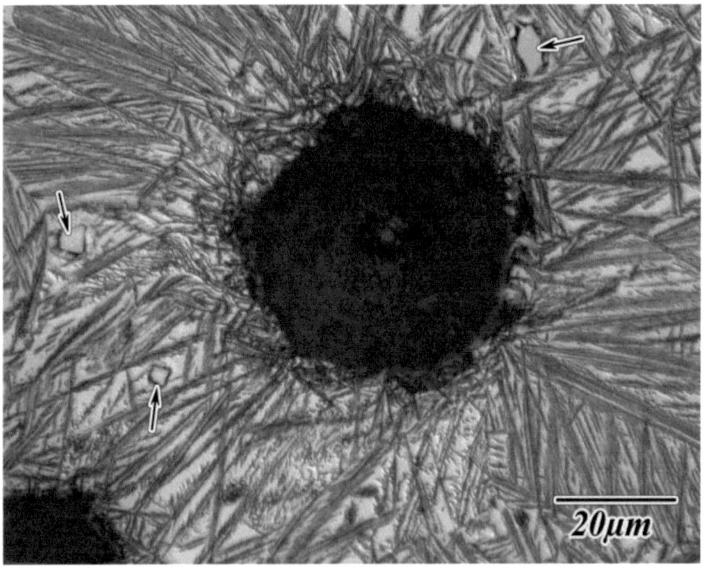

Figure 15. Austempered ductile iron with 0.3%Nb. Note the presence of NbC (arrowed) in the ausferritic structure.

Ductile iron is almost always made for applications where some toughness or ductility is required, and the presence of massive carbides is considered as a detrimental feature for this particular alloy. An increase in strength is generally associated with a decrease in ductility or toughness in an alloy. This is particularly true in ductile irons; it is always desirable to increase strength and retain a good ductility. This has led to the development of heat treatments to obtain high strength in ductile irons as well as good ductility. A particular heat treatment is the "austempering" and the heat-treated irons are called "*austempered ductile irons*" (ADI) [43]. The acicular structure obtained by this heat treatment provides high strength but low ductilities (usually below 5%). **Figure 15** shows the ausferritic structure typical of ADI; in this case, it is a ductile iron with 0.3%Nb. Note the presence of the NbC (arrowed) particles in the structure. This structure of the matrix minimises the effect of niobium in strength and wear resistance of the iron and also its effect on fatigue properties [44].

However, heat treatment always implies higher production costs and is time-consuming. Therefore, it is preferable to obtain the desirable mechanical properties for the alloy in the as-cast conditions by controlling the solidification process and/or chemical composition.

The election of heat treatments or chemical additions is a particular decision of each foundry according to a balance of production costs. In some countries, the addition of a particular alloying element may be cheaper that undertaking a heat treatment, whereas in others to undertake a heat treatment is preferable. When chemical addition is the preferred production practice, the addition of a single element may contribute to a significant refinement of the pearlite constituent, particularly when such element dissolves in austenite, favouring hardenability.

The influence of alloying elements on the mechanical properties of ductile iron is related to the effect of such elements on the ferrite and pearlite contents in the matrix. Alloying elements may increase the strength of ferrite by solid solution strengthening and also of pearlite by reducing the interlamellar spacing. They may also affect the relative amount of pearlite and ferrite as a result of their effect on the eutectoid temperature and transformation kinetics. The strengthening obtained by alloying depends on the type of matrix since the strengthening mechanisms in pearlite and ferrite are not the same. In irons with mixed ferrite–pearlite matrices, the strengthening level depends on the relative amounts of ferrite and pearlite.

The addition of considerable amounts of nickel (4%) and molybdenum (0.5%) is a common practice to produce high resistance ductile irons in the as-cast conditions due to the bainitic matrix produced by these chemical combinations. Such a matrix is obtained due to the excellent synergism of these elements in improving hardenability, as mentioned above. However, the production cost may be high since these are very expensive elements in most of the countries.

It has been reported by Kawamoto et al. [45] that the addition of carbide-forming elements (Mo, V and Ti) to ductile iron affects mechanical properties by influencing the strength of the matrix and the composition and dispersion of the carbide phase. Niobium is known to form and extremely hard MC type carbide.

Niobium has been added to refine austenite grain size during solidification. This phenomenon has been observed to take place owing to niobium carbide formation which retards austenite grain growth. These carbides form at high temperatures and may serve as nuclei for the proeutectic austenite precipitation and perhaps for the eutectic colonies as well. Niobium increases the volume fraction of eutectic austenite as this element reduces the carbon amount precipitating as graphite during eutectic solidification. However, the effect of some other elements in the alloy also alters the eutectic carbon content (silicon and phosphorous), causing a change in the amount of proeutectic austenite.

Niobium in ductile irons has been reported to be found as NbC which forms during the first stage of solidification, during proeutectic austenite formation and before the eutectic reaction [46]. In a particular research work [47], niobium additions up to 0.8% to a ductile iron, formed rectangular niobium carbides whose size and volume fraction increased as niobium content increased. Particles up to 4 μm size were found in irons with niobium additions below 0.4% and particles as large as 10 μm were observed for irons with 0.8%Nb (see **Figure 16**). The volume fraction of NbC was about 0.01 for this later amount of niobium. Niobium also contributed to an increase in the amount of pearlite in the matrix and a decrease in the graphite volume fraction. It is suggested that most of the niobium content combines to form NbC, but a small amount must be dissolved in matrix contributing to the increase in pearlite formation. The increase in pearlite was reported from 70 to 77% when the niobium amount was 0.8%. Regarding to the graphite volume fraction, it decreased from 12 to 8% for the 0.8%Nb. Since niobium carbides form before eutectic solidification, they consume some of the available carbon for graphite nodules; therefore, the volume of graphite after eutectic solidification was observed to diminish. This effect of niobium in the microstructure contributed to a small increase in yield (from 650 to 722 MPa) and tensile strength (from 866 to 930 MPa) and hardness (from 27 to 33 HRC) along with a decrease in ductility (from 12 to 5%).

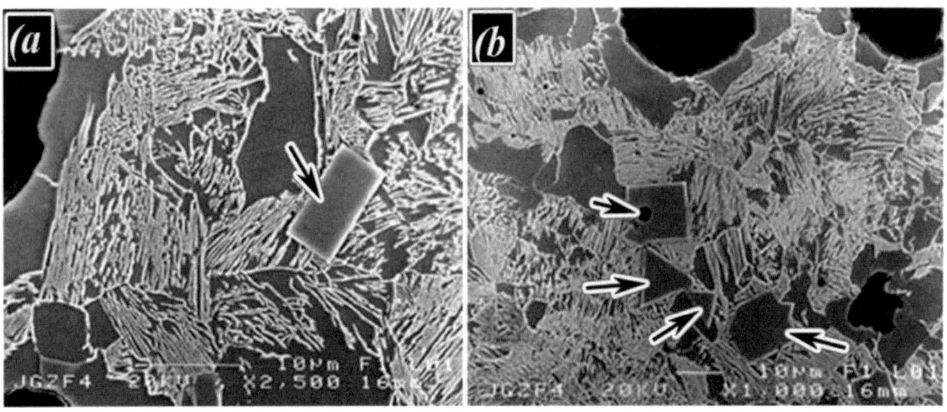

Figure 16. As-cast ductile irons (a) 0.4%Nb and (b) 0.8%Nb. Note the presence of NbC (arrowed).

2.3. Niobium in mottled iron

The main application for mottled irons with nodular graphite is for rolls in hot rolling and in other wear-resistant applications. They contain both graphite and carbide (Fe_3C or M_3C depending of the alloying elements). The presence of carbide and graphite ensures good enough wear resistance. According to Nylen [36], there are two strategies to improve wear performance for this kind of irons: (i) increase the overall hardness by increasing the proportion of cementite, but the alloy becomes more sensitive to crack and (ii) increase the total amount of carbides but with the risk to offset the graphite/carbide balance. The combination of graphite and carbide gives to the alloy other important properties for this iron: low adhesion to the trip and good thermal crack resistance as well as good elastic modulus [48, 49].

One of the strategies for improving the performance of these alloys has been focussed on the increase in the amount of the carbide phase. The AKERS group, in Sweden, has been working in the addition of a carbide-forming element to increase the carbide phase in mottled iron and other irons for rolling mills, niobium. However, the main difficulties of adding niobium to melt iron are the segregation of the particles during solidification, since these particles are the first to solidify. Under these conditions, it is difficult to obtain a good distribution of the NbC in the alloy. **Figure 17** shows the microstructure of a mottled cast iron with 1% Nb, note the agglomeration of NbC particles caused by segregation during solidification.

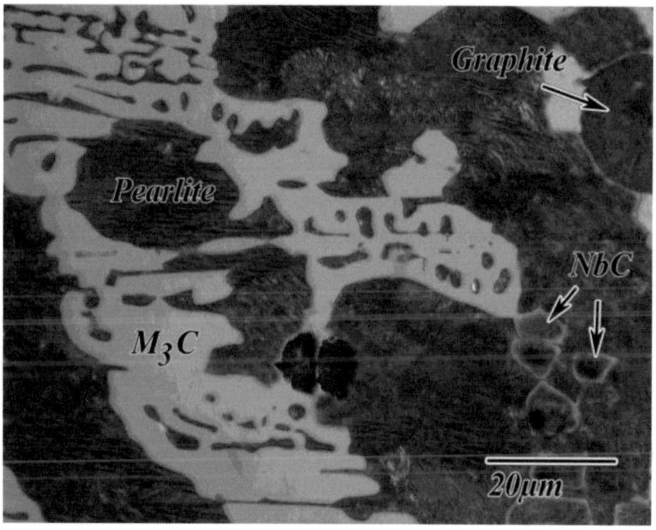

Figure 17. Mottled cast irons showing the segregation of NbC in the microstructure.

The purpose of Akers has been to develop a more wear-resistant material by adding carbide-forming elements without losing the favourable properties of the alloy. This should be done without upsetting the balance of carbide/graphite or to any major extent change the matrix composition of the base alloy. The presence of graphite in the alloy greatly improves the ability of the roll to withstand the thermal shocks associated with hot rolling. Graphite also reduces the friction between the roll and the strip and reduces the potential for welding of the trip to the roll.

In order to get the proper balance of graphite and carbides in these alloys, some requirements have been highlighted by Nylen [36], and these include as follows: (i) an extremely careful selection of melting raw materials, (ii) controlled melting conditions, (iii) adequate control of composition and (iv) efficient inoculation techniques to get the required type and distribution of graphite. According to him, this correct balance may help to avoid the use of stronger carbide-forming elements, which can promote the formation of massive carbides within the structure of the iron. The Akers group has been working hard in these actions, and they conclude that the use of niobium in mottled irons has a massive potential future.

2.4. Niobium in white iron

Undoubtedly, the major amount of work on niobium additions to cast irons has been undertaken to white cast irons, since the main application for these alloys is in the field of the wear resistance, where the NbC particles play an important role.

Since the accidental discovery of the white cast iron by Robert Ransome, an agricultural machinery manufacturer in Ipswich, England, at the turn of the eighteenth century [50], the art of manufacturing this iron has slowly become a science. At the time of Ransome's discovery, it was known only that a harder, more wear-resistant iron was formed when iron was rapidly cooled. Furthermore, when this iron was fractured, the surface appeared white and, hence, the name white cast iron. Initial applications for this iron at that time were limited to thin sections since a high freezing rate was required for its manufacture. It was not until the first part of the twentieth century that the carbide-stabilising ability of chromium was fully recognised with the advent of a 25–30% chromium-alloyed white cast iron. This new alloy could be cast white into almost any section size and was significantly more wear resistant than unalloyed white cast iron. Lower chromium irons were subsequently introduced that were alloyed with nickel mainly for hardenability [51].

Since this discovery, so many researchers have gradually contributed to the knowledge of the metallurgy of cast iron and to establish the relationship between chemical alloying, processing routes and mechanical properties, as highlighted by Tabret et al. [52] in their interesting review. In addition, since nowadays white iron alloys are highly alloyed complex materials, they suggest that this complex chemistry demands a deep understanding of the metallurgy of cast irons during their production and processing. ASTM specification A532 [53] covers the composition of the abrasion-resistant white irons in use.

The ASTM standard classifies these alloys as abrasion-resistant because in the mining industry, which is the main user, such irons are used in a range of situations which are subject to varying conditions where the main wear mechanism is the abrasive one, that is crushing, screening, and pumping [52]. Furthermore, in the steel industry, Hi-Cr irons are widely used for rolls in hot rolling mills [54], where the mechanisms of damage are abrasion of the matrix and rolling contact fatigue [55, 56]. Such deterioration of the rolls surface causes strip surface flaw [57].

Hi-Cr white irons are based on the Fe-Cr-C ternary system. Jackson [58] was one of the first researchers to study systematically this alloy and its solidification sequence. Commercial Hi-Cr alloys often contain further alloying elements such as molybdenum, nickel, copper and manganese [52], which may modify the solidification behaviour and the amount of phases present in the alloy. For example, molybdenum additions contribute to increase hardenability and avoid pearlite formation; in such case, the analysis of a quaternary system is necessary as highlighted by Qiu [59]. However, chromium and carbon contents are the main parameters describing the solidification path of the Hi-Cr irons and the ternary Fe-Cr-C system is used to describe the first stages of solidification on any commercial alloy of this type.

According to this, the common compositions go from 1.8 to 3.5%C and from 12 to 30%Cr. Under these conditions, the most common alloys are hypoeutectic and the solidification path is as follows: solidification starts with the formation of dendrites of primary austenite which grow and reject carbon towards the interdendritic regions; as the solidification progresses, temperature decreases and the remaining liquid gets richer in carbon until it reaches the eutectic composition. At this temperature, the eutectic reaction occurs

Although the austenite that forms on solidification is not stable at room temperature, a metastable austenitic matrix is commonly retained in the ambient temperature structure. More commonly, the M_S temperature is slightly above ambient temperature, while the martensite finish temperature, M_f, is below. This produces some transformation on cooling so that martensite may be present in the predominantly austenitic as-cast structure [52].

The as-cast microstructure of hypoeutectic Hi-Cr white irons typically consists of austenite dendrites (with possibly partial transformation to martensite, pearlite or bainite) and an interdendritic eutectic of carbides and partially transformed austenite. An example of the microstructure of a sand cast 17%Cr-2.6%C-2%Ni-2%Mo is given in **Figure 18**. The optical micrograph shows that the structure is mainly austenitic with some martensite along the eutectic carbides and in the eutectic matrix.

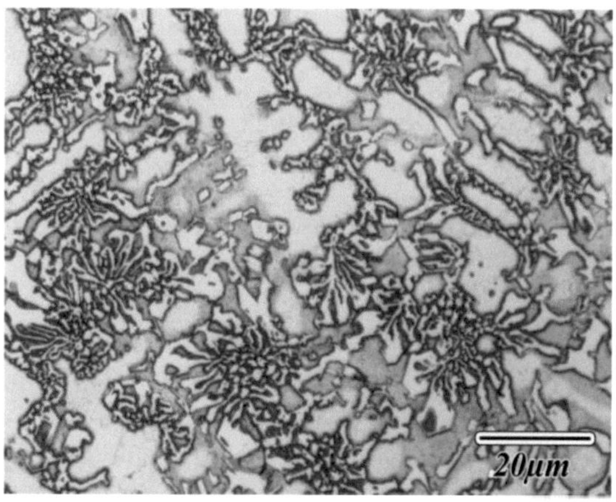

Figure 18. As-cast structure of a high-chromium white cast iron.

Although the chromium content of many Hi-Cr white iron alloys is quite high (typically between 12 and 30%), the majority of this is combined with carbon in the carbides. The chromium content remaining in the matrix is therefore quite low. For instance, Laird et al. [60] measured a matrix chromium content of only 9.5% in 17.8%Cr-3%C iron using electron microprobe analysis and De Mello et al. [61] obtained 9.5%Cr in the matrix of a 16.1%Cr-3.26%C. In the same way, Dogan et al. [62] obtained a value of 16.2%Cr in the matrix of a 26%Cr-2.76%C. Then, additional alloying elements are required to achieve sufficient hardenability, particularly with larger section sizes [63]. The most common elements used for these additions are molybdenum, nickel, manganese and copper.

Molybdenum is added to high-chromium white irons in amounts between 0.5 and 3.5%. Although it is believed that amounts greater than 1% are required to be effective [64], additions of more than 3-4% achieve little [65]. It acts to suppress pearlite formation and increase hardenability [61, 65, 66] by effectively inhibiting secondary carbide precipitation during cooling. Molybdenum also has a synergistic effect on the influence of other alloying elements such as nickel and copper, which are more effective in delaying pearlite formation if added in conjunction with molybdenum [61, 66–69]. Part of the molybdenum in the irons forms a eutectic carbide of the type M_2C at the final stage of solidification [61, 65, 66, 70–72]. Another advantage of molybdenum is that it has little effect on the Ms temperature [73], while most elements tend to decrease the Ms temperature possibly leading to over stabilisation of austenite.

Nickel, manganese and copper are added to these irons mainly to improve hardenability. They are commonly found in amounts up to 2%. Nickel and copper partitioned to the matrix [51], while manganese may be partly dissolved in matrix and partially segregated to the carbide phase, which reduces its effectiveness. These elements contribute to lower the Ms temperature of austenite during cooling and therefore increasing the alloy content often produces more austenite which in turn represents a decrease in the as-cast hardness [67]. When nickel is added to the iron along with molybdenum, the effectiveness in improving hardenability is quite considerable and the undesirable pearlite formation is avoided [67].

Less traditional alloying elements such as vanadium, tungsten, titanium and niobium have also been added to white iron alloys [45, 46, 64, 70, 74–80]. The aim of these additions is usually to achieve some modification of the eutectic carbide structure by obtaining harder carbides, though they may improve the hardenability of the matrix.

The eutectic carbide structure in Hi-Cr irons is strongly influenced by chromium as this is a strong carbide-forming element. It has been pointed out by Pearce [81] that as the chromium content increases above 10-12%, the type of eutectic carbide that forms changes from M_3C to M_7C_3. For both types of eutectic carbide, increasing the chromium content of the alloy increases the proportion of chromium to iron as the metal species in the carbides [82], while also, increasing the carbide hardness.

It has been recognised that a possible strategy for improving the toughness of white iron alloys as well as the wear resistance under sliding conditions involves the refinement of the eutectic carbide structure by producing finer, more globular carbides [81, 83, 84].

Alloying additions have also been used to modify the eutectic carbide structure [60, 85–94]. Generally, attempts to modify the eutectic carbide structure through alloying elements have had limited success. Boron has been used within the range of 0.1-0.3% [91, 92] with this purpose. The argument is the thought that the presence of boron in iron alloys decreases the solubility of carbon in austenite; which may conduct to a higher and fines carbide precipitation during solidification. The effect of silicon and rare earth elements has been investigated in these irons, in the case of silicon due to its effect on transformation behaviour, and in the case of rare earths due to its high segregation effect. Results of these elements on the carbide structure have produced inconsistent results.

The formation of other type of carbide, such as niobium carbide [78] and vanadium carbide [90], though the appropriate alloying additions, has been associated with the formation of finer rounder carbides.

The commonly applied heat treatment to destabilise the austenitic matrix, according to Pearce [81], involves holding at a temperature between 920 and 1060°C for 1–6 h. During soaking at these temperatures, secondary carbides precipitate in matrix reducing its alloy content, particularly the amount of carbon. If the alloy content is too low in austenite, the M_S temperature increases [69] and a higher amount of austenite will transform to martensite during subsequent cooling down to room temperature. For these alloys, air cooling after destabilisation heat treatments is usually a common practice to produce a predominantly martensitic structure and the risk of cracking by rapid cooling is avoided [81]. **Figure 19** shows a micrograph of the structure of a destabilised Hi-Cr iron 17%Cr-2.5%C-2%Ni-2%Mo, which was held at 1000°C for 45 min and air cooled to room temperature. The structure is composed of the eutectic carbides, and the former austenite has transformed to martensite and secondary carbides.

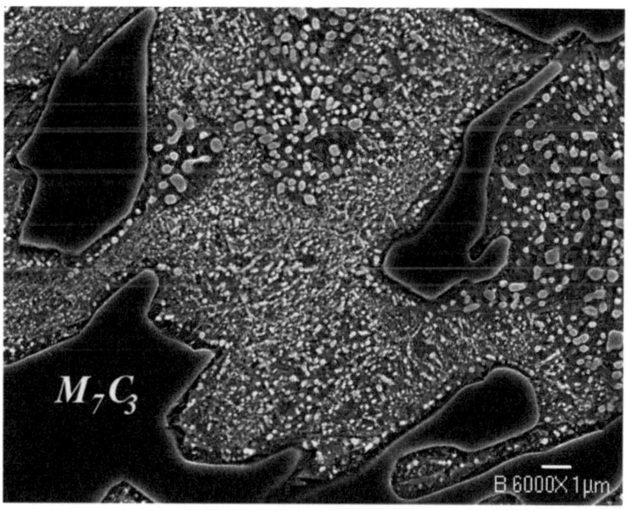

Figure 19. Microstructure of the heat-treated high-Cr white iron.

Although a predominantly martensitic matrix is formed after destabilisation treatment, retained austenite may be present in amounts up to 35% [95–97]. According to Tabret et al. [52], the retained austenite volume depends on the amount of dissolved carbon after the destabilisation heat treatment. Dissolved carbon content in austenite is in turn influenced by the composition of the alloy, the time and temperature of heat treatment and as well as the thickness of the cast. Eutectic carbides, on the other hand, are unaffected at the temperatures of the destabilisation heat treatment.

It has been widely accepted the relationship between bulk hardness and wear resistance for most alloys under abrasive conditions; according to the *Archard's Law*, the harder the material, the more the wear resistant it is. However, as Tabret et al. [52] have pointed out, there are many factors influencing the abrasive wear resistance of an alloy for a given wear system. These include the following: load, type of abrasive material, speed, etc, and then, they suggest that the microstructure of the alloy must play a strong role on the overall abrasive wear behaviour.

The eutectic carbides in white iron are main responsible for the excellent wear resistant of these irons when compared with other alloys. The eutectic carbide volume fraction is commonly between the range of 20 and 35% in the microstructure of hypoeutectic high-chromium white irons. It could be thought that increasing the carbide volume fraction would increase wear resistance; however, some other factors such as carbide type and hardness, orientation, and the role of the matrix as a carbide supporter may determine the overall wear behaviour [52].

The effect of carbide type and hardness could be considered negligible since not much difference in hardness has been observed in Hi-Cr white irons and the abrasive materials generally used in laboratory tests. High-chromium white irons commonly form the M_7C_3 carbide as the eutectic carbide. The addition of niobium and vanadium to Hi-Cr white irons to form even harder carbide types has been shown to improve the abrasion resistance [77–79] due to the increase in the relative hardness, as shown below.

On the other hand, the carbide volume fraction (CVF) clearly influences the abrasive wear resistance of these irons. Although we could expect an increase in wear resistance when increasing the amount of the carbide phase, the wear system along with the wear mechanisms taking place determines the actual wear behaviour of the alloy under abrasive conditions [52].

From the asseveration of the main role of the carbide phase on the abrasive wear resistance, it could be thought that the roll of the matrix is just to protect or provide mechanical support to the carbides. However, the actual roll of the matrix in these irons must be much more complex, according to Fulcher et al. [98]. They suggested that the roll of the matrix depends on the protection level that carbides provide to the matrix during abrasive wear. If the abrasive particles are larger than the mean free path of carbides (matrix region between carbides), the main roll of the matrix is to provide mechanical strength to the carbides. However, if the abrasive particles are small enough to abrade the matrix, it is preferentially removed and the naked carbides become unsupported and prone to crack. When this happens, the abrasion resistance of the matrix is of main importance, as this controls the rate at which the carbides become unsupported and fracture [52].

Most studies on the effect of the matrix structure have compared the as-cast, predominantly austenitic matrix with the heat treated predominantly martensitic matrix [99–101]. And it has been established that a pearlitic matrix reduced the abrasion resistance, due to poor support of the carbides, and this matrix structure is generally avoided if good abrasion resistance is required.

Under this basis, the precipitation of secondary carbides within the matrix of high-chromium white irons may also influence the abrasion behaviour. These secondary carbides strengthen the matrix by a dispersion hardening effect therefore increasing the mechanical support to the eutectic carbides and, in turn, leading to improved wear resistance [79].

Carbide-forming elements such as titanium, niobium, vanadium and tungsten have been widely used as alloying elements in high-chromium white irons. These elements either partition to the eutectic carbide phase or form their own carbides. In any case, the overall hardness of the iron is increased. In the particular case of niobium, it forms hard niobium carbides that improve hardness and abrasive resistance.

The influence of niobium (from 0 to 2.06%) on the microstructural characteristics of a 16.7%Cr white iron was examined in both as-cast and heat-treated conditions [78].The as-cast structure of the Fe-Cr-C alloy consisted of primary dendrites of austenite and the eutectic austenite-M_7C_3. Fe-Cr-C-Nb alloys also contained various amounts of NbC. In alloys containing up to 1%Nb, NbC carbides were present in the petal-like form, whereas in the alloy with 2.06%Nb, they were in the compact shaped form.

Reaustenitising the as-cast structure resulted in the precipitation of secondary carbides and depletion of carbon in the matrix, which transformed a large amount of austenite to martensite during quenching to room temperature. The amount of retained austenite was minor in the alloys containing niobium. Niobium increases the amount of NbC while decreases the amount of M_7C_3. These carbides also became elongated as the percentage of niobium increased which increases hardness and wear resistance.

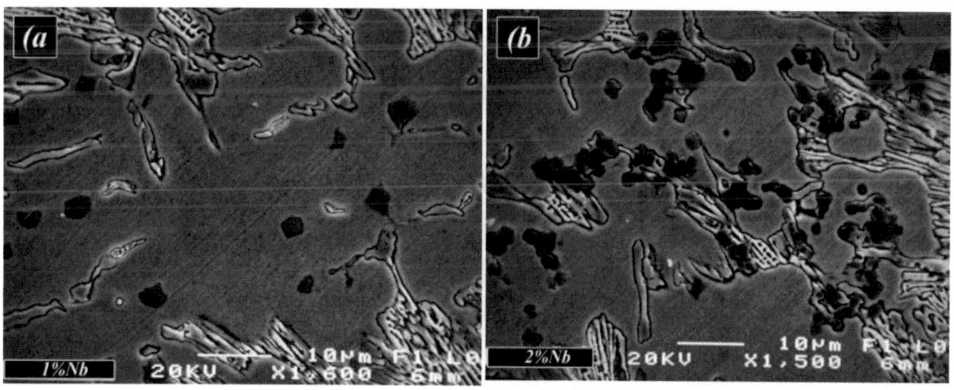

Figure 20. Niobium carbides in high-chromium white iron. (a) 1%Nb, most of the carbides are within the matrix, (b) 2%Nb, the carbides are agglomerated and segregated to the eutectic carbide/matrix interface.

Niobium in Fe-Cr-C-Nb alloys forms carbides of the type MC. According to some authors [45, 46, 102], the solubility of niobium in austenite or M_7C_3 is very low, so the majority of niobium present in the alloy is in the form of MC carbides. These carbides are formed before M_7C_3, which causes depletion of carbon in the liquid. Since carbon is the primary element that determines the amount of carbide in high-chromium irons, the amount of M_7C_3 carbide should decrease as the Nb increases. Furthermore, the decrease in the M_7C_3 carbide due to the increase in niobium reduces the amount of chromium to form such eutectic carbide and more chromium remains in austenite increasing its hardenability. They also suggest that the presence of NbC contributes to a reduction in the size of the M_7C_3 due to a change of the solidification sequence, and this increases the fracture toughness of the alloy.

The microstructure when adding niobium always shows the presence of the NbC particles embedded in the matrix or segregated to the interface matrix/eutectic carbide, depending on the amount of niobium in the alloy (see **Figure 20**).

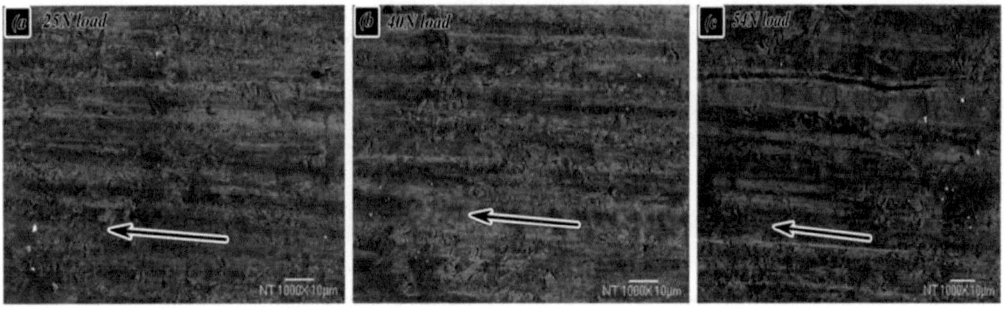

Figure 21. SEM micrographs showing the worn surface of an as-cast white iron after an abrasive wear test at three different loads, and evidencing the increase in roughness as load increases. The arrows indicate the wear direction [103].

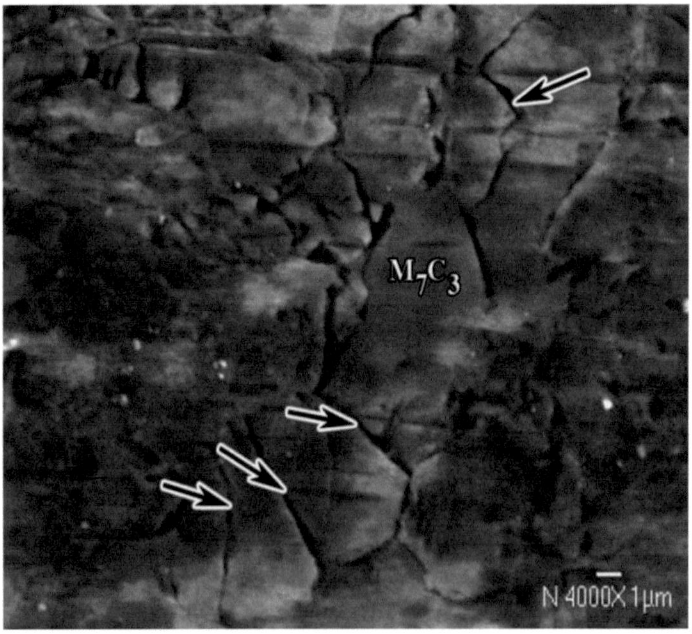

Figure 22. SEM micrograph showing carbide cracking at the surface of an as-cast iron after wear test under 54 N load [103].

The presence of such carbide particles in the structure has proven to increase wear resistance either under sliding and abrasive conditions. The effectiveness of the primary carbides and the strengthening of matrix by the heat treatment are the cause of the better wear behaviour these alloys. For example, **Figure 21** shows a series of SEM micrographs of the worn surface of a 3.1%C-17%Cr-1.1%Ni-0.98%Mo iron alloyed with 2%Nb-1.8%Ti-2%V in the as-cast conditions for different applied loads during an abrasive wear test [103]. From this figure, as the load increased, the wear groves were more evident. Also, for high loads carbide cracking commonly occurred, particularly for the as-cast iron. The soft matrix, highly deformed by friction during wear provides low support to the carbides and they crack (**Figure 22**); on the other hand, for heat-treated irons carbide cracking is less common since the strengthened matrix allows minimum deformation. These observations indicate a more stable matrix supporting the carbide phase in the alloyed iron due to the presence of MC carbides.

Figures 23 and **24** show SEM micrographs of the cross section of the iron without the alloying Nb-Ti-V elements in both as-cast and heat-treated conditions (**Figure 23**) and of the alloyed iron also in both as-cast and heat-treated conditions (**Figure 24**) after being wear tested with loads of 25 and 54 N. (analysed in Ref. [103]). From these micrographs, higher levels of deformation and carbide cracking are evident for the as-cast iron and particularly for the higher applied load; this destabilises the surface and contributes to higher wear rates. On the other hand, low deformation of matrix and no carbide cracking are observed for the heat-treated alloys, particularly for the alloyed iron (see **Figure 23d**), contributing to higher wear resistance. Although the depth of deformation was not measured for these alloys, due to the friction forces, the depth at which carbide cracking can be observed should increase with load. In the alloyed irons, one feature that highly contributes to lower eutectic carbide cracking is the presence of the hard MC carbides that strengthen matrix, even in the as-cast conditions. This phenomenon of subsurface destabilisation by carbide cracking during wear tests has been widely analysed and reported first by Fulcher et al. [98] and then by some other authors [79, 104, 105]. This is a very important factor to consider when analysing wear resistance of white irons and some alloys with a massive brittle ceramic phase in the structure.

Under these basis, for the analysed irons by Bedolla-Jauinde et al. [103], a considerable increase in wear resistance should be expected for the alloyed heat-treated iron. However, according to them, the increase was moderate and this has been attributed to the abrasive conditions used for the test. Based on the roll of the matrix during abrasive wear tests described by Fulcher et al. [98], and on the asseverations of Zum Gahr and Doane [99] that martensitic matrices obtained by heat treatments in these irons improves the wear behaviour under low stress abrasion, Bedolla-Jacuinde et al. [103] describe the behaviour of the analysed irons. The martensitic matrix reinforced with secondary carbides and with primary MC carbides reduces wear of matrix and minimises carbide cracking. This metallurgical phenomenon has been widely reported to improve the wear behaviour of high-chromium irons either under sliding and abrasion [52, 79, 98, 104, 106].

The wear behaviour of the irons analysed by Bedolla-Jacuinde et al. [103], particularly at higher loads, seems to be determined mainly by the eutectic carbide volume fraction and matrix has a minor effect. An additional important factor to consider is the eutectic carbide alignment with the surface. Dogan and Hawk [106] have reported superior wear resistance under high stress abrasion for irons that have carbides aligned parallel to the surface, compared with the same irons with carbides aligned perpendicular to the surface. They argue that long carbides perpendicular to the surface are more prone to bend and crack, particularly close to the worn surface, where the plastic deformation of matrix is considerable. Carbide cracking was observed by Bedolla-Jacuinde et al. [103] and is shown in the micrographs shown from **Figures 23** and **24**. They explain that the bending of carbides aligned perpendicular to the surface is caused by the friction generated by the surfaces in contact, and during bending, as highlighted by Fulcher et al. [98], tensile stresses are developed at the back side of the carbides. Such tensile stresses along with the matrix deformation may conduct to the carbide fracture and the consequent surface destabilisation.

A general conclusion established by Bedolla-Jacuinde et al. [103] for explaining the better wear behaviour of the iron alloyed with niobium and titanium was the higher carbide volume fraction compared with the unalloyed iron. These include hard primary MC carbides within the matrix and the eutectic M_7C_3 carbides. Furthermore, the improved wear resistance observed after heat treatment was also attribute to the strengthening of the matrix, which transformed from austenite to a complex mixture martensite plus some retained austenite a high amount of tiny secondary carbides.

As a summary, niobium in cast irons has been particularly used in low amounts as a carbide-forming element to increase hardness and basically wear resistance of these alloys.

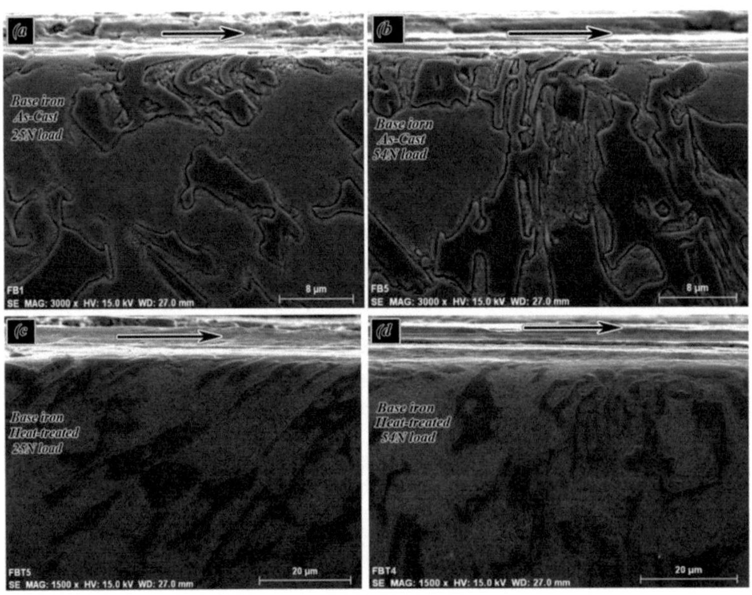

Figure 23. SEM micrographs showing the structure below the worn surface in both as-cast (a and b) and heat-treated conditions (c and d) of the unalloyed irons analysed in Ref. [103].

http://dx.doi.org/10.5772/64498

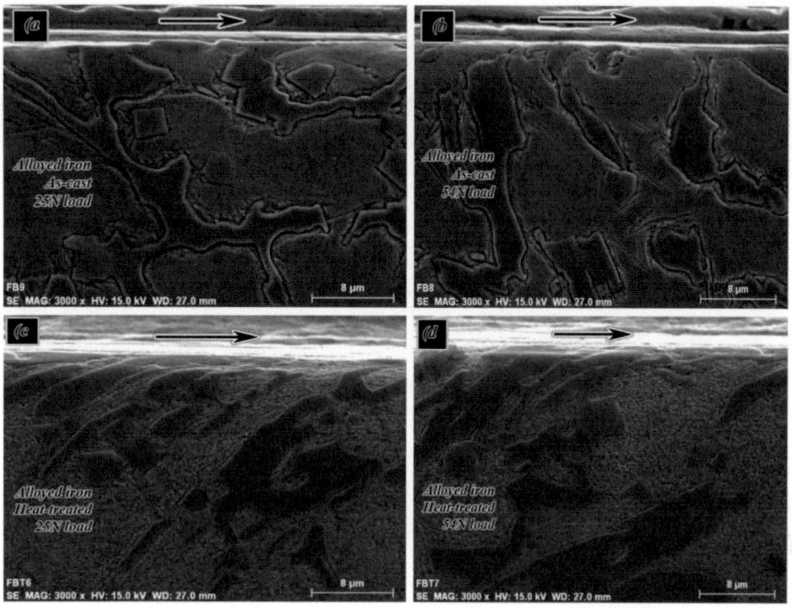

Figure 24. SEM micrographs showing the structure below the worn surface in both as-cast (a and b) and heat-treated conditions (c and d) of the alloyed irons analysed in Ref. [103].

Author details

A. Bedolla-Jacuinde

Address all correspondence to: abedolla68@gmail.com

Institute for Metallurgical and Materials Research, Michoacan University of Saint Nicholas of Hidalgo, Morelia, Michoacán, Mexico

References

[1] ASTM A247 – 10 Standard Test Method for Evaluating the Microstructure of Graphite in Iron Castings, Philadelphia, PA, ASTM.

[2] J. Krawczyk, Structural Causes of Defects in a Cast Iron Mill Roll, *Archives of Foundry Engineering*, 8(2), 2008, s.93–s.98. ISSN1897-3310.

[3] T. Takai, Present Supply – Demand Situation of Niobium, *Cryogenics*, Vol. 14 1974, 301–306.

[4] D.C. Goldberg, G. Dicker, and S.A. Worcester, Niobium and Niobium Alloys in Nuclear Power, *Nuclear Engineering and Design*, 22, 1972, 95–123.

[5] Y. Furuto, Fundamentals of Niobium Based Superconductors, *Proceedings of the International Symposium of Niobium*, The Metallurgical Society of ASME, Edited by Harry Stuart. Pittsburgh, PA, 1981, 445–494.

[6] W.K. McDonald, Niobium Based Superconductor Composites, *Proceedings of the International Symposium of Niobium*, The Metallurgical Society of ASME, Edited by Harry Stuart. Pittsburgh, PA, 1981, 495–502.

[7] E. Gregory, Applications and Secondary Fabrication of Practical Superconducting Materials, *Proceedings of the International Symposium of Niobium*, The Metallurgical Society of ASME, Edited by Harry Stuart. Pittsburgh, PA, 1981, 503–532.

[8] R.R. Rice, Niobium Compounds in Acoustics and Electro-Optics, *Proceedings of the International Symposium of Niobium*, The Metallurgical Society of ASME, Edited by Harry Stuart. Pittsburgh, PA, 1981, 579–602.

[9] B. O'Brien, J. Stinson, and W. Carroll, Development of a New Niobium-Based Alloy for Vascular Stent Applications, *Journal of the Mechanical Behavior of Biomedical Materials*, 1, 2008, 302–312.

[10] D. Lupton, F. Aldinger, and K. Schilze, Niobium in Corrosive Environments, *Proceedings of the International Symposium of Niobium*, The Metallurgical Society of ASME, Edited by Harry Stuart. Pittsburgh, PA, 1981, 533–560.

[11] B. Aronsson and L.J. Aschan, Niobium in Cemented Carbides, *Proceedings of the International Symposium of Niobium*, The Metallurgical Society of ASME, Edited by Harry Stuart. Pittsburgh, PA, 1981, 637–654.

[12] X. Liu, Y. Ding, S. Zhao, Y. Long, R. Ye, and Y. Chang, Effect of Niobium on Glass Formation Ability and Soft Magnetic Properties of Gd-doping Glassy Alloys, *Journal of Rare Earths*, 26, 2008, 89–92.

[13] F.M. Ahmed, D.S. Edgley, and I.R. Harris, Effect of Niobium Addition on the Nd-Fe-B Alloy and Magnet, *Journal of Alloys and Compounds*, 209, 1994, 363–368.

[14] S.A. Sinan, D.S. Edgley, and I.R. Harris, Effect of Additions of Nb to Sm_2Fe_{17}-Based Cast Alloys, *Journal of Alloys and Compounds*, 226, 1995, 170–173.

[15] F.M. Ahmed, A. Ataie, A.J. Williams, and I.R. Harris, Effect of Niobium on the HDDR Behavior of Near-Stoichiometric NdFeB Alloy, *Journal of Magnetism and Magnetic Materials*, 157/158, 1996, 59–60.

[16] C.G. Schon and J.A. Soares Tenorio, The Chemistry of the Iron-Niobium Intermetallics, *Intermetallics*, 4, 1996, 211–216.

[17] C.H. Tsau, J.S. Ching, and J.W. Yeh, The Microstructure and Mechanical Properties of Niobium-Modified Ni-25Al-27.5Fe Intermetallics, *Intermetallics*, 5, 1997, 433–441.

[18] N. Muruyama, R, Uemori, and M. Sugiyama, The Role of Niobium in the Retardation of the Early Stage of Austenite Recovery in Hot-Deformed Steels, *Materials Science and Engineering A*, 250, 1998, 2–7.

[19] L. Qing-You, D. Su-Huai, S. Xin-Jun, D. Han, and W. Yu-Qing, Effect of Dissolved and Precipitated Niobium in Microalloyed Steel on Deformation Induced Ferrite Transformation (DIFT), *International Journal of Iron and Steel Research*, 16(4), 2009, 67–71.

[20] L. Liang-Yun, Q. Chun-Lin, Z. De-Wen, G. Xiu-Hua, and D. Lin-Xiu, Dynamic and Static Recrystallization Behavior of Low Carbon High Niobium Microalloyed Steel, *International Journal of Iron and Steel Research*, 18(1), 2011, 55–60.

[21] B. Dutta, E.J. Palmiere. and C.M. Sellars, Modelling the Kinetics of Strain Induced Precipitation in Nb Microalloyed Steels, *Acta Materialia*, 49, 2001, 785–794.

[22] Z. Jia, R.D.K. Mirsa, R. O′Malley, and S.J. Jansto, Fine-Scale Precipitation and Mechanical Properties of Thin Slab Processed Titanium-Niobium Bearing high Strength Steels, *Materials Science and Engineering A*, 528, 2011, 7077–7083.

[23] Y. Maehara and Y. Ohmori, The Precipitation of AlN and NbC and the Hot Ductility of Low Carbon Steels, *Materials Science and Engineering*, 62, 1984, 109–119.

[24] W.M. Rainforth, M.P. Black, R.L. Higginson, E.J. Palmiere, C.M. Sellars, I. Prabst, P. Warbichler, and F. Hofer, Precipitation of NbC in a Model Austenitic Steel, *Acta Materialia*, 50, 2002, 735–747.

[25] J. Irvine and T.N. Baker, The Influence of Rolling Variables on the Strengthening Mechanisms Operating in Niobium Steels, *Materials Science and Engineering*, 64, 1984, 123–134.

[26] J.D. Boyd, The Microstructure and Properties of a Quenched and Tempered Low-Carbon-Manganese-Niobium Steel, *Metallurgical Transactions A*, 7, 1976, 1577–1586.

[27] G.D. Almeida-Soares, L.H. Alemida, T.L. Silveira, and I. Le May, Niobium Additions in HP Heat-Resistant Cast Stainless Steels, *Materials Characterization*, 29, 1992, 387–396.

[28] R.M. Brito and H.J. Kestenbach, On the Dispersion Hardening Potential of Interphase Precipitation in Micro-Alloyed Niobium Steel, *Journal of Materials Science*, 16, 1981, 1257–1263.

[29] S. Kheirandish and A. Noorian, Effect of Niobium on Microstructure of cast AISI H13 Hot Work Tool Steel, *Journal of Iron and Steel Research, International*, 15(4), 2008, 61–66.

[30] Metal Casting (Reference Book for MY4130), Textbook by Karl B. Rundman, Edited by Michigan Tech. University. http://www.mse.mtu.edu/faculty/krundman.html.

[31] E. Burgos, R. Correa, and J. Batista, Injection of Ferroniobium in Gray Cast Irons, *CBMM Report*, http://www.cbmm.com.br/portug/index.html.

[32] ASM Handbook, Vol. 15, *Casting*, ASM International, Materials Park, OH, 2010.

[33] J. James, C. Ronald, M. Leonard, M. Michael, and E. John. High Strength Gray Cast Iron Containing Niobium. European Patent WO2008105987 (A1). Appl. No: WO2008US0133420080131. Sept. 04, 2008.

[34] J. Porker and W. Lotz. Optimized Gray Cast Iron Plate Alloy for Utility Vehicle Brake Disks. US Patent 5,894,010. Appl: No 08/760,642. Apr. 13, 1999.

[35] I. Hanna, Mechanical Properties of Niobium Alloyed Gray Iron, Master Thesis, Royal Institute of Technology, Stockholm Sweden, 2011.

[36] T. Nylen, Niobium in Cast Iron, *International Symposium on Niobium 2001*, Orlando FL, USA, 2–5 Dec. 2001, 1063–1080.

[37] Z. Wenbin, Z. Hongbo, Z. Dengke, Z. Hongxing, H. Qin, and Z. Qijie, Niobium Alloying Effect in High Carbon Equivalent Gray Cast Iron, *China Foundry*, 8(1), 2011, 36–40.

[38] G. Balachandran, A. Vadiraj, M. Kamaraj, and E. Kazuya, Mechanical and Wear Behavior of Alloyed Gray Cast Iron in the Quenched and Tempered and Austempered Conditions, *Materials and Design*, 32, 2011, 4042–4049.

[39] A. Vadiraj, G. Balachandran, M. Kamaraj, B. Gopalakrishna, and K. Probhakara Rao, Studies on Mechanical and Wear Properties of Alloyed Hypereutectic Gray Cast Irons in the As-Cast and Austempered Conditions, *Materials and Design*, 31, 2010, 951–955.

[40] A. Vadiraj, G. Balachandran, M. Kamaraj, B. Gopalakrishna, and D. Venkateshwara Rao, Wear Behavior of Alloyed Hypereutectic Gray Cast Iron, *Tribology International*, 43, 2010, 647–653.

[41] Z. Daxin, Z. Yuanhao, L. Jianyong, H. Hanjun, and H. Xiaoxian, Characterization of Titanium-Containing Compounds in Gray Iron, *Tsinghua Science and Technology*, 13(2), 2008, 127–131.

[42] L. Zhong, Y. Xu, X. Liu, and F. Ye, Study of Niobium Carbide Particulate Reinforced Iron Matrix Composite Production in Situ, *Journal of Materials Science*, 46, 2011, 2814–2819.

[43] R. Elliot, Current Status of Austempered Cast Irons, *Advanced Materials Research*, 4–5, 1997, 1–16.

[44] S. Salman, F. Findik, and P. Topuz, Effect of Various Austempering Temperatures on Fatigue Properties in Ductile Iron, *Materials and Design*, 28, 2007, 2210–2214.

[45] A. Sawamoto, K. Ogi, and K. Matsuda, Solidification Structures of Fe-C-Cr-(V- Nb-W) Alloys, *AFS Transactions*, 94, 1986, 403–416.

[46] H.K. Baik and C.R. Loper Jr., The Influence of Niobium on the Solidification Structure of Fe-C-Cr Alloys, *AFS Transactions*, 96, 1988, 405–411.

[47] A. Bedolla-Jacuinde, E. Solis, and B. Hernnadez, Effect of Niobium in Medium Alloyed Ductile Cast Iron, *International Journal of Cast Metals Research*, 16(5), 2003, 481–486.

[48] J.J. Coronado and A. Sinatora, Particle Size Effect on Abrasion Resistance of Mottled Cast Iron with Different Retained Austenite Contents, *Wear*, 267, 2009, 2077–2082.

[49] J. Krawczyk and J. Pacyna, The Influence of Microstructure of Mottled Cast Iron for Mill Rolls on its Properties, *Metal 2009 Proceedings*, May 19–21, 2009, Hradec nad Moravici, 1–8.

[50] R. Elliot, Eutectic Solidification Processing Crystalline and Glassy Alloys, London, Butterworth, 1983.

[51] G. Laird II, Microstructures of Ni-Hard I, Ni-Hard IV and High-Cr White Cast Irons, *AFS Transactions*, 99, 1991, 339–357.

[52] C.P Tabrett, I.R. Sare, and M.R. Gomashchi, Microstructure-Property Relationships in High-Chromium White Iron Alloys, *International Materials Reviews*, 41(2), 1996, 59-82.

[53] ASTM A532-93 Standard Specification for Abrasion-Resistant Cast Irons, Philadelphia, PA, ASTM.

[54] H. Hiraoka, Y. Kataoka, K. Yuda, K. Taniguchi, M. Sasada, and I. Hishinuma. Application of a High Chromium Steel to Roughing Work Rolls for Hot Strip Mills, *ISIJ International*, 32(11), 1992, 1177 1183.

[55] O. Kato, H. Yamamoto, M. Ataka, and K. Nakajima, Mechanisms of Surface Deterioration of Roll for Hot Strip Rolling, *ISIJ International*, 32(11), 1992, 1216–1220.

[56] J.H. Ryu, O. Kwon, P.J. Lee, and Y.M. Kim, Evaluation of the Finishing Roll Surface Deterioration at Hot Strip Mill, *ISIJ International*, 32(11), 1992, 1221–1223.

[57] Y. Oike, J. Sato, K. Minami, K. Yoshitake, and S. Yamanaka, Influence of Rolling Conditions on Chemical Compositions of Rolled Material on Strip Surface Flaw Caused by Surface Deterioration of Hot Work Rolls, *ISIJ International*, 32(11), 1992, 1211–1215.

[58] R.S. Jackson, The Austenite Liquidus Surface and Constitutional Diagram for the Fe-Cr-C Metastable System, *Journal of the Iron and Steel Institute*, 204, 1970, 163–167.

[59] C. Qiu, An Analysis of the Cr-Fe-Mo-C System and Modification of Thermodynamic Parameters, *ISIJ International*, 32(10), 1992, 1117–1127.

[60] G. Laird II and G.L.F. Powell, Solidification and Solid-State Transformation Mechanisms in Si Alloyed High-Chromium White Cast Irons, *Metallurgical Transactions A*, 24(2), 1993, 981–988.

[61] J.D.B. DeMello, M. Duran-Charre, and S. Hamar-Thibualt, Solidification and Solid State Transformations During Cooling of Chromium-Molybdenum White Cast Irons, *Metallurgical Transactions A*, 14(9), 1983, 1793–1801.

[62] O.N. Dogan, J.A. Hawk, and G. Laird II, Solidification Structure and Abrasion Resistance of High Chromium White Irons, *Metallurgical and Materials Transactions A*, 28(6), 1997, 1315–1327.

[63] J. Dodd and J.L. Parks, Factors Affecting the Production and Performance of Thick Section High-Chromium-Molybdenum Alloy Iron Castings – Part II: Production and Performance, *AFS International Cast Metals Journal*, 1980, 15–23.

[64] P. Dupin and J.M. Schissler, Influence of Additions of Silicon, Molybdenum, Vanadium, and Tungsten Upon the Structural Evolution of the As-Cast State of a High-Chromium Cast Iron (20%Cr, 2.6%C), *AFS Transactions*, 92, 1984, 355–360.

[65] F. Maratray, Choice of Appropriate Compositions for Chromium-Molybdenum White Irons, *AFS Transactions*, 79, 1971, 121–124.

[66] J.W. Choi and S.K. Chang, Effects of Molybdenum and Copper Additions on Microstructure of High Chromium Cast Iron Rolls, *ISIJ International*, 32(11), 1992, 1170–1176.

[67] R.B. Guandlach, Microstructure, Hardness and Abrasion Resistance of As-Cast 17.5% Chromium White Iron, *AFS Transactions*, 82, 1974, 309–316.

[68] J.L. Parks, Characteristics of As-Cast and Subcritically Heat-Treated High-Chromium-Molybdenum White Irons for Thick-Section Castings, *AFS Transactions*, 86, 1978, 93–102.

[69] W.W. Cias, Austenite Transformation Kinetics and Hardenability of Heat Treated 17.5%Cr White Cast Irons, *AFS Transactions*, 82, 1974, 317–328.

[70] C.R. Loper Jr. and H.K. Baik, Influence of Molybdenum and Titanium on the Microstructures of Fe-C-Cr-Nb White Cast Irons, *AFS Transactions*, 97, 1989, 1001–1008.

[71] C. P. Tong, T. Suzuki, and T. Umeda, Eutectic Solidification of High Chromium Cast Irons, *Proceedings IV International Symposium on the Physical Metallurgy*, Tokyo, Japan, September 1989, 403–410.

[72] M. Ikeda, T. Umeda, C.P. Tong, T. Suzuki, N. Niwa, and O. Kato, Effect of Molybdenum Addition on Solidification Structure, Mechanical Properties and Wear Resistivity of High Chromium Cast Iron, *ISIJ International*, 32(11), 1992, 1157–1162.

[73] C.Y. Kung and J.J. Rayment, An Examination of the Validity of Existing Empirical Formulae for the Calculation of Ms Temperature, *Metallurgical Transactions*, 13A, 1982, 328–331.

[74] H.Q. Wu, N. Sasaguri, Y. Matsubara, and M. Hashimoto, Solidification of Multialloyed White Cast Iron: Type and Morphology of Carbides, *AFS Transactions*, 104, 1996, 103–108.

[75] R. Shanzhi, Y. Shuzhi, L. Shimin, W. Mengchun, and W. Xubao, Formation and Morphology of M6C Carbide in Tungsten-Chromium White Cast Irons, *Proceedings IV*

International Symposium on the Physical Metallurgy, Tokyo, Japan, September 1989, 381–386.

[76] R. Kesri and M. Durand-Charre, Phase Equilibria, Solidification and Solid State Transformations of White Cast Irons Containing Niobium, *Journal of Materials Science*, 22, 1987, 2959–2964

[77] J.D.B. DeMello, M. Durand-Charre, and T. Mathia, Abrasion Mechanisms of White Cast Iron II: Influence of the Metallurgical Structure of V-Cr White Cast Irons, *Materials Science and Engineering*, 78, 1986, 127–134.

[78] H.X. Chen, Z.C. Chang, J.C. Lu, and H.T. Lin, Effect of Niobium on Wear Resistance of 15%Cr White Cast Iron, *Wear*, 166, 1993, 197–201.

[79] M. Radulovic, M. Fiset, K. Peev, and M. Tomovic, The Influence of Vanadium on Fracture and Abrasion Resistance in High Chromium White Cast Irons, *Journal of Materials Science*, 29, 1994, 5085–5094.

[80] S.K. Yu and Y. Matsubara, Abrasion Wear Resistance of Alloyed White Cast Iron with Several Types of Carbides and Matrices, *AFS Transactions*, 106, 1998, 53–58.

[81] J.T.H. Pearce, Structure and Wear Performance of Abrasion Resistant Chromium White Cast Irons, *AFS Transactions*, 92, 1984, 599–621.

[82] G.L.F. Powell and G. Laird II, Structure, Nucleation, Growth and Morphology of Secondary Carbides in High Chromium and Cr-Ni White Cast Irons, *Journal of Materials Science*, 27, 1992, 29–35.

[83] S. Hua-Qin, T. Chongxi, Y. Xu-Ru, and W. Qigui, Study on Raising the Impact Toughness on Wear-Resistant High-Chromium Cast Irons, *AFS Transactions*, 99, 1991, 333–337.

[84] D.N. Hanlon, W.M. Rainforth, and C.M Cellars, The Rolling/Sliding Wear Response of Conventionally Processed and Spray Formed High Chromium Content Cast Iron at Ambient and Elevated Temperature, *Wear*, 225–229, 1999, 587–599.

[85] M.A. Qian, W. Chaochang, and S. Harada, Modification of Hypoeutectic Low Alloy White Cast Irons, *Journal of Materials Science*, 31, 1996, 1865–1871.

[86] J. Shen and Q.D. Zhou, Solidification Behaviour of Boron-Bearing High-Chromium Cast Iron and the Modification Mechanism of Silicon, *Cast Metals*, 1, 1988, 79–85.

[87] J. Li and R.W. Smith, Growth Modification of Austenite-(Fe,Cr)3C Pseudo-Binary Eutectic, *Proceedings IV Decennial International Conference on Solidification Processing*, VOL. 4 1997, 481–488.

[88] G.Y. Liang and J.Y. Su, The Effect of Rare Earth Elements on the Growth of Eutectic Carbides in White Cast Irons Containing Chromium, *Cast Metals*, 4, 1992, 83–88.

[89] G.L.F. Powell, R.A. Carlson, and V. Randle, The Morphology and Microstructure of M7C3 Carbides in Fe-Cr-C and Fe-Cr-C-Si Alloys Near the Eutectic Composition, *Journal of Materials Science*, 29, 1994, 4889–4896.

[90] W.C. Chang, H.H. Tsun, and M. Qian, Formation of Spheroidal Carbides in Vanadium White Cast Iron by Rare Earth Modification, *Materials Science and Technology*, 6(9), 1990, 905–910.

[91] N. Ma, Q. Rao, and Q. Zhou, Effect of Boron on the Structures and Properties of 28%Cr White Cast iron, *AFS Transactions*, 98, 1990, 775–781.

[92] H. Fusheng and W. Chaochang, Modifying High Cr-Mn Cast Iron with Boron and Rare Earth-Si Alloy, *Materials Science and Technology*, 5, 1989, 918–924.

[93] J.S. Park and J.D. Verhoeven, Directional Solidification of White Cast Iron, *Metallurgical and Materials Transactions A*, 27(8), 1996, 2328–2337.

[94] Y. Qingxiang, L. Bo, L. Jianhua, and Y. Mei, Effect of Rare Earth Elements on Carbide Morphology and Phase Transformation Dynamics of High Ni-Cr Alloy Cast Iron, *Journal of Rare Earths*, 16, 1998, 36–40.

[95] D.K. Subramanyam, Retained Austenite Measurements in High Chromium White Irons, *AFS Transactions*, 93, 1985, 763–768.

[96] F. Maratray and A. Poulalion, Austenite Retention in High-Chromium White Irons, *AFS Transactions*, 90, 1982, 795–804.

[97] O.N. Dogan and J.A. Hawk, Effect of Retained Austenite on Abrasion Resistance of High-Cr White Cast Irons, *AFS Transactions*, 105, 1997, 167–174.

[98] J.K. Fulcher, T.H. Kosel, and N.F. Fiore, The Effect of Carbide Volume Fraction on the Low Stress Abrasion Resistance of High Cr-Mo White Cast Irons, *Wear*, 84, 1983, 313–325.

[99] K.H. Zum Gahr and D.V. Doane, Optimizing Fracture Toughness and Abrasion Resistance in White Cast Irons, *Metallurgical Transactions A*, 11(4),1980, 613–620.

[100] I.R. Sare and B.K. Arnold, The Effect of Heat Treatment on the Gouging Abrasion Resistance of Alloy White Cast Irons, *Metallurgical and Materials Transactions A*, 26(2), 1995, 357–370.

[101] I.R. Sare and B.K. Arnold, The Influence of Heat Treatment on the High-Stress Abrasion Resistance and Fracture Toughness of Alloy White Cast Irons, *Metallurgical and Materials Transactions A*, 26(7), 1995, 1785–1793.

[102] A. Bedolla-Jacuinde, Microstructure of Vanadium, Niobium, and Titanium Alloyed High-Chromium White Irons, *International Journal of Cast Metals Research*, 13, 2001, 343–361.

[103] A. Bedolla-Jacuinde, F. Guerra, I. Mejía, J. Zuno-Silva, and W.M. Rainforth, Abrasive Wear of V-Nb-Ti Alloyed High-Chromium White Irons, *Wear*, 332–333, 2015, 1006–1011.

[104] A. Bedolla Jacuinde and M. Rainforth, The Wear Behavior of High-Chromium White Irons as a Function of Silicon and Mischmetal Content, *Wear*, 250, 2001, 449–461.

[105] A. Bedolla-Jacuinde, R. Correa, I. Mejía, J. Quezada, and W.M. Rainforth, The Effect of Titanium on the Wear Behavior of a 16%Cr White Cast Iron under Pure Sliding, *Wear*, 263, 2007, 808–820.

[106] O.N. Dogan and J.A. Hawk, Effect of Carbide Orientation on Abrasion of High Cr White Cast Iron, *Wear*, 189, 1995, 136–142.

Chapter 8

Indium Phosphide Bismide

Liyao Zhang, Wenwu Pan, Xiaoyan Wu, Li Yue and
Shumin Wang

Additional information is available at the end of the chapter

http://dx.doi.org/10.5772/64565

Abstract

Indium phosphide bismide is a new member to the dilute bismide family. Since the first synthesis by molecular beam epitaxy (MBE) in 2013, it has cut a figure for its abnormal properties comparing with other dilute bismides. Bismuth (Bi) incorporation is always a difficulty for epitaxial growth of dilute. In this chapter, it shows how to regulate MBE growth parameters and their influence on Bi incorporation in $InP_{1-x}Bi_x$. Structural, electronic and optical properties are systematically reviewed. Thermal annealing to study Bi thermal stability and its effect on physical properties is performed. $InP_{1-x}Bi_x$ shows strong and broad photoluminescence at room temperature, which is a potential candidate for fabricating super-luminescence diodes applied for enhancing spatial resolution in optical coherence tomography. Quaternary phosphide bismide, including InGaPBi and InAlPBi, is briefly introduced in this chapter.

Keywords: $InP_{1-x}Bi_x$, MBE, structure property, electronic property, optical property, thermal stability

1. Introduction

III–V compound semiconductor plays an important role in modern optoelectronic devices for its special physical properties whose band structure could be tailored to meet various device applications. While extensive researches on III-Vs have been carried out for over 60 years, bismuth (Bi) containing III-V-Bi is the least investigated compound. As GaAsBi was first successfully grown by metal organic vapour phase epitaxy (MOVPE) in 1998 [1] and by molecular beam epitaxy (MBE) in 2003 [2], people found several promising physical properties of various dilute bismides, which arouse an upsurge of research on these novel class III-Vs.

As Bi is the heaviest non-radioactive element, it is predicted [3, 4] that dilute bismides will get large spin-orbit splitting energy, which could suppress Auger recombination and improve the characteristic temperature of lasers fabricated with dilute bismide materials. When Bi is incorporated into III-Vs, Bi will form impurity levels close to the valence band (VB) of the host material [5]. If the incorporated Bi concentration increases from doping level to dilute alloy level, the Bi states will broaden and interact with the host valence band, which will cause band gap bowing effect [3]. The bandgap of InP is 1.35 eV at room temperature and will be reduced by about 106 meV for incorporating per 1% Bi [6]. This could help to regulate the material band gap, such as red shifting the bandgap to telecommunication band on GaAs, or even mid-infrared range on GaSb. Unlike dilute nitrides in which incorporating N atoms will significantly reduce the electron mobility and form a lot of non-radiative recombination centres, dilute bismides do not reduce the electron mobility for a small amount of Bi [7]. Bismuth could also act as a surfactant during III-V material growth [8], which makes the interface and/or surface much smoother, improving the optical properties of the materials.

Berding *et al.* predicted InSbBi, InAsBi and InPBi to be potential candidates for mid- and far-infrared optoelectronics application in 1988 [9]. Through a series of calculations, they pointed out that InPBi and InSbBi was the most difficult and the easiest to synthesize, respectively, because there was a larger miscibility gap in $InP_{1-x}Bi_x$ than that in the other two materials. However, once InPBi was successfully mixed, it would be the most robust among the three alloys.

$InP_{1-x}Bi_x$ was first successfully grown by MBE in our group in 2013 [10] and showed several special properties comparing with other dilute bismides, such as broad and strong photoluminescence (PL) at 1.4–2.7 μm at room temperature. This property makes $InP_{1-x}Bi_x$ very attractive for fabricating super-luminescence diodes applied in optical coherence tomography (OCT), which is a non-invasive painless diagnosis technique.

In this chapter, epitaxial growth and physical properties of $InP_{1-x}Bi_x$ and its quaternary compound InGaPBi and InAlPBi are summarized. The potential application in medical diagnosis is also described.

2. Epitaxial growth and characterization

$InP_{1-x}Bi_x$ is difficult to be synthesized as mentioned and has a narrow growth parameter window. The growth parameters, such as growth temperature, Bi flux, PH_3 pressure and growth rate, were investigated to optimize the $InP_{1-x}Bi_x$ crystal quality. $InP_{1-x}Bi_x$ epitaxial layers were grown on semi-insulating InP (100) substrates by V90 gas-source MBE (GSMBE). The beam equivalent pressure (BEP) of the In and Bi sources was measured by a retractable ion gauge. The fluxes of elemental In and Bi were controlled by adjusting the respective effusion cell temperatures while P_2 was cracked from PH_3 at 1000°C.

The growth temperature was decreased to about 275–364°C to initiate the growth of 250–420 nm thick $InP_{1-x}Bi_x$. High-resolution X-ray diffraction (HRXRD), atomic force microscopy

(AFM), energy dispersive X-ray spectroscopy (EDX), scanning electron microscope (SEM) and Rutherford backscattering spectroscopy (RBS) were used to characterize surface and structural properties of the $InP_{1-x}Bi_x$ thin films.

2.1. Effect of growth temperature

Growth temperature is a key parameter for growing dilute bismides as it directly affects the epi-growth process. In GSMBE technique for growth of $InP_{1-x}Bi_x$ alloys with smooth surface and good crystal quality, the optimized growth temperature is about 300–330°C. A set of samples with thickness of 420 nm labelled as A, B and C were grown by GSMBE at 275°C, 324°C and 364°C, respectively, all measured by a thermocouple. AFM result, as shown in **Figure 1(a)**, shows that In/Bi droplets appear in Sample A grown at 364°C with a density in the order of 10^7 cm^{-2} while smooth surface with a low root-mean-square (RMS) roughness value of only 1 nm can be achieved in Sample B when decreasing the growth temperature to 324°C. If the growth temperature is further lowered to 275°C, a whisker-like surface feature with a high RMS roughness value of 17 nm is observed in Sample C.

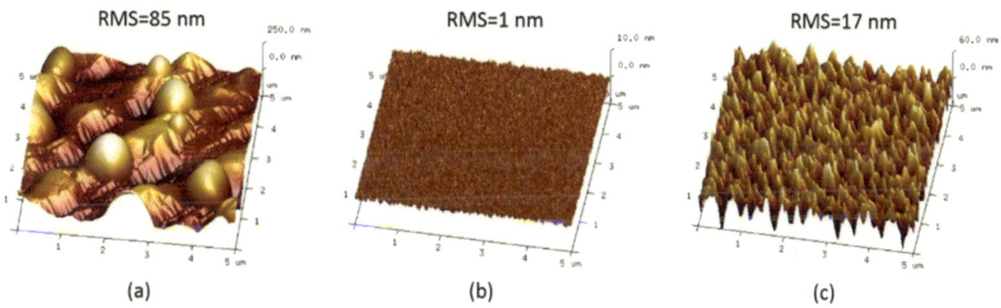

Figure 1. Surface morphology of InP(Bi) samples grown at (a) 364°C – Sample A, (b) 324°C – Sample B and (c) 275°C – Sample C.

Figure 2 shows HRXRD (004) ω/2θ rocking curves for the three samples. The left peak corresponds to the $InP_{1-x}Bi_x$ layer, which is weak and wide for Sample A while a narrow full width at half-maximum (FWHM) of 46 arcsec is obtained in Sample B in which the Bi content is estimated to be 2.4% and interference fringes can be clearly seen indicating a sharp $InP_{1-x}Bi_x$/InP interface. The $InP_{1-x}Bi_x$ peak shifts to the lower angle with decreasing the growth temperature, indicating the probability of Bi atoms sticking on surface and incorporation into lattice is enhanced at low growth temperatures. As the growth temperature is further lowered to 275°C, the $InP_{1-x}Bi_x$ peak broadens.

These results reveal that growth temperature is a critical parameter to obtain smooth $InP_{1-x}Bi_x$ single crystal. The optimized growth temperature of 324°C measured by a thermocouple is believed to be close to the Bi melting point. When the impingent Bi atoms are balanced between desorption and incorporation to lattice sites, single crystal $InP_{1-x}Bi_x$ film is achieved with no Bi droplets on the surface formed by excess Bi atoms. In addition, Bi is known to be an excellent surfactant [8], which could improve the surface and interface quality. When the growth temperature is further decreased, excess Bi and P atoms tend to accumulate on InP deteriorating the surface and the structural quality.

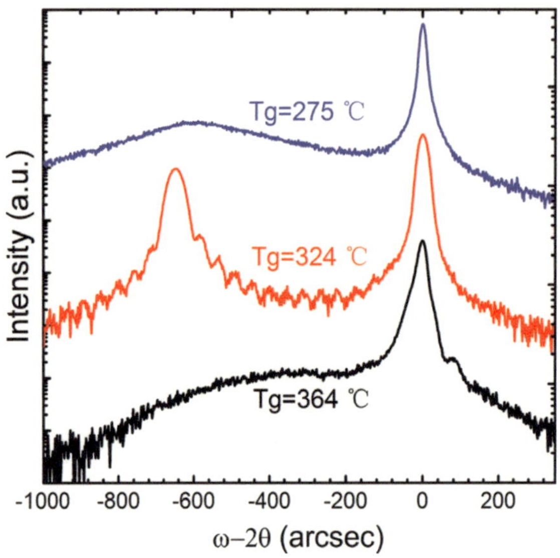

Figure 2. HRXRD (004) ω/2θ rocking curves from $InP_{1-x}Bi_x$ samples grown at different temperatures.

2.2. Effect of Bi flux

At the optimized growth temperature, we have tried to grow $InP_{1-x}Bi_x$ using different Bi flux to check the effect of Bi flux on the growth. A set of samples were grown with Bi/In BEP ratio varied from 0 to 0.16 (0.78 μm/h, PH_3 pressure = 350 Torr). The HRXRD (004) ω/2θ rocking curves are shown in **Figure 3(a)**, and the layer peak shifts to the lower angle when the Bi flux increases, indicating the increased incorporation of Bi in InP. **Figure 3(b)** shows the relationship between the Bi content and the FWHM of the InPBi layers and the Bi/In BEP ratio. It was found that the Bi incorporation increases linearly while the FWHM deceases for Bi concentration up to 2.4% and then broadens rapidly when further increasing the Bi flux. Both the samples with 1.8% and 2.4%, Bi has a narrow FWHM of ~50 arcsec and shows interference fringes, indicating improved composition uniformity and a smooth $InP/InP_{1-x}Bi_x$ interface.

Figure 4(a–c) shows AFM images of the $InP_{1-x}Bi_x$ with $x = 0\%$, 1.8% and 3.0%, respectively. The RMS roughness values of the three samples are 3.9 nm, 0.7 nm and 32.6 nm, respectively, suggesting proper Bi flux can improve surface quality. As the growth temperature is very low, excessive P atoms stick on the surface, and the surface migration length of In is reduced. Thus, the InP reference sample shows a rough texture surface. With a proper Bi flux, the growth front could be covered with Bi atoms, which are commonly considered to be surfactants to improve the surface quality. However, after further increasing the Bi flux, the sample achieving 3.0% Bi incorporation exhibits a much broader diffraction line width in **Figure 3(a)**, and micrometre-scale In/Bi droplets are occurred as shown in **Figure 4(c)**, indicating Bi atoms will be accumulated on surface forming large metallic Bi droplets and further hinder the uniform growth of $InP_{1-x}Bi_x$.

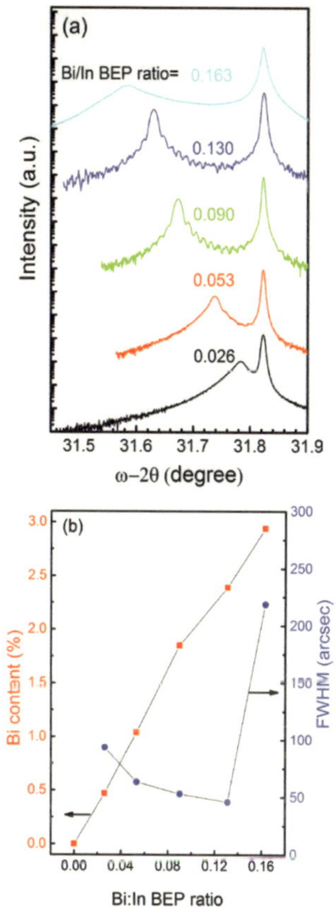

Figure 3. (a) HRXRD (004) ω/2θ rocking curves from $InP_{1-x}Bi_x$ samples grown at varying Bi flux; (b) dependence of the Bi content and the FWHM on Bi/In BEP ratio.

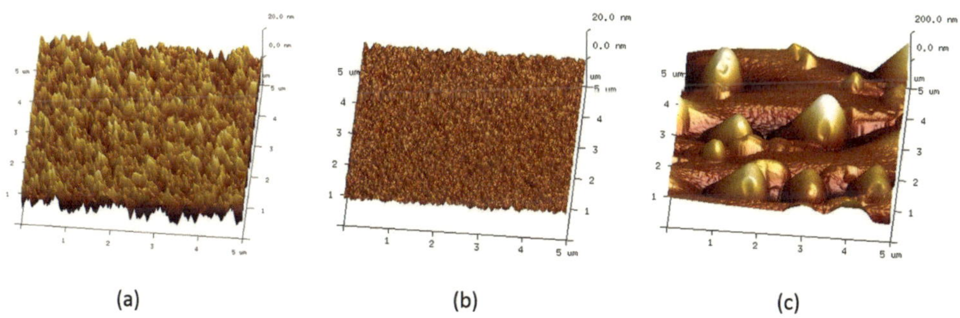

Figure 4. Surface morphology of 5 × 5 μm² InP(Bi) samples grown with different Bi flux conditions. Bi/In BEP ratio = x: 1 for (a) x = 0; (b) x = 0.090 and (c) x = 0.163.

2.3. Effect of PH_3 pressure

Normally, the growth parameter of V/III flux ratio has a direct influence on the growth of dilute bismides. In the MBE growth of a wide range of highly mismatched bismide alloys including GaAsBi, GaSbBi, InAsBi and InSbBi, a low growth temperature and near-stoichiometric growth conditions are needed to achieve efficient Bi incorporation due to the strong competition between As/Sb and Bi atoms. We have grown a set of $InP_{1-x}Bi_x$ layers with PH_3 pressure varied over a wide range from an In-rich to a P-rich condition. However, we found that the incorporation of Bi in InP is independent of the PH_3 pressure studied. High PH_3 pressure causes rough surface, and the introduction of Bi improves surface quality.

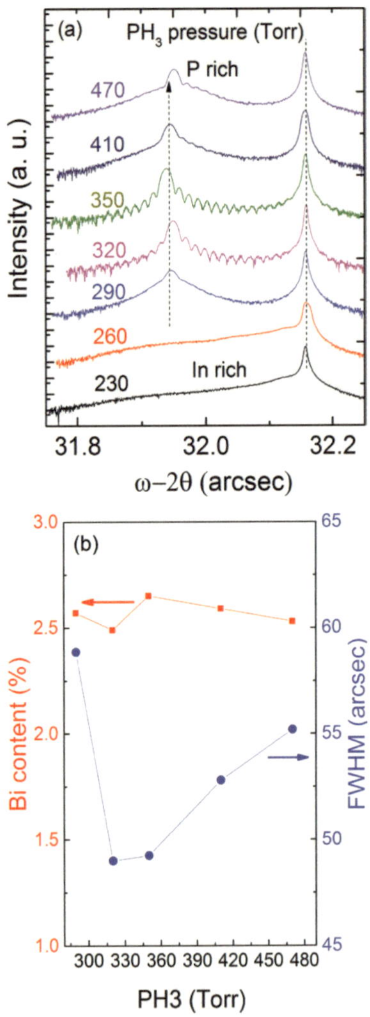

Figure 5. (a) HRXRD (004) ω/2θ rocking curves of $InP_{1-x}Bi_x$ layers grown at varying PH_3 pressures; (b) dependence of the Bi content and the FWHM on the PH_3 pressure.

The HRXRD (004) ω/2θ rocking curves are shown in **Figure 5(a)**, and the effect of the PH_3 pressure on the Bi content and the FWHM is shown in **Figure 5(b)**. The PH_3 pressure of 290 Torr is near the stoichiometry threshold condition where a transition from the In-rich to the P-rich condition is expected for these samples. Under the In-rich condition, no clear epitaxial layer peak is detected by XRD measurement expected a lot of In-rich droplets on their surfaces that were examined by the AFM and EDX, indicating excess In atoms hinder the InPBi growth. For the samples grown at $PH_3 \geq 290$ Torr, i.e., under the P-rich condition, two peaks can be well seen suggesting the success of InPBi growth. Surprisingly, with increasing the PH_3 pressure, the $InP_{1-x}Bi_x$ diffraction peak fluctuates slightly while the FWHM is nearly unchanged and the interference fringes disappear. For the sample grown with the PH_3 pressure = 350 Torr, the interference fringes and a narrow FWHM of 49 arcsec are obtained, indicating an optimized III/V flux ratio to grow $InP_{1-x}Bi_x$ with sharp interface and uniform Bi incorporation.

This growth property is quite different from that in solid source MBE as well as MOCVD growth of dilute bismides mentioned before where the V elements, such as As and Sb, are commonly consider to be strong competitor of Bi, and the V to III overpressure ratio should be carefully characterized and adjusted to achieve Bi-droplet-free surfaces. The peculiar phenomenon in $InP_{1-x}Bi_x$ may be related to the group-V source used rather than the differences between the InBi and GaBi bonds.

AFM results shown in **Figure 6** reveal that with a high P to In overpressure ratio, the roughening process of the (100) InP films is mainly controlled by the number of excessive P atoms and gives rise to mountain chain-like features aligned in the (0–11) direction, leading to a relative high RMS value of 5.3 nm.

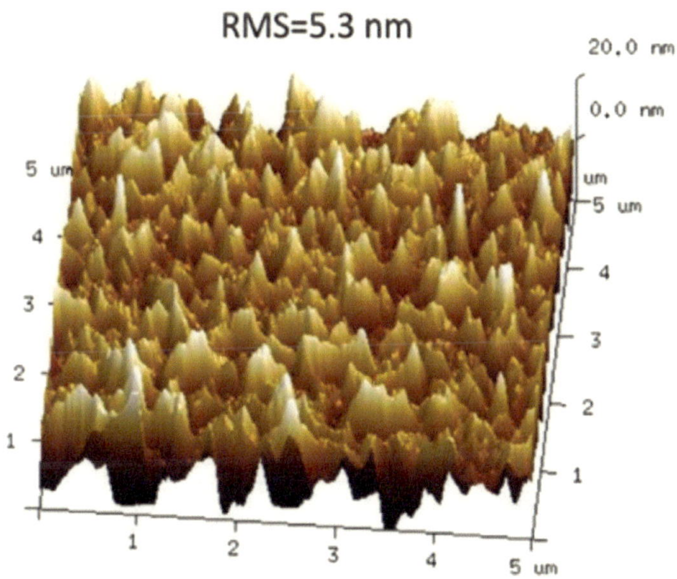

Figure 6. Surface morphology of the InP(Bi) sample grown at a relative high PH_3 pressure of 470 Torr.

2.4. Effect of growth rate

A set of 250–405 nm thick $InP_{1-x}Bi_x$ layers were grown to check the effects of growth rate varied from 0.50 to 0.81 μm/h while both the Bi flux and the PH_3 pressure were fixed. The HRXRD (004) ω/2θ rocking curves are shown in **Figure 7**. The Bi content as a function of the inverse InP growth rate is shown in **Figure 8(a)**. It is shown that the change in the Bi concentration is inversely proportional to the InP growth rate. These results indicate that Bi atoms are incorporated into the $InP_{1-x}Bi_x$ layer like dopant atoms, such as Be or Si. Be or Si has a unity sticking coefficient and the non-sensitive to the V supplies, and the doping concentration is inversely proportional to the growth rate of the host material. The $InP_{1-x}Bi_x$ sample with Bi incorporation of up to 3.7% was demonstrated at a relatively low InP growth rate of 0.50 μm/h and shows free of droplet surface (as shown in **Figure 8(b)**) and a smaller FWHM of ~105 arcsec than the sample with a Bi content up to 3.0% grown at a higher Bi flux as mentioned before, indicating the low growth rate can suppress the possibility of droplet formation and improve the uniformity of the Bi distribution. In addition, the observation of periodic elongated terraces aligned in the (0–11) direction in **Figure 8(b)** is in expectation because the P to In overpressure ratio is relatively high due to the low growth rate.

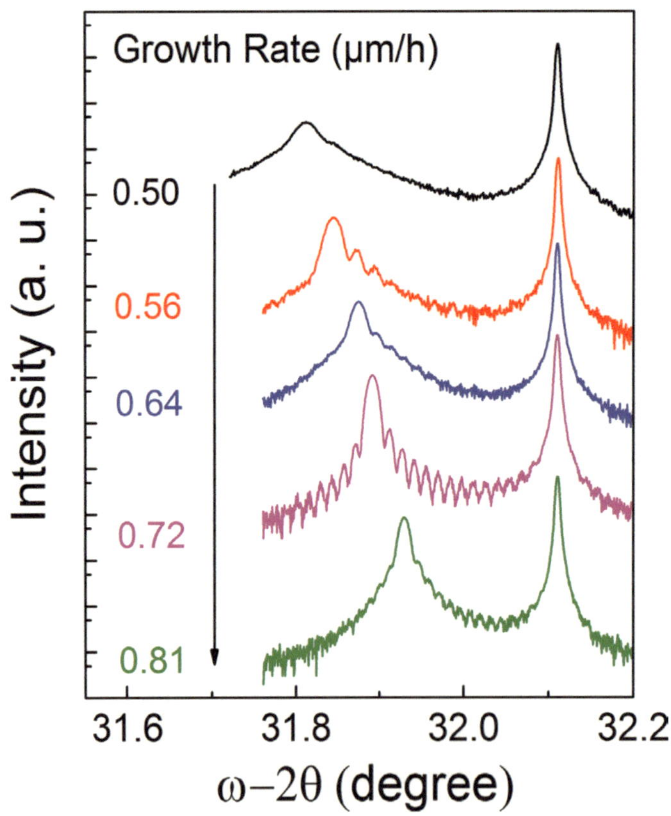

Figure 7. HRXRD (004) ω/2θ rocking curves of $InP_{1-x}Bi_x$ layers grown at varying growth rate.

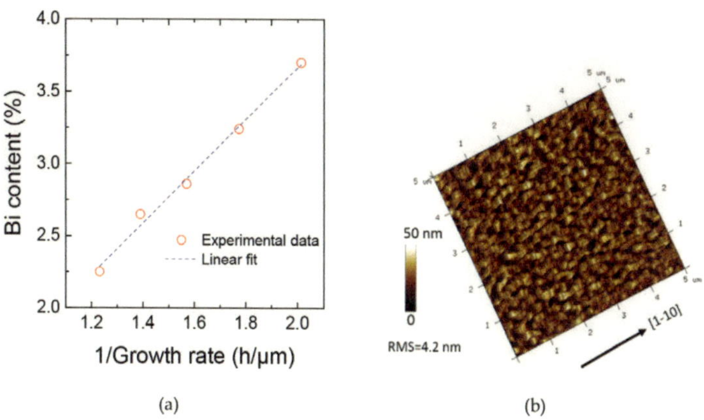

Figure 8. (a) The Bi content as a function of the inverse InP growth rate; (b) surface morphology of the InP(Bi) sample grown at a relative low InP growth rate of 0.50 μm/h.

3. Physical properties

3.1. Surface and structure properties

3.1.1. Lattice expansion

For $InP_{1-x}Bi_x$ alloy, the incorporated Bi atoms are believed to replace P atoms. Because the atom size of Bi is larger than that of P, the Bi_P replacement will cause local strain and change the lattice constant. As zinc-blende InBi has a lattice constant of 6.52 Å that is larger than InP (5.868 Å), the lattice constant of $InP_{1-x}Bi_x$ is expected to be between InBi and InP.

A 420 nm thick $InP_{1-x}Bi_x$ epilayer was grown on (100) InP substrate by GSMBE, with a 70 nm InP buffer layer grown in between to improve the initial growth surface. High-resolution XRD was performed to characterize the structural properties of $InP_{1-x}Bi_x$ films with Bi content varied from 0.47% to 3%. HRXRD (004) rocking curves in **Figure 9** show clear Pendellösung fringes, indicating the $InP_{1-x}Bi_x$ films with excellent crystal quality. All curves contain two peaks, in which the right narrow peak is from the InP substrate and the broad $InP_{1-x}Bi_x$ layer peak is located at the left side, indicating a larger lattice constant, which is in accordance with theoretical prediction. With increasing the Bi content from 0.47% to 3%, the $InP_{1-x}Bi_x$ layer peak moves to lower angles gradually. The higher the Bi content, the larger the lattice constant of $InP_{1-x}Bi_x$.

High-resolution TEM and geometric phase analysis were performed to investigate the out-of-plane (along the growth direction) strain and the in-plane (in the growth plane) strain of the epilayers to the substrate, which was performed by Dr. Minjian Wu at the Paul-Drude-Institute in Germany. **Figure 10(a)** shows the out-of-plane strain distribution. The colours in the map represent the stress intensity, in which the redder indicates a larger strain and the greener means a smaller strain. The average strain of the $InP_{1-x}Bi_x$ epilayer is about 0.60 ± 0.4%, while the strain of InP buffer layer is 0.00 ± 0.4%. The in-plane strain is 0.00 ± 0.4%, which is not shown here. These results indicate that the $InP_{1-x}Bi_x$ epilayer is psudomorphically strained.

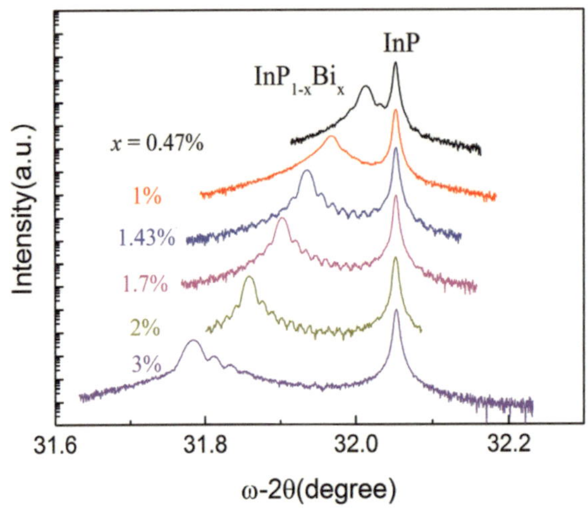

Figure 9. XRD (004) rocking curves of $InP_{1-x}Bi_x$ thin films with different Bi contents.

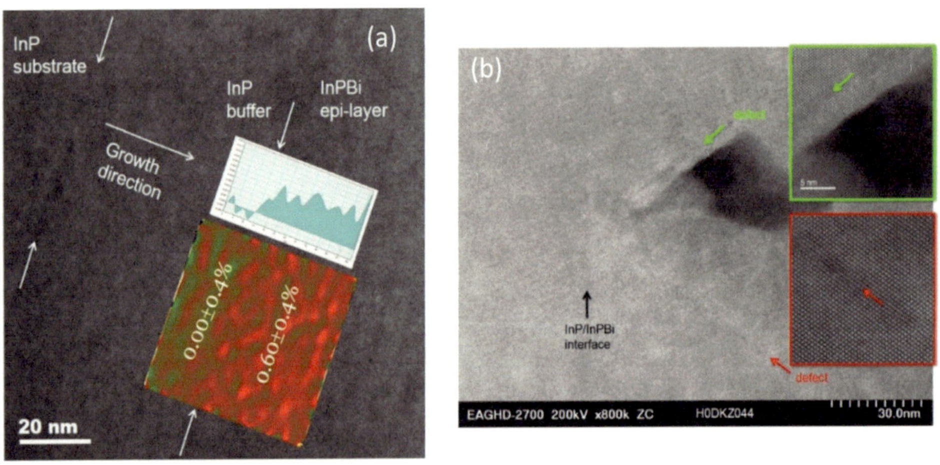

Figure 10. (a) High resolution TEM and geometric phase analysis of $InP_{1-x}Bi_x$ layers grown on InP. (b) Cross-sectional TEM of $InP_{1-x}Bi_x$ film. The arrows in the green and red boxes point out dislocation in $InP_{1-x}Bi_x$ layer.

Figure 10(b) shows the cross-sectional TEM image of $InP_{1-x}Bi_x$ films, which were grown at a temperature of 50°C lower than that of the samples measured in **Figure 2(a)**. Dislocations are obviously seen in the region pointed out by arrows in the green and red boxes. Because of the atomic size difference between Bi and P atoms, when Bi replaces P, the surrounding atoms will be compressed. Once the stress is too large, it will relief through dislocation formation. Lattice distortion may be caused by atom substitution, vacancy and interstitial atoms.

http://dx.doi.org/10.5772/64565

3.1.2. Segregation

Phase separation is a general phenomenon in III-V compounds. With a large atomic radius, a large atomic mass and a low bonding energy of Bi-III, Bi segregation occurs easily in dilute bismides. This adds difficulty in MBE growth of $InP_{1-x}Bi_x$, in which a low growth temperature should be used to ensure Bi incorporation, resulting in a high tendency to defect formation.

Atom probe tomography was performed to examine the Bi distribution in $InP_{1-x}Bi_x$ thin films. A 70 nm InP buffer layer was first grown on the InP (001) substrate to improve the interface and then 420 nm $InP_{1-x}Bi_x$ was deposited on it. For sample preparation, we first coated the sample with a protective layer and pulled a wedge out, then FIB was used to form the final tip. **Figure 11** shows In, P and Bi atom distribution in the as-grown $InP_{1-x}Bi_x$ alloys. We can see four different layers. The top layer where no In, P and Bi atoms were detected was the protective layer. The lower layers are $InP_{1-x}Bi_x$ epi-layer, InP buffer layer and InP substrate in turn. The distributions of In and P atoms are quite uniform; however, Bi atoms are found to distribute somehow non-uniformity in the $InP_{1-x}Bi_x$ layer. The red arrow in **Figure 11** points out a relatively higher Bi region than the surroundings, which is an obvious evidence of Bi segregation.

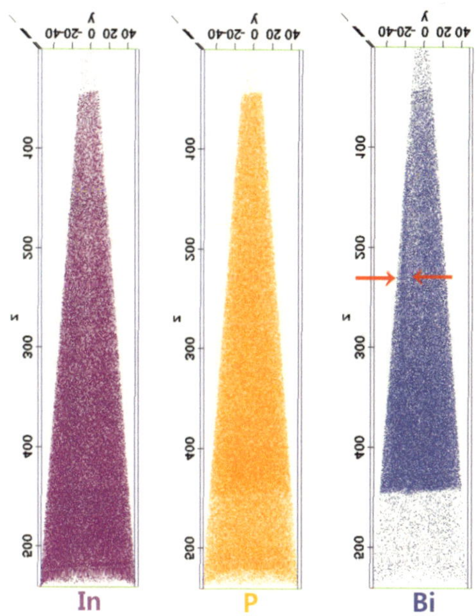

Figure 11. Distribution map of In, P and Bi atoms in $InP_{1-x}Bi_x$ thin film deposited on an InP (001) substrate observed by atom probe tomography.

3.2. Electronic property

InPBi was predicted to be difficult to fabricate partially due to the large lattice mismatch [9]. In 2013, the success growth of $InP_{1-x}Bi_x$ has stimulated theoretical studies of this novel compound further. Kopaczek *et al.* [6] have studied the band gap (E0) and spin-orbit splitting (SO) in $InP_{1-x}Bi_x$ alloys with $0 < x < 0.034$ by contactless electro-reflectance. The reduction coefficient of E0 was deduced to be 83 meV/% Bi while that of SO was 13 meV/% Bi. These results were close to the estimation in the valence band anti-crossing model and *ab initio* calculations where the shifts of E0 and SO are 106 and 20 meV/%Bi, respectively. Similarly, Bi-related band parameters were deduced theoretically by Polak *et al.* [11].

Zhang *et al.* has performed first-principle calculations systematically to investigate structural and electronic properties of the normal $InP_{1-x}Bi_x$ [12] and $In_{1-x}PBi_x$ [13] alloys, respectively. It is found that the hetero-antisite defect Bi_{In} is energetically easier than Bi_P. In $InP_{1-x}Bi_x$, the electronic states around Fermi level mainly consist of Bi-6p states, which contribute to the band gap reduction. While for $In_{1-x}PBi_x$, the Bi_{In} defect is a deep-level donor, which does not contribute to the reduction in the band gap of InP:Bi alloy but may be related to the reported strong RT PL.

3.3. Optical property

Photoluminescence (PL) and absorption measurement were carried out employing a Fourier transform infrared (FTIR) spectrometer in the rapid rather than the step-scan mode, in which a liquid-nitrogen cooled InSb detector and a CaF_2 beam splitter were used. A 532 nm laser was used as the excitation. The samples were mounted into a close-cycle refrigerator for the low temperature PL measurements.

$InP_{1-x}Bi_x$ samples of 390 nm thick and with $0\% \le x \le 2.49\%$ were grown directly on semi-insulating (100) InP substrates without InP buffer by V90 GSMBE system. An InP reference sample was grown under the same growth condition (LT InP) for comparison.

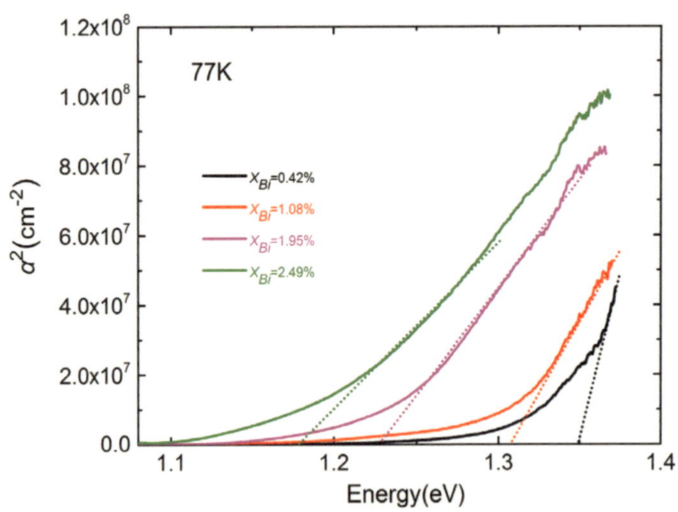

Figure 12. Square of absorption coefficient of $InP_{1-x}Bi_x$ samples with various Bi compositions as a function of photon energy at 77 K.

http://dx.doi.org/10.5772/64565

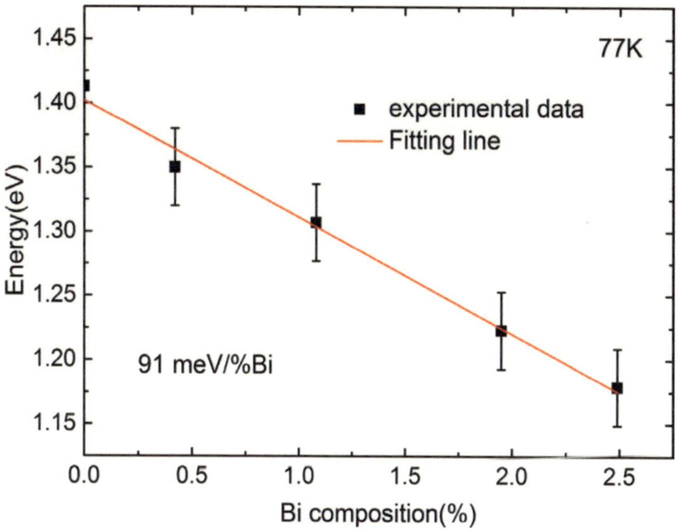

Figure 13. Band gap energy of $InP_{1-x}Bi_x$ measured from absorption spectra as a function of Bi composition. The error bars of the experimental data are labelled. The solid line is the linear fitting line of the experimental data.

Figure 12 shows square of absorption coefficient of $InP_{1-x}Bi_x$ films with various Bi compositions as a function of photon energy at 77 K. The band gap value is obtained from the linear extrapolation of the rising part for each sample. As Bi composition increases, the band edge of $InP_{1-x}Bi_x$ shifts to longer wavelengths revealing a reduction in the band gap energy. As shown in **Figure 13**, the Bi-induced band gap reduction is about 91 meV/% Bi, which is close to the measured value of 83 meV/% Bi by photoreflectance and the theoretically predicted value of 106 meV/% Bi [6].

Figure 14(a) shows PL spectra of LT InP reference sample and $InP_{1-x}Bi_x$ samples with various Bi concentrations at 10.5 K. The spectra are magnified by particular factors for visually suitable in height. Strong and broad PL peaks are observed for the samples, except for the sample with the highest Bi composition. Red arrows point to the theoretical band gap values of $InP_{1-x}Bi_x$ following the 106 meV/% Bi [6]. Blue arrows point to the band gap measured from absorption spectra as shown in **Figure 13**. In general, the PL emission features shift to lower energy with increasing Bi concentration as expected, and the spectral line shape also changes. The LT InP reference spectrum manifests two sharp peaks at about 1.42 eV and 1.39 eV, respectively, superimposed by a broad peak centred at about 1.3–1.35 eV. Compared to that of the InP grown at the same condition, the $InP_{1-x}Bi_x$ samples have totally different PL characteristic. In all cases, the observed PL signals show peak energies much smaller than the band gap of $InP_{1-x}Bi_x$ as shown by the arrows in **Figure 14(a)**. When only doping with 0.42% Bi, a strong and broad feature appears at about 1.05 eV and becomes a dominant signature in the PL spectra. As Bi content increases to 1.95%, the PL intensity begins to decrease, and a new feature at around 0.8 eV emerges. Overall, the $InP_{1-x}Bi_x$ PL spectral evolution versus Bi concentration shows that features at high energy gradually quench and features at low energy emerge. Such anomalous

PL behaviours are different from those found in GaAsBi [4] and GaSbBi [14] where PL peak follows the band gap reduction for a small amount of incorporated Bi composition.

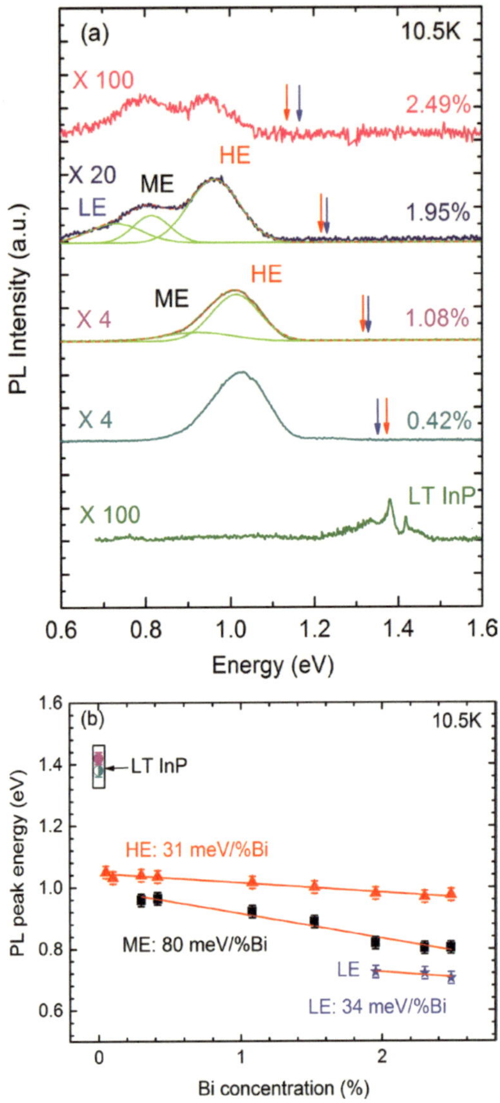

Figure 14. (a) PL spectra of $InP_{1-x}Bi_x$ samples with various Bi concentrations and LT InP reference sample at 10.5 K. Red arrows point to the theoretical band gap values of $InP_{1-x}Bi_x$ following the 106 meV/% Bi. Blue arrows point to the band gap measured from absorption spectra following the 91 meV/% Bi. The green curves are the Gaussian peak fitting. The red solid curve is the sum of the fitted lines. (b) PL peak energy evolution versus Bi concentration. The red solid line is a linear fit of HE (slope = 31 meV/% Bi); the black solid line is a linear fit of ME (slope = 80 meV/% Bi); the blue solid line is a linear fit of LE (slope = 34 meV/% Bi).

The broad and asymmetry PL peaks of $InP_{1-x}Bi_x$ can be well fitted by three features and labelled by low energy (LE), medium energy (ME) and high energy (HE). **Figure 14(a)** shows a typical fitted PL spectrum of $InP_{1-x}Bi_x$ with $x = 1.95\%$. Both HE and ME peaks exist almost in all $InP_{1-x}Bi_x$ PL spectra and are dominant, while the LE peak is resolved only for samples with x ≥ 1.95%. As shown in **Figure 14(b)**, the HE emission is found to monotonically decrease with Bi concentration from 1.05 eV down to 0.98 eV and shifts at ~31 meV/% Bi. The ME emission is found to monotonically decrease with Bi concentration from 0.95 eV down to 0.80 eV and shifts at ~80 meV/% Bi. Recent theoretical calculations by Kopaczek *et al.* reveal that the conductive band (CB) shift is 27 meV/% Bi and the VB shift is 79 meV/% Bi in $InP_{1-x}Bi_x$ [6]. Considering the HE and ME shifts versus Bi concentration are close to the CB and VB shifts, respectively, we attribute the HE transition to electrons in the CB of $InP_{1-x}Bi_x$ recombining with holes trapped at a deep level, and the ME transition to electrons trapped at a deep level to holes in the VB of $InP_{1-x}Bi_x$.

To determine the nature of $InP_{1-x}Bi_x$ PL and verify the existence of deep levels, we performed deep level transient spectroscopy (DLTS) measurements on the LT InP reference and $InP_{1-x}Bi_x$ ($x = 2.49$ %) grown on n + InP substrates. From the DLTS results, two deep levels with the ionization energy of the n-type deep level to be 0.38 eV below the CB edge and the p-type deep level to be 0.31 eV above the VB maximum of InP, respectively, are identified. Based on the PL spectra and the DLTS results, the physical origins of HE, ME and LE can be explained as follows: (1) HE is related to the transition between conductive band (CB) and the p-type deep level. (2) ME is related to the transition between the n-type deep level and valence band (VB). (3) LE is related to the transition between the n-type deep level and the p-type deep level. The 0.38 eV donor-like deep level is likely related to the intrinsic deep level in LT grown InP [15]. This level is considered to originate from the intrinsic antisite of P_{In}. While the VB of $InP_{1-x}Bi_x$ results from the anti-crossing of the Bi impurity level with the VB of InP, we attribute the p-type deep level to the formation of Bi pairs or complex Bi-related clusters. It has been theoretically shown that Bi pairs in GaP:Bi can form a series of energy levels above the VB [16]. The energy difference from the VB edge increases with decrease in the Bi pair distance and can reach as deep as 0.5 eV [16]. This is also expected in InP:Bi because the Bi impurity level is slightly below the VB edge of InP [3].

The PL intensity is found to significantly increase by Bi incorporation due to the effective spatial trapping of holes, which is a positive indication for light-emitting devices. Detailed analysis of $InP_{1-x}Bi_x$ PL has explained the origin of the strong light emission.

3.4. Electrical and transport properties

Electrical and transport measurements and first-principle calculations have been combined to investigate the dependence of electric properties of $InP_{1-x}Bi_x$ on Bi content [17]. Data of Hall effect measurements at variable temperatures for undoped $InP_{1-x}Bi_x$ show a high electron concentration in the order of 10^{18} cm^{-3}. The interesting thing is that both the electron concentration and the mobility have a nonlinear relationship with the Bi content. The electron concentration first decreases with increasing incorporated Bi content from 0% to 1.83% and then increases with keep on increasing Bi after its content exceeds to 1.83%. However, the electron mobility shows quite an inverse variation in that of the electron concentration as a function of Bi content.

P_{In} antisites and V_P vacancies are dominant donors in InP and $InP_{1-x}Bi_x$. Chen *et al.* found a level of 0.23 eV below conduction band minimum (CBM) and a level 0.12 eV above CBM hypothetically corresponding to P_{In} level and V_P level [18], respectively, in which the P_{In} level does not accord with our experimental results as shown in Section 3.3. V_P and P_{In} concentration dominates the electron concentration at low Bi content ($x < 1.83\%$) and high Bi content ($x > 1.83\ \%$), respectively. When Bi content is below 1.83%, the increasing incorporated Bi will occupy more V_P vacancies, decreasing the V_P concentration and impurity scattering, thus causing the decrease in electron concentration and the increase in electron mobility. After Bi content exceeds to 1.83%, both P_{In} and V_P density increases with increase in the Bi content, enlarging the overall electron concentration and reducing the electron mobility (**Figure 15**).

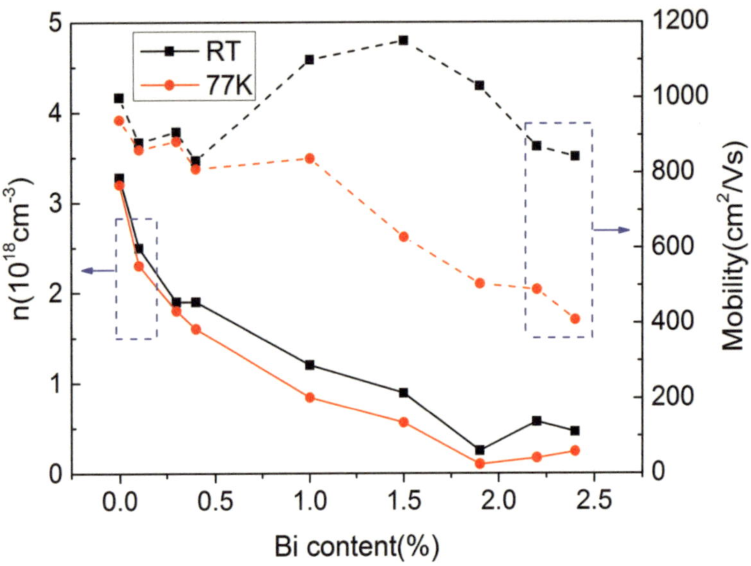

Figure 15. The relationship of electron concentration and mobility with Bi content at room temperature and 77 K.

4. Annealing

In this section, a systematic study on the effect of annealing on the structural and optical properties of $InP_{1-x}Bi_x$ is presented [19]. The 390 nm thick $InP_{1-x}Bi_x$ sample was diced into small pieces and annealed for 2 min at different temperatures ranging from 400°C to 800°C. To prevent P decomposition during annealing, the samples were capped with an InP substrate.

Figure 16(a) and **(b)** shows HRXRD (115) ω/2θ rocking curves of the as-grown and annealed $InP_{0.981}Bi_{0.019}$ layer and the extracted Bi content along with strain relaxation from four asymmetric (11±5) rocking curves, respectively. The rocking curve of the as-grown $InP_{1-x}Bi_x$ shows a narrow peak with a FWHM of only 71 arcsec, revealing high crystal quality. As the annealing temperature is raised to 400°C, the FWHM slightly increases to 84 arcsec. Further increasing annealing temperature broadens the line width. This is attributed to enhanced strain relaxation as shown in **Figure 16(b)**. The Bi content hardly changes with the annealing temperature rising up to 500°C, indicating a good stability of this material. However, when the annealing temperature increases to 600°C, the FWHM suddenly deteriorates to 161 arcsec and the diffraction peak in HRXRD shifts to a higher angle. The Bi content of InPBi annealed at 600°C is distinctly decreased comparing to that annealed at 500°C, which is similar to InGaAsBi annealed up to 650°C [20]. As the annealing temperature keeps on ascending to 700°C, the FWHM increases to 215 arcsec. Meanwhile, the diffraction peak further moves towards higher angles with the intensity dropping drastically. From **Figure 16(b)**, the Bi content decreases to 0.0165. As the annealing temperature increases to 800°C, no accurate peak position can be extracted owing to a very broad signal. This suggests the deterioration of crystal structure under high annealing temperatures. The trend of peak shifting towards high angles implies that the average Bi content continuously decreases, which is indicated by the black dashed line in **Figure 16(b)**.

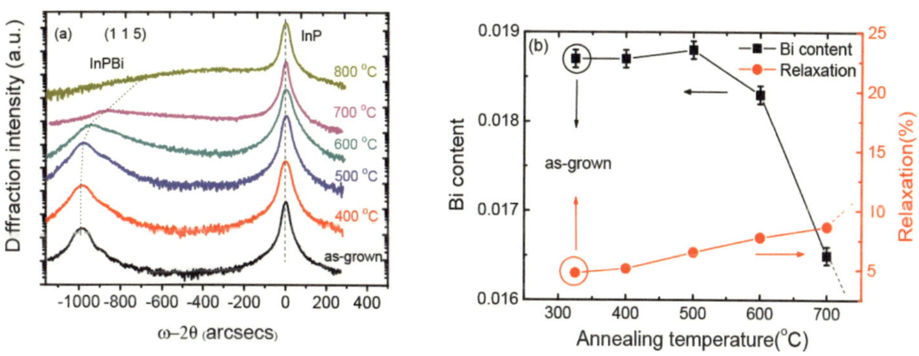

Figure 16. (a) HRXRD (004) ω/2θ rocking curves of $InP_{1-x}Bi_x$ samples with various annealing temperatures. (b) The extracted Bi content of $InP_{1-x}Bi_x$ before and after annealing at different temperatures.

The III-Bi bonding energy is small because Bi atom has a very large atomic radius compared with other group V elements. Thus, annealing at high temperatures likely causes Bi outdiffusion from $InP_{1-x}Bi_x$ leading to decrease in the Bi content. However, the measured temperature for initial Bi outdiffusion of $InP_{1-x}Bi_x$ is about 600°C, which is lower than GaAsBi (800°C) [21]. Considering the bond energy of Ga-Bi (158.6 ± 16.7 kJ/mol) is larger than In-Bi (153.6 ± 1.7 kJ/mol) [22], Bi is expected to be more stable in GaAsBi.

Room temperature PL spectra of $InP_{0.981}Bi_{0.019}$ and LT InP before and after annealing at different temperatures are shown in **Figure 17(a)**. For a visually suitable PL peak height, the spectra were magnified by particular factors. The wide PL signals can be fitted very well by three peaks, which are marked by LE, ME and HE, respectively. When the annealing temperature is below 500°C, broad PL spectra with similar figures are observed in the energy range from 0.45 to 1.1 eV. The dominant peak is ME with relative weak intensities for both LE and HE peak. As shown in **Figure 17(b)**, the normalized integrated PL intensity increases when compared with the as-grown one. When the annealing temperature rises to 400°C, a maximum PL enhancement by 20% is obtained and the PL spectrum is essentially unchanged. This result is encouraging as optical quality can be improved without compromising the PL peak energy shift. As shown in **Figure 17(b)**, when the annealing temperature is below 500°C, the integrated PL intensity of the LT InP reference sample decreases with increasing annealing temperature. This suggests that the PL improvement is mainly attributed to the removal of Bi-related defects similar to the circumstance of annealing on InPN [23].

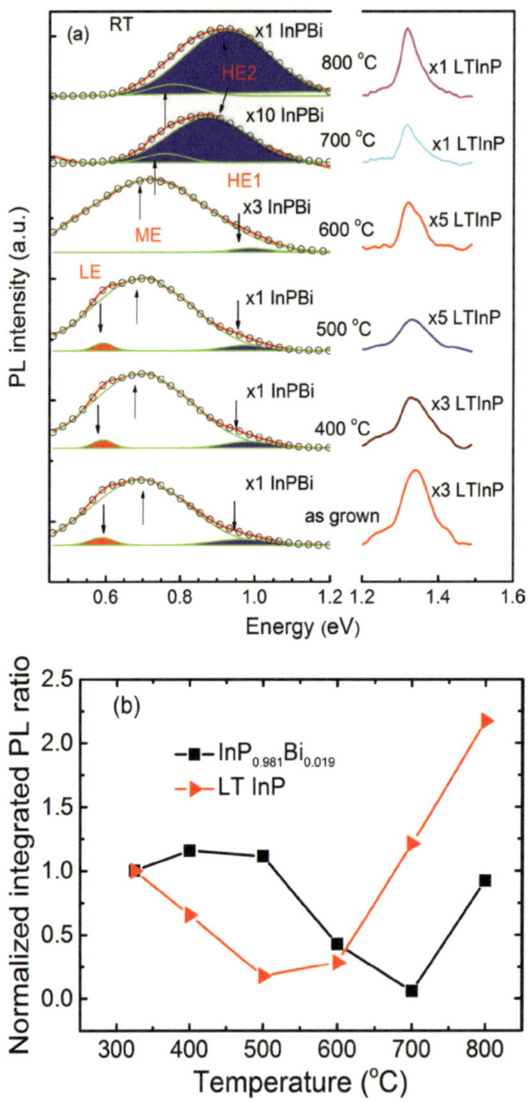

Figure 17. (a) PL spectra of $InP_{1-x}Bi_x$ (x = 0.019) and LT grown InP as reference at room temperature. The dotted lines are measured data. The green and red lines are the Gaussian peak fitting and the sum of fitted lines, respectively. (b) The normalized integrated PL intensity in respect of the intensity from the non-annealed sample as a function of the annealing temperature.

5. Quaternary phosphide bismides

Ternary alloy InGaP is an essential material for red laser diodes [22], while InAlP is an active material for nitride-free yellow-green photoelectric devices and has potential application in laser diodes, light-emitting diodes, displayers and multi-junction solar cells [24–26]. Incorporating Bi into InGa(Al)P is expected to reduce the band gap to tune the emission wavelength to longer range and improve the temperature sensitivity of InGa(Al)P lasers. Meanwhile, the strain of InGa(Al)PBi can be tuned freely from tensile to compressive by adjusting Bi and In/Ga(Al) contents.

5.1. InGaPBi thin films

InGaPBi thin films have been grown on GaAs and InP substrates using GSMBE [27, 28].The growth temperature is crucial for Bi incorporation, and the optimized temperature for InGaPBi growth on InP and GaAs substrates is 280°C and 300°C, respectively. The Bi content increased from 0.5% to 2.2% by increasing the Bi flux. A 330 nm thick $In_{0.468}Ga_{0.532}P_{1-x}Bi_x$ films were grown on GaAs substrates with Bi content of 0.5%, 1.0% and 1.8%, respectively. A 370 nm thick $In_{0.975}Ga_{0.025}P_{1-x}Bi_x$ films were grown on InP substrates with Bi content of 0.42%, 1.6% and 2.2%, respectively. InGaP reference samples were also grown for comparison.

HRXRD (004) rocking curves show clear Pendellösung fringes as shown in **Figure 18(a)** and **(b)**, indicating excellent crystal and interface quality. The strain of InGaPBi films can be tuned from tensile to compressive by increasing the Bi content. The HRXRD $\omega/2\theta$ rocking curves on the (004) crystal plane and asymmetric (115) reflections were performed to obtain the exact lattice mismatch. The FWHM of the epilayer peak for all the InGaPBi films on GaAs substrate is very small, 54.2, 49.2 and 49.9 arcsec for $In_{0.468}Ga_{0.532}P_{1-x}Bi_x$ films with 0.5%, 1.0% and 1.8% Bi content, respectively, indicating good crystal quality. For the reference $In_{0.468}Ga_{0.532}P$ film on GaAs substrate, the lattice constant is 5.646 Å with −1378 ppm negative mismatch to the GaAs substrate. As expected, with increasing the Bi content from 0.5% to 1.8%, the diffraction peak moves to smaller angles. The lattice mismatch between the InGaPBi sample with Bi content of 0.5% and 1.0% with the GaAs substrate is −810 and −314 ppm, respectively. As the Bi content is increased up to 1.8%, the InGaPBi epilayer is compressively strained to the GaAs substrate with a mismatch of 598 ppm. Through this method, InGaPBi films have been successfully grown on GaAs with controllable strain from tensile to compressive. For the $In_{0.975}Ga_{0.025}P_{1-x}Bi_x$ films grown on InP substrate with $x = 0\%$, 0.42%, 1.6% and 2.2%, the mismatch to InP is tuned from −1776 to −1189, 338 and 1093 ppm, respectively. The strain of $In_{0.975}Ga_{0.025}P_{1-x}Bi_x$ films on InP can also be tuned from tensile to compressive.

Surface morphology characterized by AFM [27, 28] indicates that Bi has little effect on the InGaPBi surface grown on a GaAs substrate but can smoothen the InGaP(Bi) samples grown on an InP substrate. Weak room temperature PL was observed at 1.78 eV for the InGaPBi films on GaAs with a Bi content of 0.5%, featuring near band optical transition with 122 meV smaller than that of InGaP. For all the InGaPBi films on InP, room temperature PL shows strong and broad light emission at energy levels much smaller than the InP band gap.

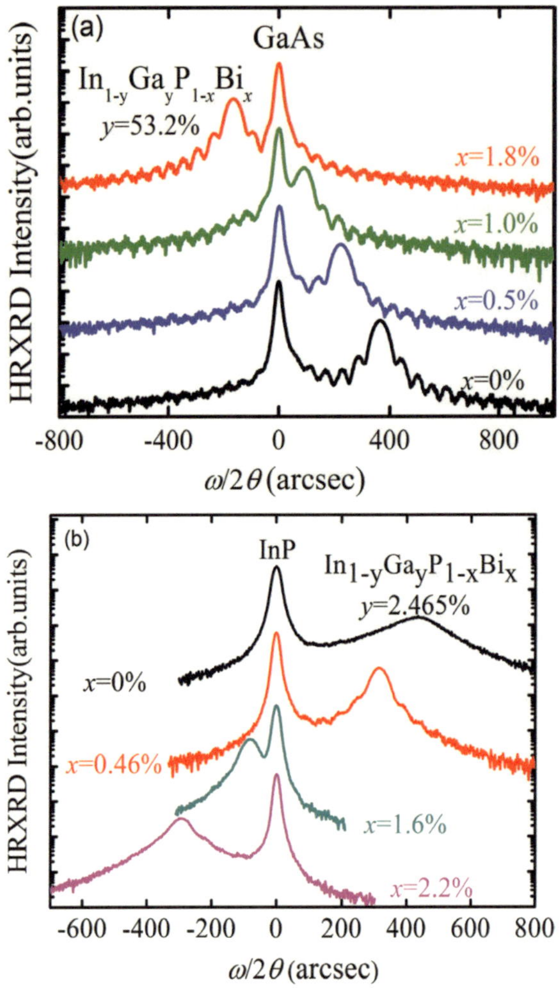

Figure 18. XRD rocking curves of InGaPBi films on (a) GaAs and (b) InP.

5.2. InAlPBi thin films

$In_{1-y}Al_yP_{1-x}Bi_x$ thin films on GaAs with Bi concentration up to $x = 1.6\%$ were grown by GSMBE. The growth temperature is critical for Bi incorporation, and the growth temperature window is quite small around the optimal temperature at 300°C. After surface deoxidation, a 100 nm undoped GaAs buffer layer was grown at 580°C and then decreases the substrate temperature to 300°C to grow the InAlPBi epilayer with a thickness of 360 nm. The pressure of the P_2 flux was kept at 350 Torr during the InAlPBi epilayer growth. Keeping the same In/Ga ratio, the temperature of the Bi cell was set between 427°C and 462°C, respectively.

Rutherford backscattering spectrometry (RBS) was performed for the InAlPBi sample with 1.6% Bi content to determine Bi composition and confirms that 93% of Bi atoms are located at

http://dx.doi.org/10.5772/64565

substitutional sites. The surface morphology was characterized using AFM. From the 2 × 2 μm² scan image of the InAlPBi samples, the RMS roughness is only 0.5 nm, revealing that all the samples have smooth surfaces.

The lattice quality was characterized using the HRXRD ω/2θ rocking curves both on the (004) crystal plane and on the asymmetric (115) reflections. $In_{1-y}Al_yP_{1-x}Bi_x$ thin films reveal excellent structural quality and perfect interfaces. The FWHMs of the epilayer peak of the InAlPBi samples are 48.2, 46.5 and 49.2 arcsec, respectively, for the samples with Bi content of 0.5%, 0.8% and 1.6%, indicating good crystal quality. For the $In_{0.467}Al_{0.533}P$ reference sample and the InAlPBi samples with 0.5% and 1.6% Bi contents, the lattice mismatch with GaAs substrate is −695, −318 and 1041 ppm, respectively. The strain can be tuned from tensile to compressive by increasing Bi contents. The corresponding strain relaxation is 0.6%, 1.6% and 1.1%, respectively. That is, all the InAlPBi films are almost strained.

6. Potential device application

Modern medical diagnostic technology tends to adopt non-invasive painless technique, such as ultrasound examination, X-ray photography, X-ray computer tomography and nuclear magnetic resonance technology. Some diagnosis technology can offer 3D images; however, the spatial resolution is limited in hundreds of micro-metres. Optical coherence tomography (OCT) is one of the non-invasive painless diagnosis techniques developed in recent years [29–31], which could offer higher resolution comparing with other diagnosis techniques. Basically, it is an interferometric technique. A typical OCT system uses a Michelson interferometer with a white light or low coherence light source. Light is split into the reference mirror and sample, respectively, and reclaimed to interfere. The interference signals will be collected by the photo detector and analysed by computer to extract the information out of the biological tissues (**Figure 19**).

The spatial resolution (Δl) of OCT technique is inversely proportional to the line width of the spectrum of the light source in the following formula [29]:

$$\Delta l = 0.44\lambda_0^2 / \Delta\lambda \tag{1}$$

λ_0 and $\Delta\lambda$ are the centric wavelength and FWHM of the light source, respectively. The wider the FWHM, the higher the spatial resolution. So, a wide-spectrum light source is preferred in the OCT system. Currently, the most popular light source used in a commercial OCT system is super-luminescence diode [29], with which the best spatial imaging resolution is around 10 μm. To get even higher resolution, we need to use a light source with a wider spectrum.

$InP_{1-x}Bi_x$ shows strong room temperature PL, which is eight times stronger in the peak intensity and 80 times stronger in the integrated PL intensity than that of InP grown under the same conditions. The spectrum is broad from 1.4 to 2.7 μm, and the widest shows a FWHM of 690 nm. If we use $InP_{1-x}Bi_x$ to fabricate super-luminescence diodes and apply them in the OCT system, the spatial resolution could be improved to 1.5 μm theoretically. It should be noted that $InP_{1-x}Bi_x$ thin films were first realized by MBE just in 2013 [6], the physical properties are still under investigation. If we further optimize the epitaxial growth parameters and tailor the material structure, it is promising to further broaden the FWHM of the room temperature PL or to blue shift the central PL wavelength while keeping the current FWHM. This will improve the spatial resolution of OCT system beyond a psychological value of 1 μm, thus significantly advancing the development of OCT diagnosis technique.

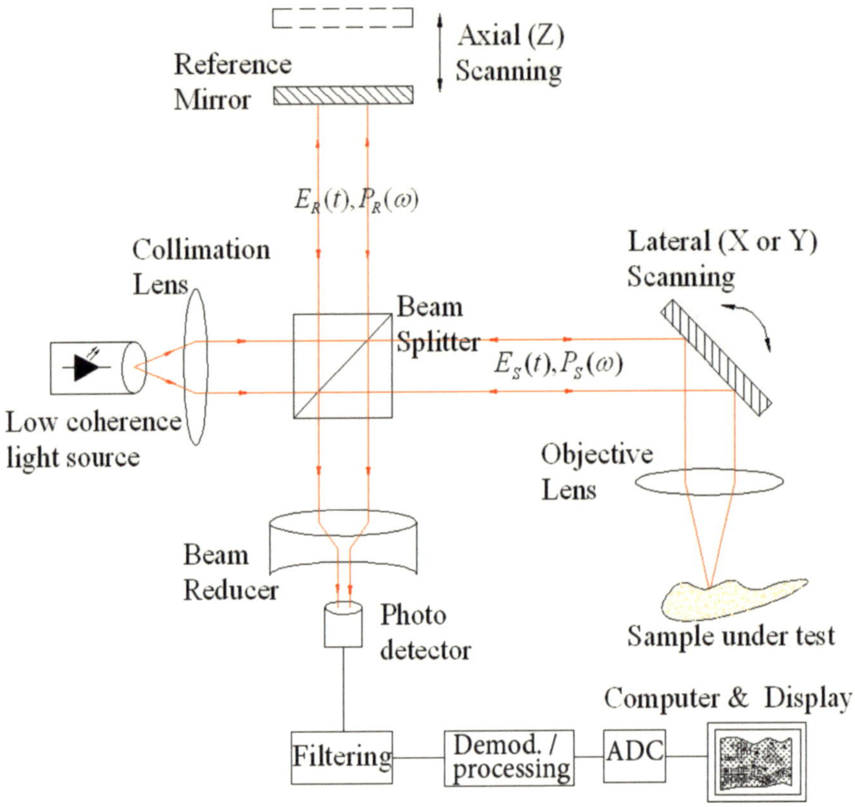

Figure 19. A typical optical setup of single point OCT system [32].

7. Summary

In this chapter, we have reviewed the MBE growth and characterization of $InP_{1-x}Bi_x$ and its quaternary alloys InGaPBi and InAlPBi. To get a high Bi content while maintaining good crystal quality is challenging because of the small growth window for $InP_{1-x}Bi_x$. Growth parameters were investigated in detail to find an appropriate growth window. Annealing is performed to study Bi thermal stability and influence on the physical properties of $InP_{1-x}Bi_x$. The optical

http://dx.doi.org/10.5772/64565

property of $InP_{1-x}Bi_x$ is distinctive comparing with other dilute bismides and shows broad and strong PL at room temperature. This makes $InP_{1-x}Bi_x$ a potential candidate for fabricating super-luminescence diodes applicable in OCT with spatial resolution in the nanometre regime.

Author details

Liyao Zhang[1], Wenwu Pan[1], Xiaoyan Wu[1], Li Yue[1] and Shumin Wang[1,2*]

*Address all correspondence to: shumin@mail.sim.ac.cn

1 State Key Laboratory of Functional Materials for Informatics, Shanghai Institute of Microsystem and Information Technology, CAS, Shanghai, China

2 Department of Microtechnology and Nanoscience, Chalmers University of Technology, Gothenburg, Sweden

References

[1] Oe, K. and H. Okamoto, New semiconductor alloy $GaAs_{1-x}Bi_x$ grown by metal organic vapor phase epitaxy. Japanese Journal of Applied Physics, 1998. 37(11A): p. L1283.

[2] Tixier, S., M. Adamcyk, T. Tiedje, S. Francoeur, A. Mascarenhas, P. Wei, et al., Molecular beam epitaxy growth of $GaAs_{1-x}Bi_x$. Applied Physics Letters, 2003. 82(14): p. 2245–2247.

[3] Alberi, K., J. Wu, W. Walukiewicz, K. Yu, O. Dubon, S. Watkins, et al., Valence-band anticrossing in mismatched III-V semiconductor alloys. Physical Review B, 2007. 75(4): p. 045203.

[4] Fluegel, B., S. Francoeur, A. Mascarenhas, S. Tixier, E. Young, and T. Tiedje, Giant spin-orbit bowing in $GaAs_{1-x}Bi_x$. Physical Review Letters, 2006. 97(6): p. 067205.

[5] Francoeur, S., M.-J. Seong, A. Mascarenhas, S. Tixier, M. Adamcyk, and T. Tiedje, Band gap of $GaAs_{1-x}Bi_x$, $0< x < 3.6\%$. Applied Physics Letters, 2003. 82(22): p. 3874–3876.

[6] Kopaczek, J., R. Kudrawiec, M. Polak, P. Scharoch, M. Birkett, T. Veal, et al., Contactless electroreflectance and theoretical studies of band gap and spin-orbit splitting in $InP_{1-x}Bi_x$ dilute bismide with $x \leq 0.034$. Applied Physics Letters, 2014. 105(22): p. 222104.

[7] Pan, W., P. Wang, X. Wu, K. Wang, J. Cui, L. Yue, et al., Growth and material properties of InPBi thin films using gas source molecular beam epitaxy. Journal of Alloys and Compounds, 2016. 656: p. 777–783.

[8] Naceur, H.B., T. Mzoughi, I. Moussa, L. Nguyen, A. Rebey, and B. El Jani, Surfactant effect of bismuth in atmospheric pressure MOVPE growth of InAs layers on (100) GaAs

substrates. Physica E: Low-dimensional Systems and Nanostructures, 2010. 43(1): p. 106–110.

[9] Berding, M.A., A. Sher, A.B. Chen, and W. Miller, Structural properties of bismuth-bearing semiconductor alloys. Journal of Applied Physics, 1988. 63(1): p. 107–115.

[10] Wang, K., Y. Gu, H. Zhou, L. Zhang, C. Kang, M. Wu, et al., InPBi single crystals grown by molecular beam epitaxy. Scientific Reports, 2014. 4.

[11] Polak, M., P. Scharoch, and R. Kudrawiec, First-principles calculations of bismuth induced changes in the band structure of dilute Ga–V–Bi and In–V–Bi alloys: chemical trends versus experimental data. Semiconductor Science and Technology, 2015. 30(9): p. 094001.

[12] Zhang, X., P. Lu, L. Han, Z. Yu, J. Chen, and S. Wang, Structural and electronic properties of InPBi alloys. Modern Physics Letters B, 2014. 28(17): p. 1450140.

[13] Wu, L., P. Lu, C. Yang, D. Liang, C. Zhang, and S. Wang, The effect of Bi In hetero-antisite defects in $In_{1-x}PBi_x$ alloy. Journal of Alloys and Compounds, 2016. 674: p. 21–25.

[14] Rajpalke, M.K., W.M. Linhart, M. Birkett, K.M. Yu, J. Alaria, J. Kopaczek, et al., High Bi content GaSbBi alloys. Journal of Applied Physics, 2014. 116(4): p. 043511.

[15] Liang, B.W., P.Z. Lee, D.W. Shih, and C.W. Tu, Electrical-properties of inp grown by gas-source molecular-beam epitaxy at low-temperature. Applied Physics Letters, 1992. 60(17): p. 2104–2106.

[16] Usman, M., C.A. Broderick, A. Lindsay, and E.P. O'Reilly, Tight-binding analysis of the electronic structure of dilute bismide alloys of GaP and GaAs. Physical Review B, 2011. 84(24).

[17] Wei, G., D. Xing, Q. Feng, W. Luo, Y. Li, K. Wang, et al., The identification of the dominant donors in low temperature grown InPBi materials. arXiv preprint arXiv: 1603.09015, 2016.

[18] Chen, W.M., P. Dreszer, A. Prasad, A. Kurpiewski, W. Walukiewicz, E.R. Weber, et al., Title: origin of [ital n]-type conductivity of low-temperature grown InP. Journal of Applied Physics, 1994. 76: p. 1.

[19] Wu, X.Y., K. Wang, W.W. Pan, P. Wang, Y.Y. Li, Y.X. Song, et al., Effect of rapid thermal annealing on $InP_{1-x}Bi_x$ grown by molecular beam epitaxy. Semiconductor Science and Technology, 2015. 30(9).

[20] Ye, H., Y.X. Song, Y. Gu, and S.M. Wang, Light emission from InGaAs:Bi/GaAs quantum wells at 1.3 mu m. Aip Advances, 2012. 2(4).

[21] Mohmad, A.R., F. Bastiman, C.J. Hunter, R. Richards, S.J. Sweeney, J.S. Ng, et al., Effects of rapid thermal annealing on $GaAs_{1-x}Bi_x$ alloys. Applied Physics Letters, 2012. 101(1).

[22] Ishikawa, M., Y. Ohba, H. Sugawara, M. Yamamoto, and T. Nakanisi, Room-temperature cw operation of InGaP/InGaAlP visible light laser diodes on GaAs substrates

grown by metalorganic chemical vapor deposition. Applied Physics Letters, 1986. 48(3): p. 207–208.

[23] Umeno, K., Y. Furukawa, N. Urakami, S. Mitsuyoshi, H. Yonezu, A. Wakahara, et al., Growth and luminescence characterization of dilute InPN alloys grown by molecular beam epitaxy. Journal of Vacuum Science & Technology B: Microelectronics and Nanometer Structures, 2010. 28(3): p. C3B22.

[24] Bour, D., J. Shealy, G. Wicks, and W. Schaff, Optical properties of $AlxIn_{1-x}P$ grown by organometallic vapor phase epitaxy. Applied Physics Letters, 1987. 50(10): p. 615–617.

[25] Mukherjee, K., D. Beaton, T. Christian, E. Jones, K. Alberi, A. Mascarenhas, et al., Growth, microstructure, and luminescent properties of direct-bandgap InAlP on relaxed InGaAs on GaAs substrates. Journal of Applied Physics, 2013. 113(18): p. 183518.

[26] Schneider, Jr, R. and J. Lott, InAlP/InAlGaP distributed Bragg reflectors for visible vertical cavity surface-emitting lasers. Applied Physics Letters, 1993. 62(22): p. 2748–2750.

[27] Wang, K., P. Wang, W. Pan, X. Wu, L. Yue, Q. Gong, et al., Growth of semiconductor alloy InGaPBi on InP by molecular beam epitaxy. Semiconductor Science and Technology, 2015. 30(9): p. 094006.

[28] Yue, L., P. Wang, K. Wang, X. Wu, W. Pan, Y. Li, et al., Novel InGaPBi single crystal grown by molecular beam epitaxy. Applied Physics Express, 2015. 8(4): p. 041201.

[29] Fercher, A.F., W. Drexler, C.K. Hitzenberger, and T. Lasser, Optical coherence tomography-principles and applications. Reports on Progress in Physics, 2003. 66(2): p. 239.

[30] Huang, D., E.A. Swanson, C.P. Lin, J.S. Schuman, W.G. Stinson, W. Chang, et al., Optical coherence tomography. Science, 1991. 254(5035): p. 1178–1181.

[31] Schmitt, J.M., Optical coherence tomography (OCT): a review. IEEE Journal of Selected Topics in Quantum Electronics, 1999. 5(4): p. 1205–1215.

[32] Pumpkinegan. Optical coherence tomography. 2006 [An image depicting a single point optical coherence tomography system]. Available from: https://en.wikipedia.org/wiki/Optical_coherence_tomography.

Chapter 9

Selecting Appropriate Metallic Alloy for Marine Gas Turbine Engine Compressor Components

Injeti Gurrappa, I.V.S. Yashwanth and A.K. Gogia

Additional information is available at the end of the chapter

http://dx.doi.org/10.5772/64329

Abstract

Metallic alloys with excellent structural and chemical properties play a significant role in a variety of applications. The selection of suitable metallic materials for marine gas turbine engine components is a real challenge as the surrounding environment is highly corrosive and the components have to function for a designed period at varied elevated temperatures. This chapter explains the selection of a suitable metallic alloy for marine gas turbine engine compressor section components based on extensive experimental data on two titanium-based alloys 6242 and IMI 834 under simulated marine gas turbine engine environmental conditions. The results revealed that 6242 exhibits superior performance over IMI 834. Therefore, the titanium alloy 6242 in association with appropriate protective coating is recommended to fabricate components intended to use for marine gas turbine engine compressor section applications.

Keywords: titanium alloys, marine gas turbine engines, hot corrosion, degradation mechanism, smart coatings

1. Introduction

Metallic alloys consist of one or more elements including small amounts of non-metals to improve their structural, corrosion, biomedical, magnetic properties, etc., which play a significant role in a variety of applications and environments. They are used extensively for commercial applications due to their cost-effective performance. Brass, bronze and various types of steels are the common metallic materials. However, selecting appropriate metallic materials for strategic applications such as defence, aerospace, and gas turbines of different types is a challenge. Excellent strength and lightweight properties made titanium-based alloys attractive candidate materials to fabricate components for aerospace applications. Enhanced strength of titanium alloys is obtained by solid solution alloying and stabilization of two-phase structures. The titanium alloys exhibit good corrosion resistance due to the formation of titania on their surface. However, the nature, composition and thickness of the protective oxide scales depend on the environmental conditions as well as nature and concentration of alloying elements. Due to the combined effects of elevated temperatures and exposure to harsh fuel and marine environmental conditions, hot corrosion has become a growing obstacle in meeting the desired efficiency of marine gas turbines. Therefore, it is essential to understand the corrosion characteristics of the titanium alloys under simulated marine engine environmental conditions [1].

Substantial amount of work was carried out by principal author and his co-workers on titanium-based alloys for their high-temperature oxidation and hot corrosion resistance in a variety of environments for aerospace applications, and the mechanisms under which they degrade were proposed [2–13]. He also developed an oxidation model to predict the life of titanium alloy components under gas turbine engine applications and demonstrated the validity of model with experimental data [14]. Furthermore, he successfully developed high-performance protective coatings for titanium alloy components for gas turbine engine applications [15–18]. Apart from aerospace applications, the utility of IMI 834 alloy for biomedical applications was also studied [19]. Recently, highly ordered titania nanotubes were successfully synthesized on different types of titanium alloys for a variety of applications including the solarization of a variety of systems starting from household to space vehicles and prevention of global warming, which is a major challenge for the world at present [20–23].

For obtaining higher efficiencies, the materials should satisfy both mechanical and corrosion resistance for industrial applications. It is pertinent to mention the fact that corrosion determines the life of components in a variety of industries. Therefore, material selection based on their corrosion resistance (if selected) enhances the efficiency, reduces down time, which in turn improves the production significantly and subsequently profitability [13]. In the present investigation, an attempt has been made to understand the properties of 6242 and IMI 834 and select a suitable alloy for fabrication of components used in marine gas turbine industries by carrying out a systematic corrosion study under marine gas turbine engine environmental conditions (hot corrosion) at different elevated temperatures.

2. Hot corrosion of titanium alloys

The chemical composition of selected titanium alloys is presented in **Table 1**.

2.1. Weight change measurements

Figure 1 shows the weight change as a function of exposure time for the titanium alloy IMI 834 in hot corrosion environment at three different temperatures. It can be observed that the weight change is increasing with exposure time and temperature. The degree of weight change at 700°C is significant compared with the remaining temperatures. After 40 h of exposure time,

http://dx.doi.org/10.5772/64329

the weight change becomes constant at 500 and 600°C. However, at 700°C, the weight change increases rapidly with time and reaches 16 mg/cm^2, which is about four and six times more when compared with 500 and 600°C, respectively. It indicates that higher temperatures are much severe for the alloy under hot corrosion conditions.

Alloy	Al	Sn	Zr	Mo	Nb	Si	C	O	Ti
IMI 834	5.8	4.06	3.61	0.54	0.7	0.32	0.5	0.105	Balance
Ti 6242	6	2	4	2	-	0.1	-	-	Balance

Table 1. The chemical composition of selected titanium alloys (wt.%).

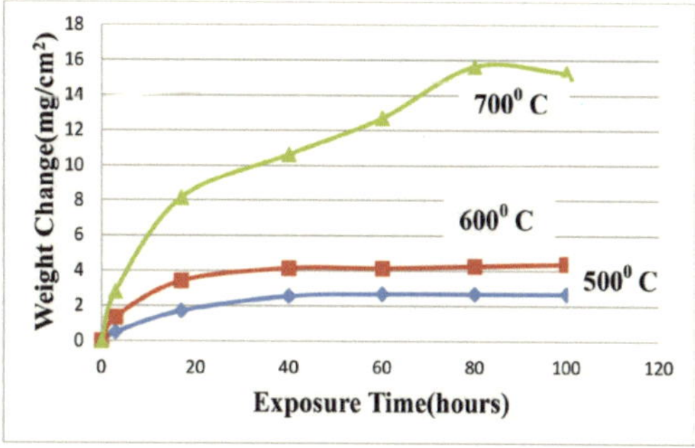

Figure 1. Weight change as a function of exposure time for IMI 834 at varied temperatures.

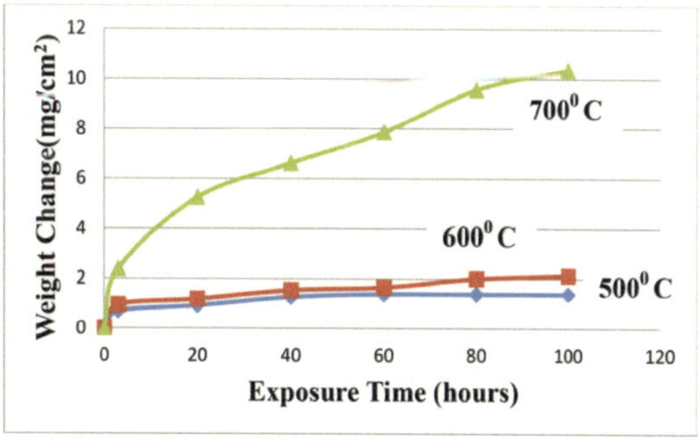

Figure 2. Weight change as a function of exposure time for Ti 6242 at varied temperatures.

Figure 2 shows the weight change as a function of exposure time for the titanium alloy 6242 in hot corrosion environment at 500, 600 and 700°C. From the data, it can be observed that the weight change is increasing with exposure time and temperature. The weight change for the alloy at 700°C is increasing rapidly compared to lower temperatures, that is, 500 and 600°C. The weight change patterns are almost similar at 500 and 600°C and the change is only about 2 mg/cm^2. It is important to mention that unlike other alloy IMI 834, the weight change is less for 6242 at all the studied temperatures.

Temperature (°C)	IMI 834	6242
500	1.68×10^{-2}	1.6×10^{-4}
600	1.81×10^{-2}	1.6×10^{-3}
700	6.5	1.52

Table 2. Rate constants of titanium alloys at different temperatures ($mg^2/cm^4/h$).

Alloy	Activation energy (KJ/mole)
IMI 834	-587.41
6242	-286.31

Table 3. The activation energies of IMI 834 and 6242 alloys.

Based on the weight change data at different temperatures for both the alloys, the rate constants were evaluated and are presented in **Table 2**. The rate constant is almost twice for IMI 834 compared with 6242 at 500°C, one order more at 600°C and four times at 700°C. The activation energies evaluated for both the alloys are shown in **Table 3**. It is important to mention that the activation energy for IMI 834 is about twice as that of titanium alloy 6242, indicating that it is more reactive with the marine environment.

2.2. X-ray diffraction (XRD)

The XRD results revealed that the hot corrosion products for both the alloys studied at different temperatures are predominantly the oxides of titanium, aluminium and tin.

2.3. Scanning electron microscopy/electron dispersive X-ray spectroscopy (SEM/EDS)

Figures 3 and **4** show the surface morphologies of titanium alloys IMI 834 and 6242, respectively, at 500, 600 and 700°C. The formed scale on IMI 834 was cracked and intact at 500°C and spalled at 600°C. While at 700°C, the scale was not only spalled but also the depth of the crack was significantly deep. In the case of 6242, the scale was intact at 500°C, cracked at 600°C and spalled at 700°C (**Figure 4**). However, the cracks are not as deep as that of IMI 834 at similar environmental conditions. The EDS results of hot-corroded IMI 834 and 6242 at various temperatures are shown in **Figures 5** and **6**. The EDS data of IMI 834 at all the temperatures show the formation of oxides of titanium, aluminium, tin and silicon. The peaks corresponding to sodium and chlorine are due to the presence of residual salt on the samples. In the case of 6242, the EDS data show the formation of oxides of titanium, aluminium, molybdenum and small amount of silicon at 500°C. When the temperature increased to 600°C, a very small amount of molybdenum in association with titanium and aluminium oxides was formed. While at 700°C, only the oxides of titanium and aluminium were observed and there was no presence of molybdenum or silicon. The formation of oxide scale clusters on the surface of titanium alloys at 500, 600 and 700°C, which gradually increased with increase in temperature, is responsible in degrading the alloys by affecting the mechanical properties that make them

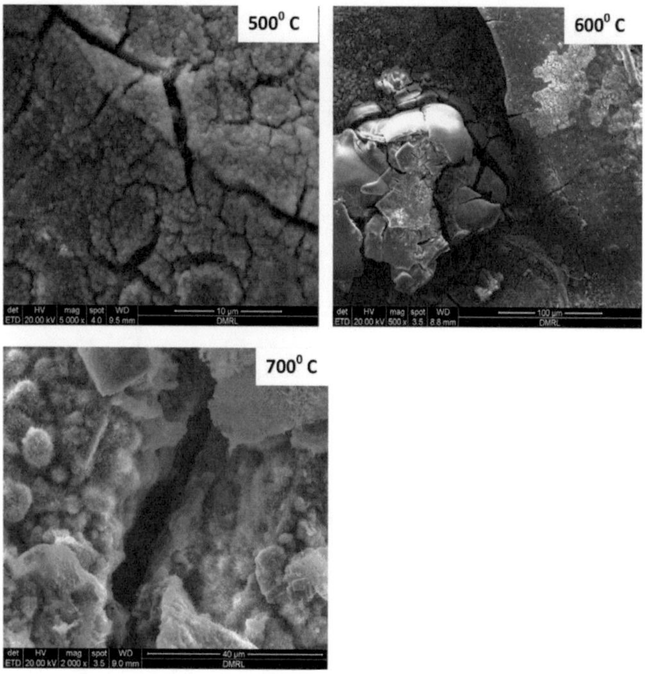

Figure 3. Surface morphologies of hot-corroded IMI 834 alloy at various temperatures.

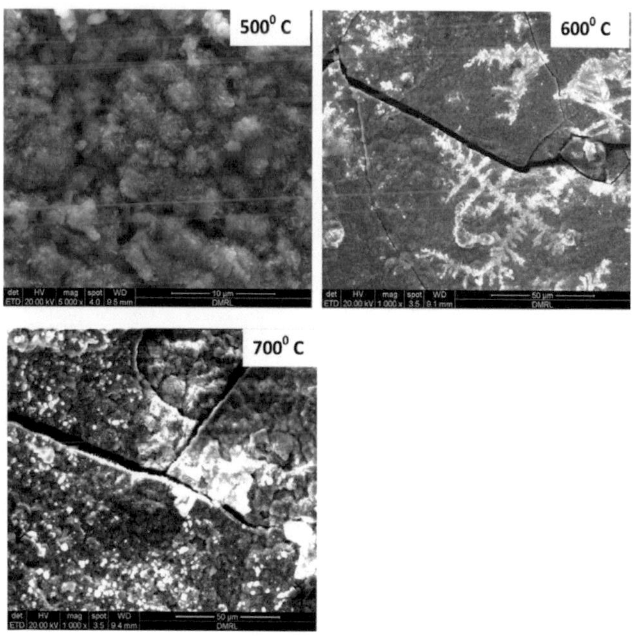

Figure 4. Surface morphologies of hot-corroded Ti 6242 alloy at different temperatures.

unsuitable for marine gas turbine engine applications. However, the 6242 showed better performance when compared with IMI 834.

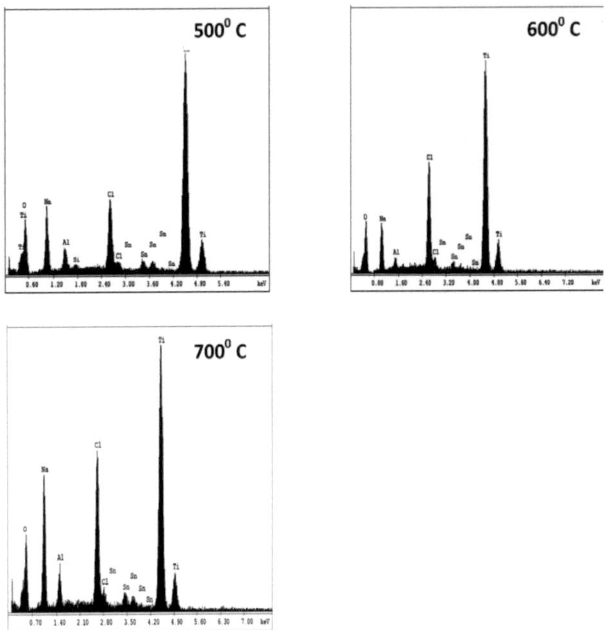

Figure 5. Electron dispersive spectroscopy (EDS) patterns of hot-corroded IMI 834 at different temperatures.

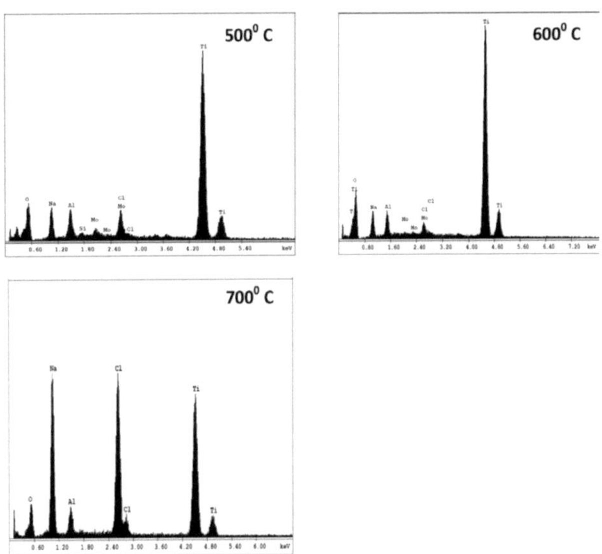

Figure 6. Electron dispersive spectroscopy results of hot-corroded 6242 at different temperatures.

http://dx.doi.org/10.5772/64329

2.4. Micro-hardness measurements

Micro-hardness profiles as a function of depth from the surface for hot-corroded IMI 834 and 6242 are presented in **Figures 7** and **8**, respectively. It is observed that the hardness is higher at the surface of the alloys than in the substrate. This is attributed to the corrosion that takes place at the surface. The high hardness is due to the dissolution of oxygen in the titanium alloys because of exposure to high temperatures in marine environmental conditions. The oxygen-dissolved region is also known as alpha case. It is also observed that the hardness gradually decreases from the surface to the substrate and finally becomes constant. The depth at which the hardness becomes a constant provides the depth of the alpha case. It is seen from the micro-hardness data that the depth of alpha case is more for higher temperatures due to increased corrosion rates at elevated temperatures. It is known that the corrosion rate increases with increase in temperature (**Table 2**). It is also evident from **Figures 7** and **8** that the depth of alpha case is higher for IMI 834 compared to 6242 at all the studied temperatures.

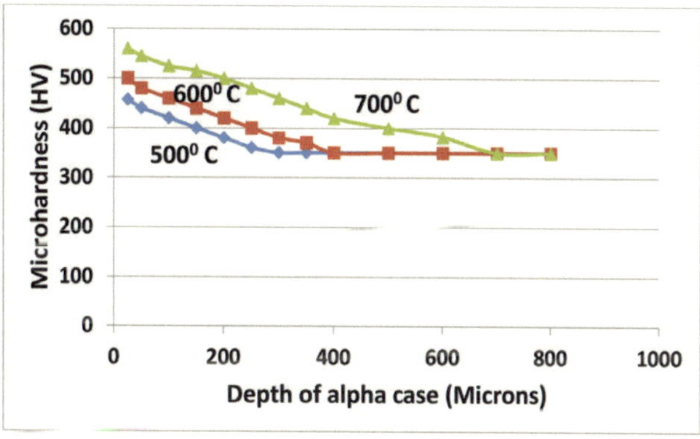

Figure 7. Micro-hardness as a function of depth of α case for hot-corroded IMI834 at varied temperatures.

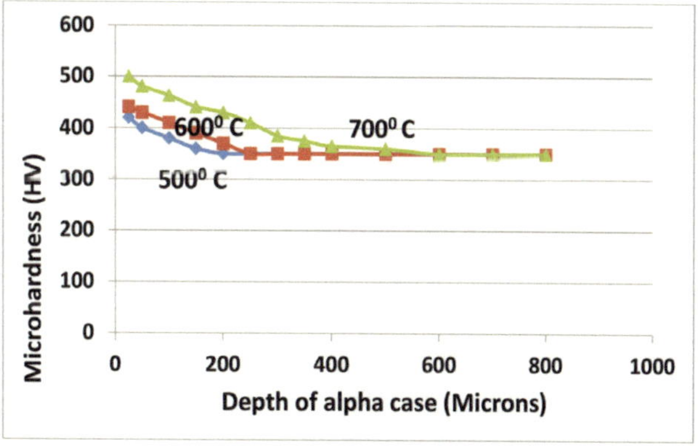

Figure 8. Micro-hardness as a function of depth of α case for hot-corroded 6242 at different temperatures.

2.5. Cross sections

The cross sections of hot-corroded titanium alloys IMI 834 and 6242 exposed at 500, 600 and 700°C are shown in **Figures 9** and **10**, respectively. The figures reveal that the corrosion-affected region increases with increase in temperature for both the alloys. However, the titanium alloy IMI 834 has more oxide scale at all the studied temperatures compared to Ti 6242. It is also clearly observed that the corrosion-affected zone is more for the hot-corroded IMI 834 when compared to Ti 6242 at all the temperatures. It indicates that Ti 6242 alloy provides good corrosion resistance and is suitable for marine applications.

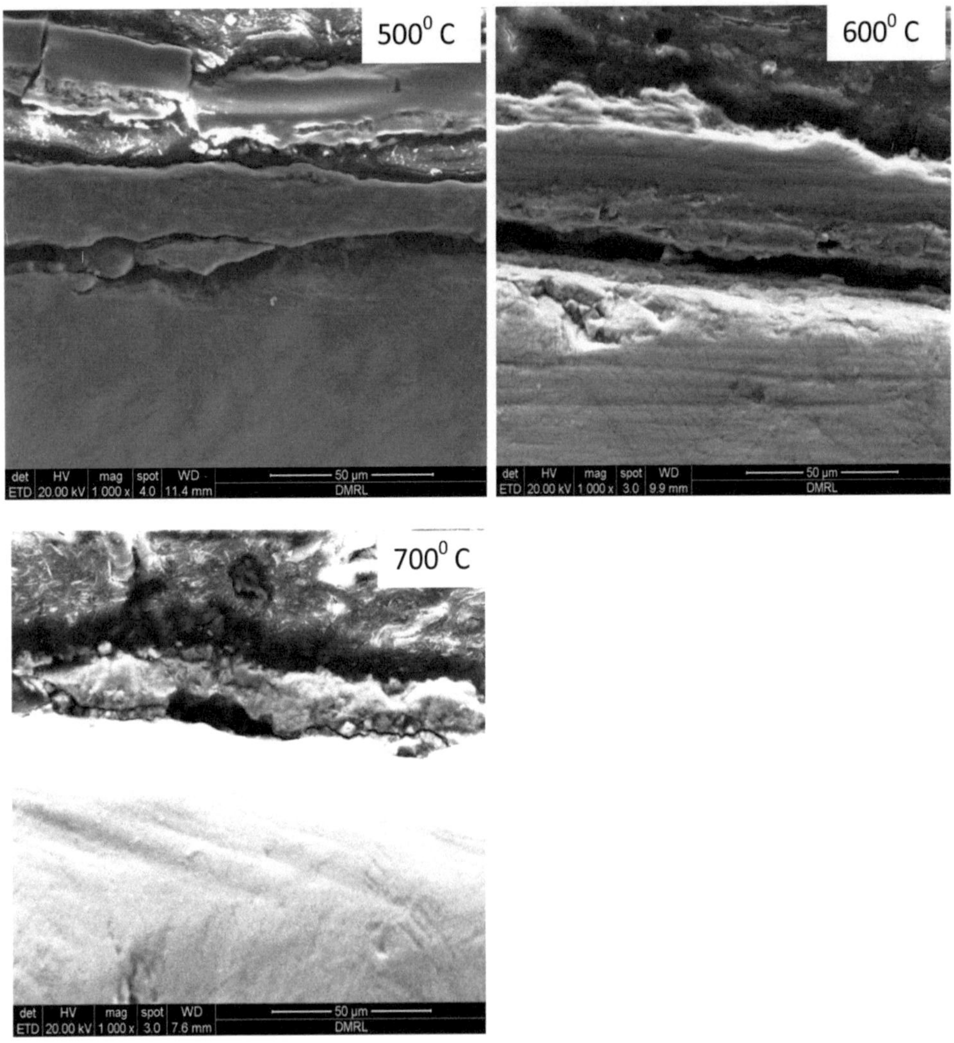

Figure 9. Cross sections of IMI 834 after exposure to hot corrosion environment at various temperatures.

http://dx.doi.org/10.5772/64329

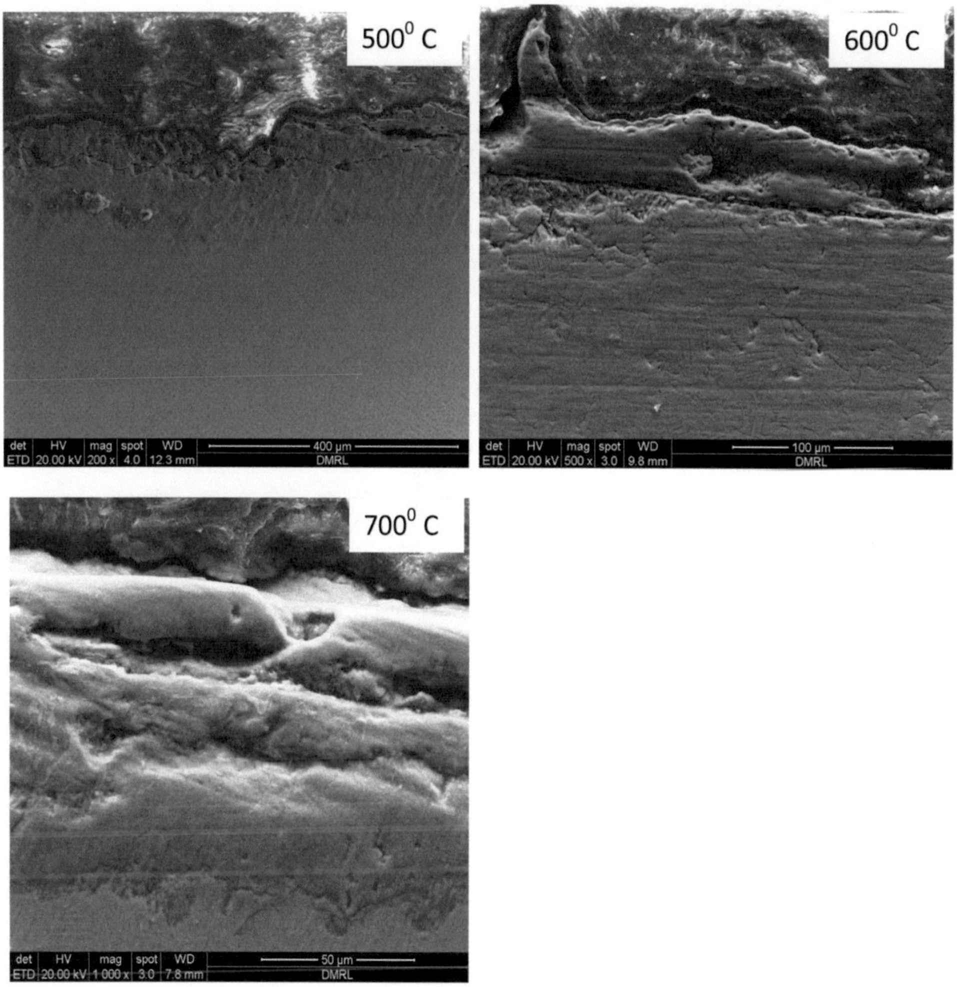

Figure 10. Cross sections of Ti 6242 after exposure to hot corrosion environment at 500, 600 and 700°C.

3. Comparative evaluation

Any material that is to be chosen for marine gas turbine engine applications should perform its function at service environmental conditions for a designed period. Therefore, suitable selection of a material is mandatory. It can be done based on extensive laboratory experimentation only because it is not possible to test the materials directly in the field. Therefore, comparative evaluation of titanium alloys IMI 834 and 6242 has been carried out under simulated marine gas turbine engine conditions. In order to compare the results precisely for both the alloys, the data were redrawn at different temperatures. The variation of weight change as a function of time for IMI 834 and Ti 6242 at temperatures of 500, 600 and 700°C is

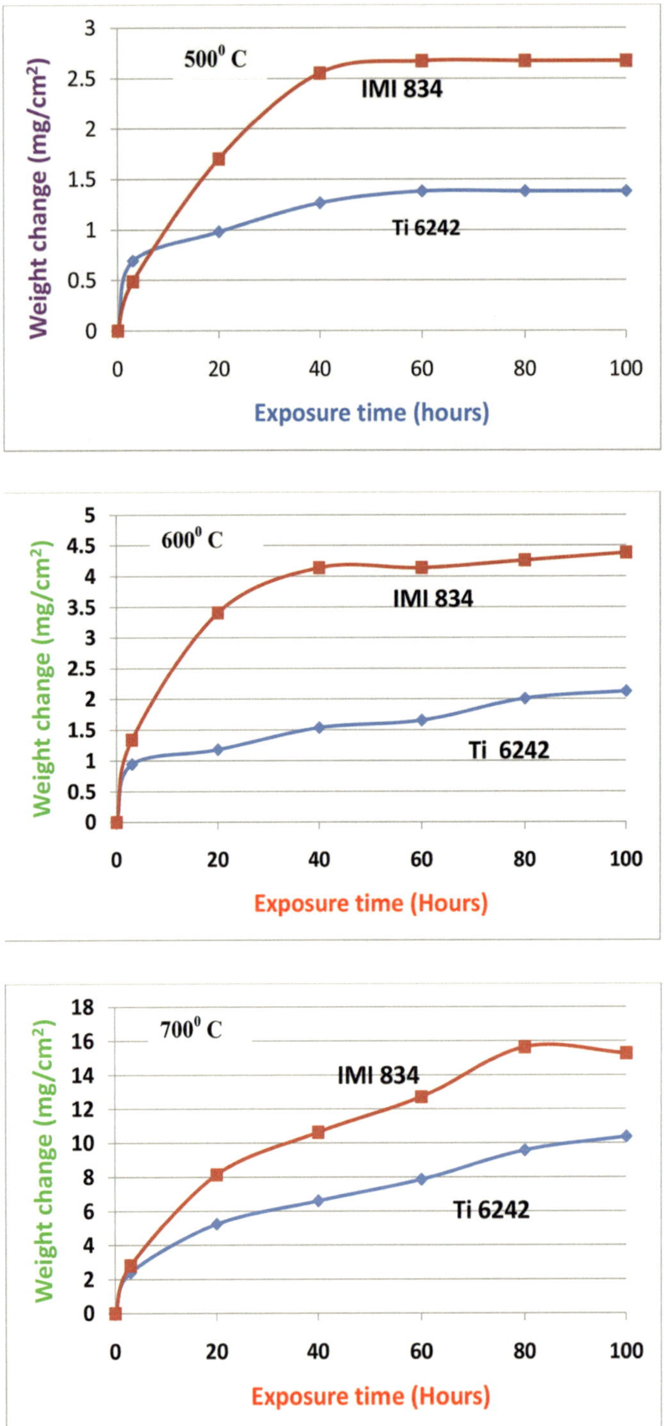

Figure 11. Comparative weight change as a function of exposure time for Ti 6242 and IMI 834 at different temperatures.

http://dx.doi.org/10.5772/64329

compared as shown in **Figure 11**. There is a large difference in weight change as the time progresses till 100 h. For example, after 20 h the weight change for IMI 834 is 1.7 mg/cm^2 and it is 1 mg/cm^2 for 6242. As the time increased to 40 h, the weight change is increased to 2.5 mg/cm^2 for IMI 834 and 1.25 mg/cm^2 for 6242. With further increase in exposure time, the weight change increased marginally for both the alloys. After 60 h, the weight change remains constant for both the alloys. In essence, the weight change is almost doubled for IMI 834 compared to 6242 at 500°C indicating that IMI 834 is highly vulnerable to hot corrosion and corrodes at a faster rate. Similar behaviour is clearly observed at 600 as well as at 700°C (**Figure 11**) although the difference is less at 700°C. It is to be noted that the weight change is considerably more at 700°C for both the alloys. However, it is less for 6242 compared to IMI 834. This observation clearly shows that Ti 6242 is more preferable than IMI 834 for marine applications.

SEM results of IM 834 and 6242 show that the oxide scale cracked in the presence of marine environment. This indicates that the chloride ions present in the environment promote oxide scale cracking and facilitate oxygen diffusion into the alloy, which is the reason for observing significant weight change. Cross sections of hot-corroded specimens revealed the presence of an oxide scale and alpha case (**Figures 9** and **10**). The oxide scale consists of granular rutile and alumina. The region just below the oxide scale is the oxygen-dissolution region (α case). As expected, the thickness of the oxide scale increases with increase in temperature.

In order to understand the effect of hot corrosion of IMI 834 and 6242 more clearly, the microhardness data were redrawn at various temperatures. The depth of alpha case at 500°C is about 175 μm for 6242 and 300 μm for IMI 834. At 600°C, the depth of alpha case is about 200 μm for 6242 and 400 μm for IMI 834. While at 700°C, it is 500 μm for 6242 and 700 μm for IMI 834 (**Figure 12**). The results clearly indicate that the depth of alpha case is considerably high for IMI 834 than 6242 at all the temperatures. It indicates that the corrosion rate is much faster for IMI 834 compared to 6242. It is further evidenced by observing low corrosion-affected zone from the cross sections of both the alloys. The hardened zone due to diffusion of oxygen into the alloys exhibits a considerable surface hardness, which is sufficient to affect the mechanical properties of the titanium alloys by forming a highly brittle zone in which cracks initiate during service. The total depth of α-case formed for IMI 834 in marine environment is several times more than that of specimens oxidized in air at the same temperature. This clearly demonstrates the aggressiveness of marine environment compared to oxidizing conditions. Analysis of scales by XRD revealed the presence of $TiCl_2$, TiO_2, $ZrCl_2$, $AlCl_3$ and Al_2O_3 indicating that oxygen and chlorine present in the marine environment react with the alloy constituents to form their corresponding chlorides and oxides.

The present results suggest that the titanium alloy 6242 can be used for the fabrication of components intended for marine gas turbine engines. The investigations further suggest that effective protective measures are essential as the life of alloy improves significantly with appropriate protective coatings. Therefore, it is recommended to use titanium alloy 6242 instead of IMI 834 with appropriate protective coatings for marine gas turbine engine applications for their enhanced efficiency.

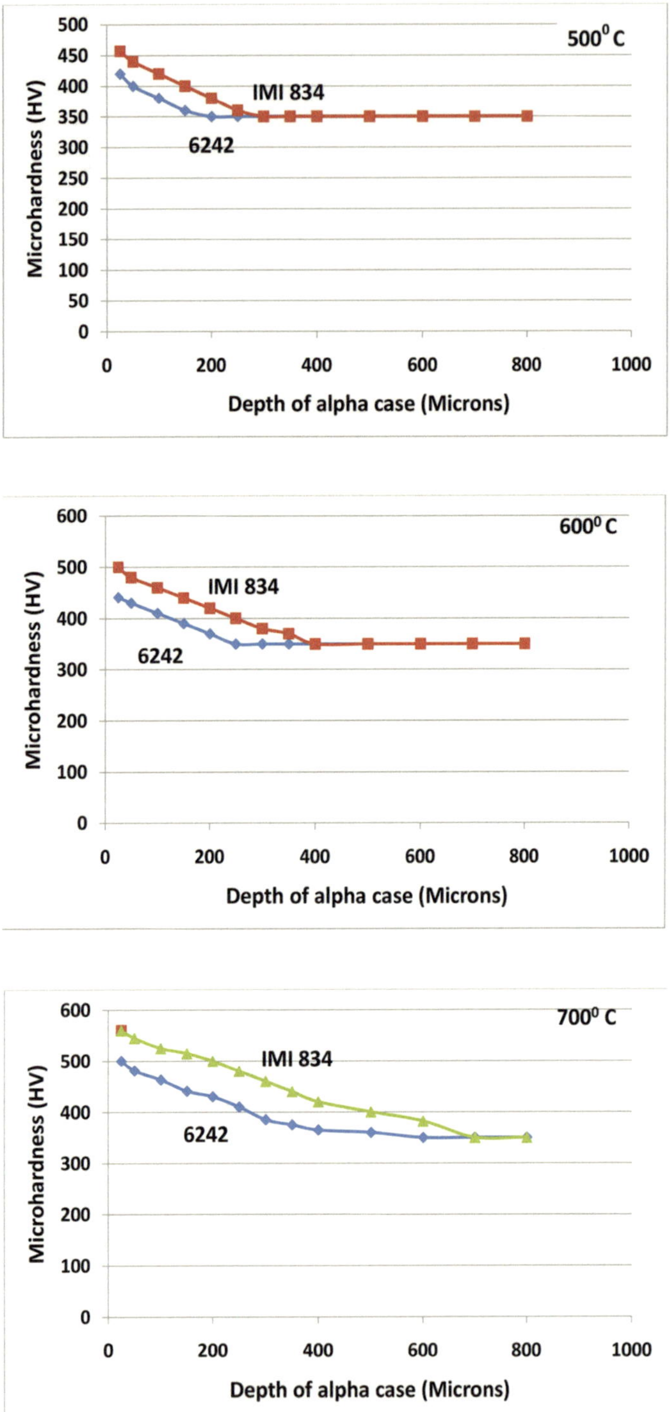

Figure 12. Comparative depth of alpha case for IMI 834 and 6242 at different temperatures.

http://dx.doi.org/10.5772/64329

4. Degradation mechanism

In general, titanium alloys containing less than 50 at.% aluminium do not form a continuous, dense and protective alumina scale upon exposure to elevated temperatures due to thermodynamic considerations [24]. IMI 834 and 6242 alloys containing only about 6 wt.% aluminium were not able to form a continuous, protective alumina scale. The oxide scale formed on the surface of IMI 834 and 6242 started spalling after exposure in marine environment indicating the formation of a poorly adherent oxide scale after reacting with chloride ions present in the environment. The previous results revealed that rutile formation is predominant followed by alumina on rutile at elevated temperatures [2, 3]. The XRD results show the formation of titanium and aluminium chlorides due to the fact that the titania readily reacts with chloride ions present in the environment to form chlorides at elevated temperatures. Given below are the proposed mechanistic steps that degrade IMI 834 and 6242 under marine environmental conditions:

1. The oxide scale that forms on the surface of IMI 834 and 6242 are predominantly TiO_2 in association with Al_2O_3. The TiO_2 reacts with chloride ions present in the marine environment to form volatile $TiCl_2$

$$TiO_2 + 2Cl^- = TiCl_2 + 2O^{2-} \tag{1}$$

The $TiCl_2$ dissociates at elevated temperatures to form Ti^{2+} and Cl^- ions

$$TiCl_2 = Ti^{2+} + 2Cl^- \tag{2}$$

The titanium ions then react with oxygen ions present in the environment to form a non-adherent and non-protective TiO_2 scale, which spalls very easily. Chloride ions present in the marine environment penetrate into the alloy to form volatile chlorides. This process continues until titanium in the alloys is consumed. In other words, the reaction is auto-catalytic. The oxygen ions that form in the above reaction diffuse into the alloy and form an oxygen-dissolution region due to more solubility of oxygen in titanium alloys.

2. Al_2O_3 reacts with Cl^- ions to form aluminium chloride

$$Al_2O_3 + 6Cl^- = 2AlCl_3 + 3O^{2-} \tag{3}$$

The $AlCl_3$ that formed in the above reaction dissociates to form Al^{3+} and Cl^- ions

$$AlCl_3 = Al^{3+} + 3Cl^- \tag{4}$$

The Al^{3+} ions react with oxygen ions to form a loose and non-protective alumina scale, which spalls very easily, as in the case of titania

$$Al^{3+} + 3Cl^{-} = Al_2O_3 \quad (5)$$

Here again, the chloride ions penetrate into the titanium alloy to form volatile chlorides. As mentioned above, the reaction is autocatalytic. The oxygen ions that formed in reaction (3) diffuse into the alloy and react with the alloys. Both the above processes (reactions (1) and (3)) contribute to the formation of α case in the titanium alloys subsurface.

As a result of the above reactions, the degradation of titanium alloys takes place at a faster rate. The results of Du et al. [25] provide evidence for the proposed mechanism. This mechanism is also supported by the significant depth of α-case formation (**Figure 12**) as well as weight change (**Figure 11**) in the marine environment. Further, SEM results clearly show cracking of the oxide scale in marine environments (**Figures 3** and **4**). Once the cracks are formed in the oxide scale, because of the formation of volatile chlorides, oxygen can penetrate more easily into the alloy and form the oxygen-dissolution region at a faster rate. This is the reason for the significant depth of α case in marine environments. The proposed mechanism is supported by the reported literature [26, 27]. This cracking situation easily makes the components fabricated from IMI 834, susceptible to failure under normal service conditions of the compressor section of gas turbines. This mechanism is evidenced by observing stress-corrosion cracking in titanium alloys in the presence of halides and stress [26, 27]. It was also reported that the extent of attack depends on temperature, exposure time and tensile stress [28, 29]. Cracking in titanium alloys was also reported by Bauer [30] who traced the crack origins. It is further supported by Beck, who reported cracking in stressed Ti-8Al-1Mo-1V alloy when immersed in anhydrous LiCl-KCl eutectic at 707 °F [31]. Even in actual jet engines, cracking was observed on salted Ti-6Al-4V alloy discs, indicating comprehensive evidence for the suggested mechanism [32]. It is well agreed by researchers in the field that NaCl is the more severe environment [33]. It was also mentioned that a liquid phase is not essential for causing cracking. Chemical periodicity is also not a factor. What appears pertinent is the reactivity of the salt in penetrating titanium oxide and in developing brittle products. The cracking mechanism is well supported by Logan et al. [34]. They proposed that oxygen ions from the scale and chlorine ions from the NaCl diffuse into the titanium alloy, react with alloy constituents to destroy atomic-binding forces and cause cracking. In essence, the reported literature clearly provides sufficient evidence for the proposed mechanism. It suggests the necessity of protecting titanium alloy components from hot corrosion to eliminate failures during service and thereby enhance their life. The studies also stress the need to develop coatings to protect titanium alloys from both oxidation and hot corrosion that are experienced by gas turbine engine compressor section components.

5. Smart coatings development

Various smart coatings containing iridium, tungsten, palladium, tantalum, aluminium and their combination were designed and developed on titanium alloy, IMI 834, using various surface engineering techniques and investigated systematically under both oxidation and hot corrosion conditions [35]. The results revealed that the smart coatings based on aluminium that were developed by innovating a new-pack composition showed an excellent resistance under both oxidation and hot corrosion conditions [35, 36]. The elemental distribution showed a protective, continuous and adherent alumina scale over the coating during both oxidation and hot corrosion conditions [36]. The XRD patterns of oxidized and hot-corroded smart coatings revealed that the major phase in the scale was primarily alumina [36]. No outward diffusion of alloying elements such as Sn, Zr, Mo and Nb was noticed. In addition, there was no inward diffusion of oxygen or nitrogen observed. Further, micro-hardness measurements made on oxidized and hot-corroded smart coatings indicated that there was no hardened zone formed, whereas the oxidized and hot-corroded uncoated alloys showed a significant hardened zone [36]. It indicates that an excellent protection was provided by the designed and developed smart coating based on aluminium to the titanium alloys from oxidation, alpha-case formation as well as hot corrosion. Modern gas turbines need to operate at high temperatures for their enhanced efficiency and the high-temperature oxidation; α-case formation and hot corrosion are the major concerns as mentioned above. The developed coating exhibited an excellent oxidation and hot corrosion resistance to the titanium alloys, prevented alpha-case formation effectively, and thus the smart coated titanium alloy components can be used safely at higher temperatures for achieving enhanced efficiency. Further, the developed coatings can be prepared by a simple technique, easy to coat large components and moreover highly economical. Hence, it is recommended to use the designed and developed smart coatings for the modern gas turbine engine titanium alloy components.

6. Summary

Hot corrosion problems in marine gas turbine engine compressor section components fabricated from titanium-based alloys along with their degradation mechanisms have been explained in detail by systematically studying the titanium alloys 6242 and IMI 834 in simulated marine gas turbine engine conditions. The ingress of oxygen increases with time and temperature, which causes to form hardened zone that initiates crack formation and ultimately fails the components. The importance of selecting suitable titanium alloy, which exhibits good hot corrosion resistance for fabrication of compressor components, has been stressed. The development of smart coatings based on newer composition of elements and their combination has been explained and showed that among the developed coatings, a smart coating based on aluminium that was deposited by innovating new-pack composition exhibits an excellent protection to the titanium alloys against oxidation, alpha-case formation and hot corrosion, which in turn enhances the efficiency of modern gas turbine engines of various types, that is, aero, marine and industrial. Therefore, it is recommended to use the titanium alloy 6242 in association with developed smart coating for advanced marine gas turbine engine compressor section components for their enhanced service life and maximum efficiency.

Author details

Injeti Gurrappa[1*], I.V.S. Yashwanth[1,2] and A.K. Gogia[1]

*Address all correspondence to: igp1@rediffmail.com

1 Defence Metallurgical Research Laboratory, Kanchanbagh, Hyderabad, India

2 Texas University, El Paso, TX, USA

References

[1] I. Gurrappa and D. Venugopala Reddy, J. Alloys. Comp. 390 (2005) 270.

[2] I. Gurrappa and A.K. Gogia, Proceedings of the 5th National Convention on Corrosion, vol. I, New Delhi, India, November 1999, pp. 210–219.

[3] I. Gurrappa, Corr. Prev. Control, 49 (2002) 79.

[4] I. Gurrappa, Oxid. Met. 59 (2003) 321.

[5] I. Gurrappa, D. Manova, J.W. Gerlach, S. Mandl and B. Rauschenbach, J. Alloys. Comp. 426 (2006) 375.

[6] I. Gurrappa, D. Manova, J.W. Gerlach, S. Mandl and B. Rauschenbach, Surf. Coat. Technol. 201 (2006) 3536.

[7] I. Gurrappa, D. Manova, S. Mandl and B. Rauschenbach, J. Adv. Mater. 41 (2009) 40.

[8] I. Gurrappa, J. Mater. Sci. Lett. 22 (2003) 771.

[9] I. Gurrappa, Proceedings of the National Symposium on Electrochemical Science and Technology, Bangalore, India, July 2002, p. 11.

[10] I. Gurrappa, Platinum Met. Rev. 45 (2001) 124.

[11] I. Gurrappa, Oxid. Met. 56 (2001) 73.

[12] I. Gurrappa, I.V.S. Yashwanth and A.K. Gogia, Design and Development of Smart Coatings for Titanium Alloys in Gas Turbines, in "Gas Turbines: Technology, Efficiency and Performance" ISBN: 978-1-61122-842-7, D.J. Ciafone (Editor), (2011), Nova Publishers, USA, pp. 59–69.

[13] I. Gurrappa, I.V.S. Yashwanth, A.K. Gogia and J.S. Burnell-Gray, The Significance of Titanium Based Alloys for Industrial Applications, in "Titanium Alloys: Formation, Characteristics and Applications" ISBN: 978-1-62808-329-3, L.C. Zhang and C. Yang (Editors) (2013), Nova Publishers, USA, pp. 1–31.

[14] I. Gurrappa, J. Alloys. Comp. 389 (2005) 190.

[15] I. Gurrappa, A. Wilson and P.K. Datta, J. Coat. Tech. Res. 6 (2009) 257.

[16] I. Gurrappa and A.K. Gogia, Surf. Coat. Technol. 139 (2001) 216.

[17] I. Gurrapa and A.K. Gogia, Mater. Sci. Technol. 17 (2001) 581.

[18] I. Gurrappa, Proceedings of the GLUCORR 2002, New Delhi, India, February 2002, IL24.

[19] I. Gurrappa, Mater. Charac. 49 (2002) 54.

[20] I. Gurrappa and I.V.S. Yashwanth, The Significance of Electrochemical Techniques in Nanocoatings Technology, in "Comprehensive Guide for Nanocoatings Technology", ISBN: 978-1-63482-447-7, Nova Publishers, USA, 2015, pp. 389–426

[21] I.V.S. Yashwanth and I. Gurrappa, Mater. Lett. 142 (2015) 328–331.

[22] I.V.S. Yashwanth and I. Gurrappa, J. Nanomater. Mol. Nanotechnol. 3(2) (2014) (doi:http://dx.doi.org.10.4172/2324-8777,1000141).

[23] I.V.S. Yashwanth and I. Gurrappa, Adv. Sci. Eng. Med. 6 (2014) 608.

[24] R.A. Perkins, K.T. Chiang and G.H. Meier, Scripta. Met. 21 (1987) 1505.

[25] H.L. Du, P.K. Datta, D.B. Lewisand and J.S. Burrell-Gray, Mater. Sci. Eng. A. 205 (1996) 1999.

[26] 26. R.F. Simenz, J.M. Van Orden and G.G. Wald, Spec. Tech. Publ. No. 297, ASTM, Materials Park, OH, 1996, p. 53.

[27] H.B. Bomberger, Titanium and Its Alloys, Source Book, American Society for Metals, 1982, p. 161.

[28] D. Singagia, G. Taccani and B. Vicentine, Corros. Sci. 18 (1978) 781.

[29] M.W. Mahoney and A.S. Tetelman, Met. Trans. 7A (1976) 1549.

[30] G.W. Bauer, Phys. Metall. Symp. Watertown Arsenal, September, 1955.

[31] T.R. Beck et al., Quart. Progr. Rept. No. 14, Contract NAS 7 489, Boeing Sci. Res. Labs., 1969.

[32] Pratt and Whitney, TML Report 88, Battelle Memorial Institute, 1957.

[33] H.H. Uhlig, Mater. Prot. Perf. 12 (1976) 42.

[34] H.L. Logan et al., Spec. Tech. Publ. No. 397, ASTM Materials Park, OH, 1966, p. 215.

[35] I. Gurrappa, A. Wilson and P.K. Datta, J. Coat. Tech. Res. 6 (2009) 257.

[36] I. Gurrappa, European Commission Final Report, July 2008.

Chapter 10

Magnetocaloric and Magnetic Properties of Meta-Magnetic Heusler Alloy $Ni_{41}Co_9Mn_{31.5}Ga_{18.5}$

Takuo Sakon, Takuya Kitaoka, Kazuki Tanaka,
Keisuke Nakagawa, Hiroyuki Nojiri,
Yoshiya Adachi and Takeshi Kanomata

Additional information is available at the end of the chapter

http://dx.doi.org/10.5772/64375

Abstract

$Ni_{41}Co_9Mn_{31.5}Ga_{18.5}$ is a magnetic Heusler alloy, which indicates metamagnetic transition at the reverse martensite transition. In this paper, caloric measurements were performed and discussed about magnetocaloric effect. We also performed magnetization measurements around Curie temperature T_C in the martensite phase and analyzed by means of the spin fluctuation theory of itinerant electron magnetism. From the differential scanning calorimetry (DSC) measurements in zero fields, the value of the latent heat λ was obtained as 2.63 kJ/kg, and in magnetic fields the value was not changed. The entropy change ΔS was − 7.0 J/(kgK) in zero fields and gradually increases with increasing magnetic fields. The relative cooling power (RCP) was 104 J/kg at 2.0 T, which was comparable with In doped $Ni_{41}Co_9Mn_{32}Ga_{16}In_2$ alloy.

The results of magnetization measurement were in agreement with the theory of Takahashi, concerning itinerant electron magnetism. From the M^4 vs H/M plot, the spin fluctuation temperature T_A can be obtained. The obtained T_A was 703 K and which was smaller than Ni (1.76×10^4 K). The value was comparable to that of UGe_2 (493 K), which is famous for the strongly correlated heavy fermion ferromagnet.

Keywords: shape memory alloys, differential scanning calorimetry, magnetocaloric effect, magnetization, itinerant electron magnetism

1. Introduction

Recently, ferromagnetic shape memory alloys (FSMA) have been studied extensively as candidates for highly functional materials. Between FSMA, Ni_2MnGa is the most renowned alloy [1]. The alloy has a cubic $L2_1$ Heusler structure, with a space group of $\boldsymbol{Fm\bar{3}m}$, and ferromagnetic transition occurs at the Curie temperature, T_C, 365 K [2, 3]. Cooling from ordinary temperature, a martensite transition occurs at the martensite transition temperature, T_M, 200 K. Below T_M, a superstructure state is realized as a result of lattice modulation [4, 5]. For Ni_2MnGa type alloys, $\boldsymbol{T_M}$**'s** are varied from 200 to 330 K by non-stoichiometrically varying the concentration of constituent elements of the alloys. Sakon et al. studied about the magnetic properties of $\mathbf{Ni_{50+x}Mn_{27-x}Ga_{23}}$ [6]. The martensite transition and the ferromagnetic transition occurred at the same temperature at the martensite transition temperature T_M for $x = 2.5, 2.7$. The T_M shift in magnetic fields around a zero magnetic field was estimated to be $dT_M/dB = 1.1 \pm 0.2$ K/T, thus indicating that magnetization influences martensite transition.

New alloys in the Ni–Mn–In, Ni–Mn–Sn, and Ni–Mn–Sb Heusler alloy systems that are expected to be ferromagnetic shape memory alloys have been studied [7, 8]. A metamagnetic transition from paramagnetic martensite phase to ferromagnetic austenite phase was observed, and reverse martensite transition induced by magnetic fields was occurred under high magnetic fields [9, 10]. These alloys are promising as a metamagnetic shape memory alloys with a magnetic field-induced shape memory effect (MSIF) and as magnetocaloric materials. Ni–Co–Mn–In alloys, in which Co is substituted for Ni in Ni–Mn–In alloys to increase the Curie temperature, indicate shape memory behaviors in compressive **stress–strain** measurements. It is noticeable that 3% MFIS has been observed for $Ni_{45}Co_5Mn_{36.7}In_{13.3}$ [11].

The substitution of Co for Ni or Ga atoms in Ni_2MnGa type alloys has turned the magnetic ordering of the parent phase from partially antiferromagnetic or paramagnetic to ferromagnetic, resulting in a large magnetization change across the transformation, which dramatically enhances the magnetic field driving force [12–40]. The phase diagram in the temperature-concentration plane is determined on the basis of the experimental results. The determined phase diagram is spanned by a paramagnetic austenite (Para-A) phase, paramagnetic martensite phase, ferromagnetic austenite phase, ferromagnetic martensite (Ferro-M) phase of $\mathbf{Ni_{50-x}Co_xMn_{31.5}Ga_{18.5}}$ [41]. The measurements indicated that a magnetostructural transition between the Para-A and Ferro-M phases. As for $Ni_2MnGa_{1-x}Co_x$ ($0 \leq x \leq 0.20$), which was substituted of Co for Ga, the measurements showed that a magnetostructural transition between the Para-A and Ferro-M phases does not appear in this alloy system [21]. Therefore, the experimental studies of the Ni_2MnGa type Heusler compounds, which were substituted of Co for Ni atoms, are important to clarify the nature of the magnetostructural interactions, which is one of the important problems in the physics of Heusler compounds. The transformation temperature can be downshifted by magnetic field at a rate $\boldsymbol{dT_R/\mu_0 dH}$ or $\boldsymbol{dT_M/\mu_0 dH}$ up to 14 K/T in $Ni_{37}Co_{13}Mn_{32}Ga_{18}$ [39]. The aging effect is also important. Segui et al. measured the thermo-magnetization, *M-T*, of $Ni_{43}Co_7Mn_{32}Ga_{18}$ at the constant magnetic fields of **5** mT [13, 15, 42]. As longer aging time, the larger value of the magnetization was obtained at ferromagnetic austenite phase around 420 K. This indicates that the improvement of atomic

http://dx.doi.org/10.5772/64375

order enhances the ferromagnetic character. Castillo-Villa et al. reported about the elastocaloric effect in a $Ni_{50}Mn_{25-x}Ga_{25}Co_x$ [25]. They also studied the influence of applied magnetic fields on this effect experimentally, and a comparative investigation with the magnetocaloric effect, which was exhibited by the alloy, was performed. Both of the elastocaloric and magnetocaloric effects are a result of the martensite transition incurred by the alloy. The influence of a compressive stress and of a magnetic field is to gain the stability of the martensite phase. It leads to an increase of the transition temperature with applied stress and magnetic field. The magnetic properties of $Ni_{33.0}Co_{13.4}Mn_{39.7}Ga_{13.9}$ were investigated by Xu *et al.* [28]. It is noticeable that the kinetic arrest phenomenon was observed at about 120 K by thermomagnetization measurements. Magnetic field-induced transformation was also detected at the temperatures between 4.2 and 300 K. In this phenomenon, martensite transition is interrupted at certain temperatures (kinetic arrest temperature, T_{KA}) during field cooling and does not proceed with further cooling. It was confirmed that the transformation entropy change below 120 K becomes almost zero, which results in the kinetic arrest phenomenon by evaluation of the equilibrium magnetic fields and temperatures based on the transformation fields and temperatures.

Albertini *et al.* has been performed experimental studies about the composition dependence of the structural and magnetic properties of the Ni–Mn–Ga ferromagnetic shape memory alloy substituted of Co for Ni atoms around the composition of $Ni_{50}Mn_{30}Ga_{20}$ [12, 31]. The magnetic and structural properties displayed remarkable discontinuities across the martensite transition. There was a clear jump (ΔM) in the saturation magnetization at the transformation, which indicates that a metamagnetic transition appeared in the magnetic field. The field dependence of the martensite transition temperature, $dT_M/\mu_0 dH$, and that of the crystalline volume change, $\Delta V/V$, was reported. The most notable alloy is $Ni_{41}Co_9Mn_{32}Ga_{18}$. When cooling from 500 K, a ferromagnetic transition in the austenite phase at $T_C{}^A$ = 456 K. At the martensite transition temperature, T_M = 420 K, the AC susceptibility decreased drastically. Below 300 K, the AC susceptibility gradually increased and a distinct peak was found at the Curie temperature $T_C{}^M$ = 257 K in the martensite phase. Heating from 200 K, the Curie temperatures $T_C{}^M$ and $T_C{}^A$ were just the same with the temperatures in the cooling procedure. The reverse martensite temperature T_R was 436 K. Therefore, the AC susceptibility indicates re-entrant magnetism, ferromagnetic-paramagnetic state, which should be released to the crystal structures. The $dT_M/\mu_0 dH$ and $\Delta V/V$ were found to be considerably enhanced by the additional in-doping of the **Ni–Co–Mn–Ga** alloy [12]. The entropy change is larger than that of **Ni–Co–Mn–Ga,** which suggests large relative cooling power (RCP).

We studied about the physical properties and magnetism of $Ni_{50-x}Co_xMn_{31.5}Ga_{18.5}$ [41]. Crystallographic, thermal strain, and magnetic properties of $Ni_{50-x}Co_xMn_{38.5}Ga_{18.5}$ $(0 \leq x \leq 9)$ were investigated across the martensite transition temperature T_M and the reverse martensite transition temperature T_R at atmospheric pressure above $T_C{}^M = 263$ K. These transition temperatures increased gradually with increasing Co component x. Moreover, temperature hysteresis in the thermal cycles of the magnetization across the T_R and T_M, became larger with increasing x. As for $x = 9$, $Ni_{41}Co_9Mn_{31.5}Ga_{18.5}$, extensive temperature hysteresis of $T_M - T_R = 65$ K was found in the thermal strain measurement. The metamagnetic transition was observed from the paramagnetic martensite phase to the ferromagnetic austenite phase between 330 and 390 K. Under atmospheric pressure, the observed magnetostriction value of this alloy at 350 K was 0.11%. This value is larger than that of Tb-Dy-Fe single crystal [43]. In this study, $Ni_{50-x}Co_xMn_{38.5}Ga_{18.5}$ was polycrystal, and it is easy for processing and handling. Moreover, the magnetostriction effect occurred at a temperature between room temperature and 400 K, which was suggested that it is useful in the high temperature region, *for example,* apparatus in the engine room of the motorcar.

The effects of Co addition on the properties of $Ni_{8-x}Mn_4Ga_4Co_x$ ($x = 0, 1, 2$) ferromagnetic shape memory alloys are systematically investigated by first-principles calculations by Bai et al. [26]. The results of formation energy indicate that the added Co preferentially occupies the Ni sites in Ni_2MnGa alloy. The total energy difference between the paramagnetic and the ferromagnetic austenite plays an important role on the Curie transformation. Increasing Co content, electron density of states, DOS, of down spins around the Fermi level gradually decreases. On the contrary, those of the up spin almost remain without changing. This causes enhancement of the magnetic moments in these alloys. This result indicates that the investigation of the itinerant electron magnetism is important to understand the physics and magnetism of these alloys more deeply.

In this paper, caloric measurements were performed. On the basis of the experimental results, magnetocaloric effect was discussed. We also studied about the itinerant electron magnetic properties of $Ni_{41}Co_9Mn_{31.5}Ga_{18.5}$. We performed the magnetization measurements by means of the pulsed magnetic fields. The M^4 vs H/M plot crossed the coordinate axis at the Curie temperature in the martensite phase, T_C^M, and the plot indicates a good linear relation behavior around the T_C^M. The magnetization results were analyzed by means of the theory of Takahashi, concerning itinerant electron magnetism [44, 45] and the spin fluctuation temperature T_A can be obtained. We discussed about the itinerant magnetism of $Ni_{41}Co_9Mn_{31.5}Ga_{18.5}$ by means of the magnetization measurements.

2. Experimental procedures

The sample used in this study was synthesized at Yamagata University. The $Ni_{41}Co_9Mn_{31.5}Ga_{18.5}$ alloy was prepared by arc melting 4N Ni, 4N Co, 4N Mn, and 6N Ga in an argon atmosphere. The sample annealed at 1123 K for three days to homogenize the sample in a double evacuated silica tube, and then quenched in cold water. The obtained sample was polycrystalline. DSC measurements were performed by means of Helium-free magnet at High Field Laboratory for Superconducting Materials, Institute for Materials Research, Tohoku University. The bore of this magnet is 100 mmϕ in the air and which installed the factory-made DSC equipment.

Magnetization measurements were performed by means of the pulsed field magnet at Ryukoku University. The absolute value was adjusted by Ni. The diamagnetism of the sample was also concerned to analyze the field dependence of the magnetization.

http://dx.doi.org/10.5772/64375

3. Results and discussions

3.1. Crystallography of $Ni_{41}Co_9Mn_{31.5}Ga_{18.5}$

From X-ray powder diffraction shown in **Figure 1**, the sample was confirmed as a single phase with a tetragonal $D0_{22}$ structure at 298 K, as shown in **Figure 2**. The lattice parameters of the tetragonal structure were $a = 3.8794$ Å and $c = 6.6247$ Å. The size of the sample was 0.8 mm × 3.0 mm × 4.0 mm.

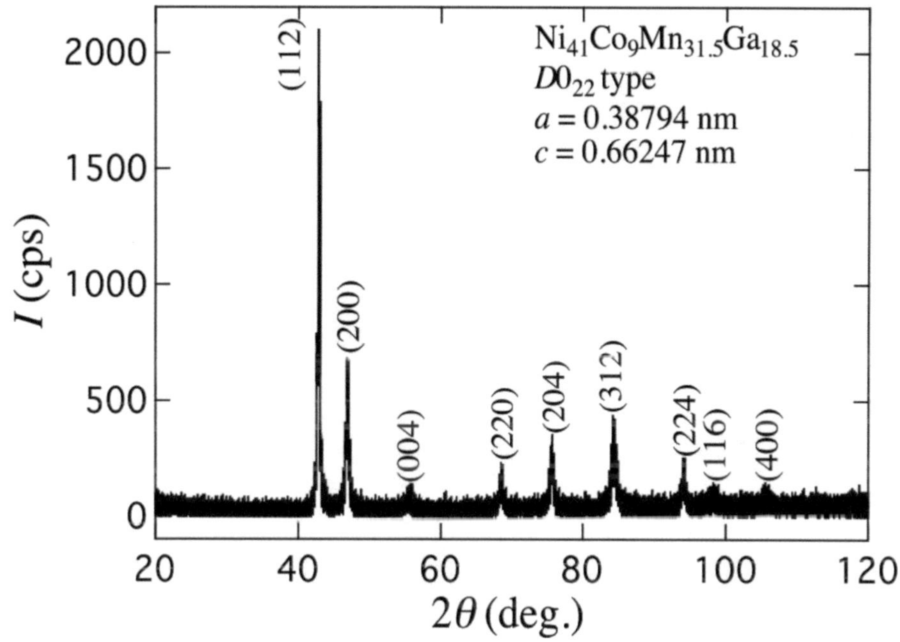

Figure 1. X-ray powder diffraction pattern of $Ni_{41}Co_9Mn_{31.5}Ga_{18.5}$ at 298 K, which indicates $D0_{22}$ type martensite phase.

The final compositions of the grown sample were verified by energy dispersive spectroscopy and were close to the nominal values with a deviation of <1%.

The Scanning Electron Microscope (SEM) image of $Ni_{41}Co_9Mn_{31.5}Ga_{18.5}$ at 298 K by means of FE-SEM (JSM6300F, JEOL Co. Ltd.) shown in **Figure 3** indicates that there are macroscopic twin variants on a scale of a few micrometers. The twins were arranged neatly in the domains. A single martensite phase characterized by typical lamellar twin substructures was observed, agreeing well with the X-ray diffraction results. This result is well agree with the optical micrographs of microstructure of $Ni_{56-x}Co_xMn_{25}Ga_{19}$ ($x = 2, 4, 8$) [32].

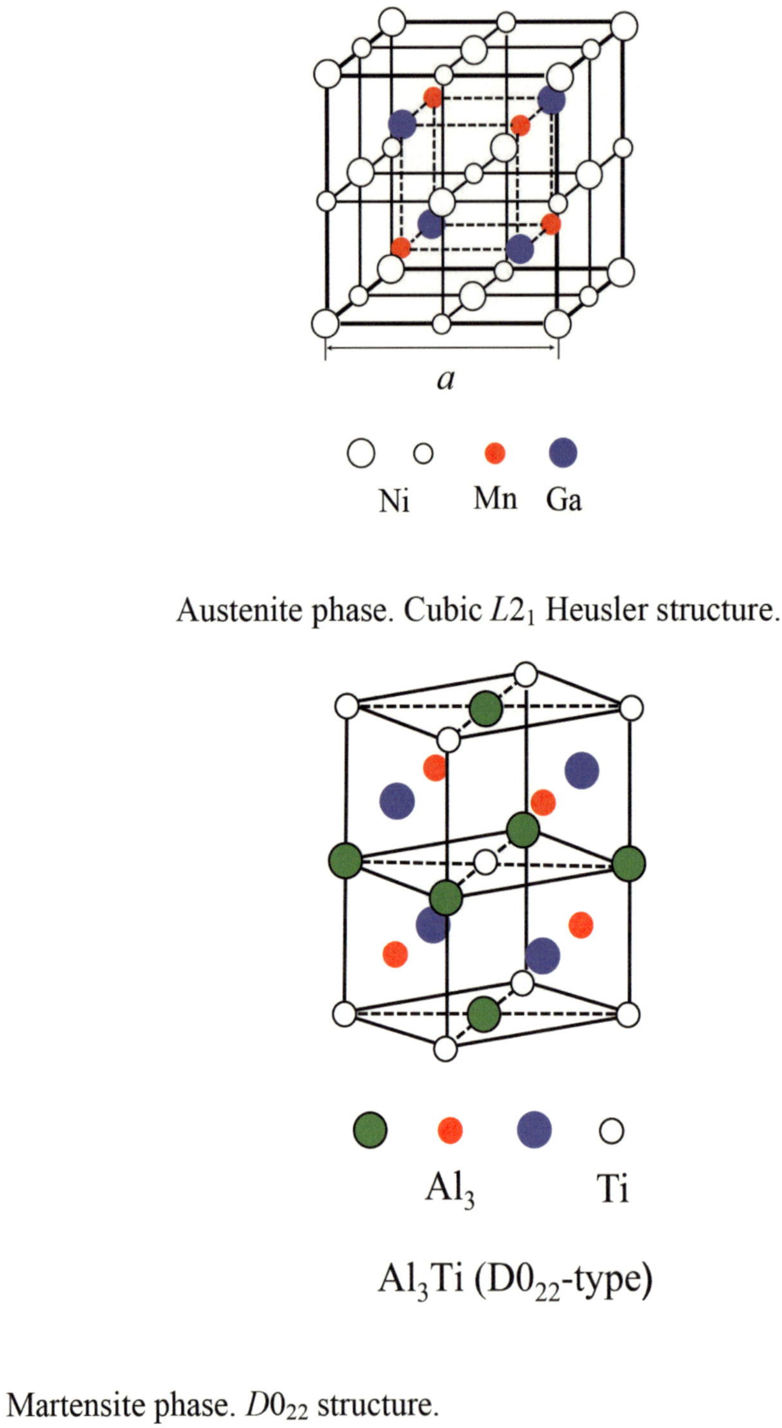

Figure 2. Schematic drawings of crystal structures in austenite phase and martensite phase.

Figure 3. Scanning electron microscope i_m age of $Ni_{41}Co_9Mn_{31.5}Ga_{18.5}$ at 298 K by means of FE-SEM.

The calorimetric measurements, which allowed for the estimation of the latent heat and magnetocaloric analysis, were performed with factory-made differential scanning calorimeter able to work up to 6 T. This setup exploits Peltier cells in order to measure heat flow of the sample. Calorimetric measurements of $Ni_{41}Co_9Mn_{31.5}Ga_{18.5}$ polycrystalline ferromagnetic shape memory alloy (FSMA) were performed across the T_R. As for the reference value of the latent heat λ at zero magnetic fields, calorimetric measurements were performed by the commercial DSC system, DSC3300 (Materials Analysis and Characterization Co., Ltd.). The measured value of the λ at zero fields was 2.63 kJ/kg around $T_R = 380$ K.

3.2. Magnetocaloric effect of $Ni_{41}Co_9Mn_{31.5}Ga_{18.5}$

The thermodynamic properties of the presented sample in magnetic fields were studied experimentally by measuring the heat flow by means of the DSC equipment. The four panels of **Figure 4** show the heat flow of $Ni_{41}Co_9Mn_{31.5}Ga_{18.5}$ in zero and magnetic fields. The endothermic reaction was occurred around the reverse martensite temperature T_R. These calorimetry experiments confirm the values of the ratio of field change of the transition temperature $dT_R/\mu_0 dH$ deduced from thermal expansion measurements [46]. The latent heat λ of the fully transformed reverse martensite transition, as shown in the four panels of **Figure 5** was calculated by integrating the heat flow profile in **Figure 4** for each magnetic field. The entropy S was calculated as $S = \lambda/T$. The entropy of $Ni_{41}Co_9Mn_{31.5}Ga_{18.5}$ at 0 and 2 T are shown in **Figure 6**.

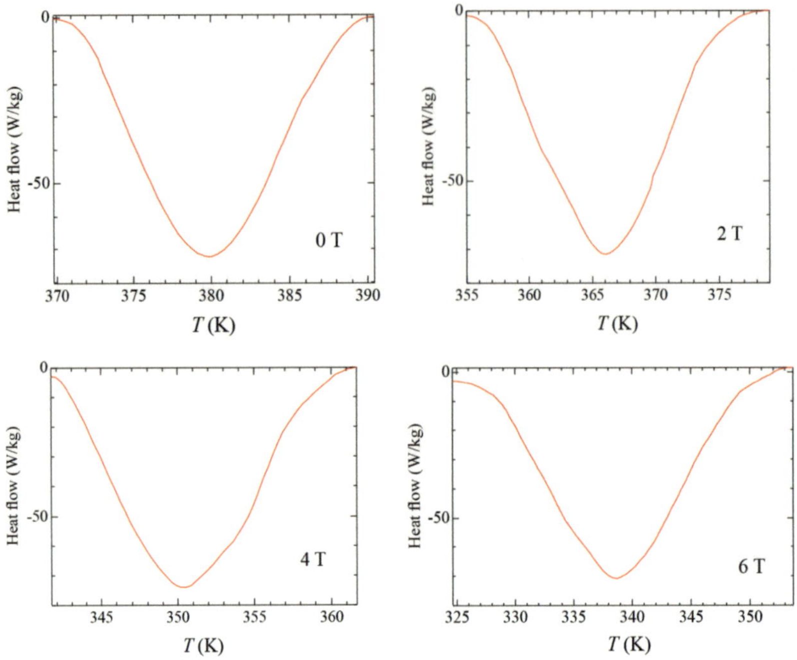

Figure 4. Heat flow of $Ni_{41}Co_9Mn_{31.5}Ga_{18.5}$ in zero and magnetic fields.

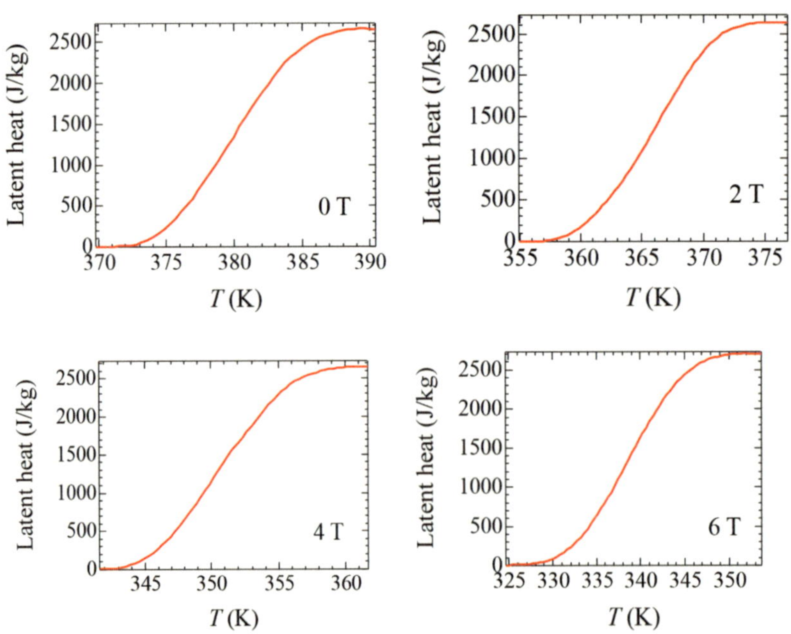

Figure 5. Latent heat of $Ni_{41}Co_9Mn_{31.5}Ga_{18.5}$ in zero and magnetic fields.

http://dx.doi.org/10.5772/64375

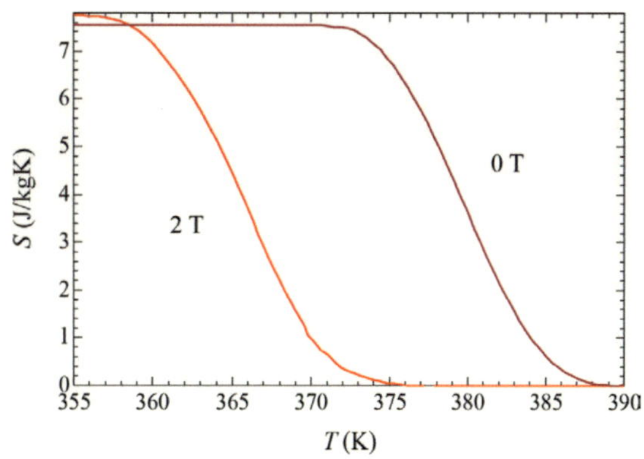

Figure 6. Entropy of $Ni_{41}Co_9Mn_{31.5}Ga_{18.5}$ at 0 and 2 T.

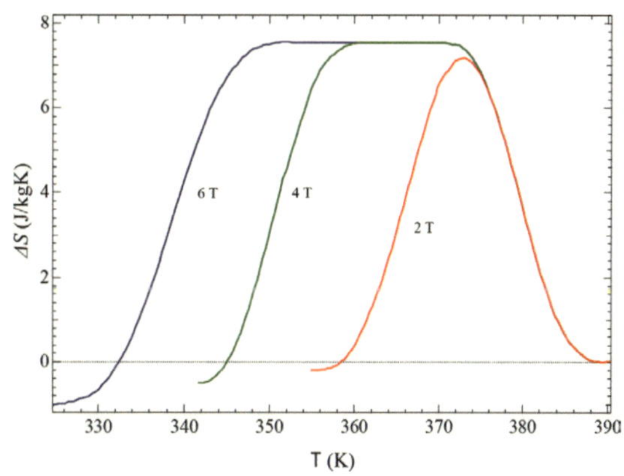

Figure 7. Entropy change ΔS of $Ni_{41}Co_9Mn_{31.5}Ga_{18.5}$.

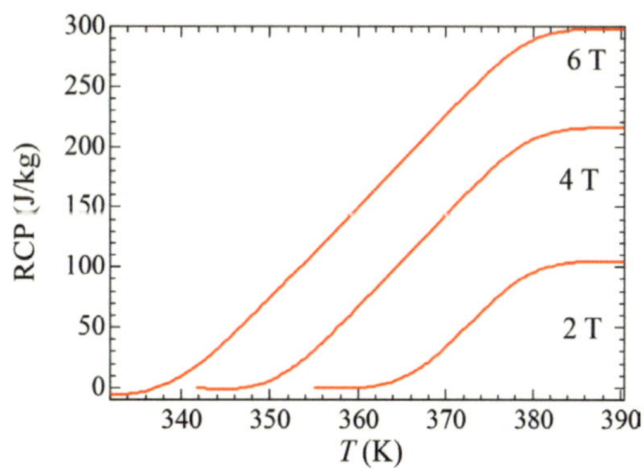

Figure 8. Relative cooling power of $Ni_{41}Co_9Mn_{31.5}Ga_{18.5}$.

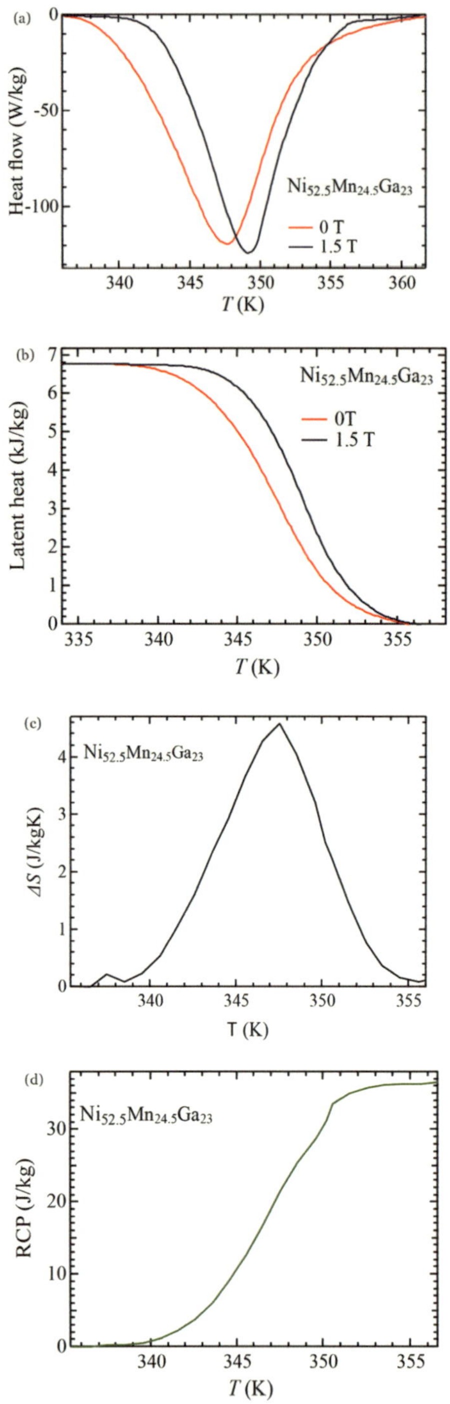

Figure 9. (a) Heat flow of $Ni_{52.5}Mn_{24.5}Ga_{23}$; (b) latent heat of $Ni_{52.5}Mn_{24.5}Ga_{23}$; (c) ΔS of $Ni_{52.5}Mn_{24.5}Ga_{23}$; and (d) RCP of $Ni_{52.5}Mn_{24.5}Ga_{23}$.

http://dx.doi.org/10.5772/64375

Figure 7 shows the entropy change $\Delta S = S(\mu_0 H) - S(0)$ of $Ni_{41}Co_9Mn_{31.5}Ga_{18.5}$. The relative cooling power (RCP) was calculated by integrating the ΔS in the temperature, as shown in **Figure 8**. The calculated RCP was 104 J/kg at 2.0 T, which was comparable with $Ni_{50}Mn_{35}In_{14}Si_1$ and $Ni_{41}Co_9Mn_{32}Ga_{16}In_2$ alloy [47].

We also performed the DSC measurement of $Ni_{52.5}Mn_{24.5}Ga_{23}$ in zero and magnetic fields by means of the water-cooled electromagnet in Ryukoku University. **Figure 9a** shows the heat flow of $Ni_{52.5}Mn_{24.5}Ga_{23}$ in a heating process. The endothermic reaction was occurred around $T_R = 348$ K. In the external magnetic field of 1.5 T, the reaction was also occurred. The $dT_R/\mu_0 dH$ obtained from DSC measurements in **Figure 9a** were 1.1 K/T, which is comparable to the results of the thermal strain measurements [6]. **Figure 9b** shows the latent heat λ of $Ni_{52.5}Mn_{24.5}Ga_{23}$. The λ is larger than that of $Ni_{41}Co_9Mn_{31.5}Ga_{18.5}$. **Figure 9c** shows the entropy change ΔS of $Ni_{52.5}Mn_{24.5}Ga_{23}$. The value was 4.6 J/kg K, which is comparable to the value of $Ni_{52.6}Mn_{23.1}Ga_{24.3}$ [48]. **Figure 9d** shows the RCP of $Ni_{52.5}Mn_{24.5}Ga_{23}$. The value was 36 J/kg.

Table 1 shows the T_M, ΔS, δT, and RCP at 2 T. δT indicates the half width of the ΔS. The ΔS, δT, and RCP of $Ni_{52.5}Mn_{24.5}Ga_{23}$ were estimated from the DSC result at zero field and 1.5 T in this study. The three alloys of the beginning cause martensite phase transition at temperature of the T_M in paramagnetic austenite from ferromagnetic martensite. Four last alloys cause re-entrant magnetism at the temperature of T_M. The $dT_M/\mu_0 dH$ of the four last alloys is larger than that of the three alloys of the beginning. Therefore, the RCP is relatively larger than former alloys.

Sample	T_M (K)	ΔS (J/kg K)	δT (K)	RCP (J/kg)	Reference
$Ni_{52.6}Mn_{23.1}Ga_{24.3}$	297	-6	1.8	11	[48]
$Ni_{52.5}Mn_{24.5}Ga_{23}$	348	-6.1	8.0	48	This work
$Ni_{55.4}Mn_{20}Ga_{24.6}$	313	-41	1.1	45	[49]
$Ni_{45}Co_5Mn_{38}Sb_{12}$	264	26	2.8	73	[50]
$Ni_{50}Mn_{35}In_{14}Si_1$	288	36	2.6	94	[51]
$Ni_{43}Co_7Mn_{31}Ga_{19}$	420 (T_R 433)	13.3 (5 T)	–	188 (5 T)	[24]
$Ni_{41}Co_9Mn_{32}Ga_{18}$	421 (T_R 456)	17.8 (5 T)	12 (5 T)	156 (5 T)	[24]
$Ni_{45}Co_5Mn_{37.5}In_{12.5}$	355	7	16	112	[52]
$Ni_{41}Co_9Mn_{31.5}Ga_{18.5}$	348 (T_R 380)	7.2	14	104	This work

Table 1. The martensite transition temperature T_M, the maximum value of the entropy change ΔS, the half width of the entropy change δT, and the relative cooling power (RCP) at 2 T.

The magnetostructural transformation in this system can be described, in the frame of a simple geometrical model, by a relation linking the field-induced adiabatic temperature change ΔT_{ad} with $dT_M/\mu_0 dH$, with the martensite specific heat value $C_p^{Mart} = 490$ J/kg K, which was obtained by means of DSC in zero fields, the transformation temperature T_M and the entropy change ΔS [47], as the formula of,

$$\Delta T_{ad} = \frac{\Delta S \cdot \Delta T_M}{\Delta S + \Delta T_M \cdot \frac{C_p^{Mart}}{T_M}} \qquad (1)$$

Here, $\Delta T_M = (dT_M/\mu_0 dH) \cdot \mu_0 \Delta H$ is the effective transformation shift in temperature induced by a magnetic field variation $\mu_0 \Delta H$. Fabbrici et al. commented about the relation between ΔT_{ad} and $dT_M/\mu_0 dH$. Eq. (1) provides important information about the relation between ΔT_{ad}, $dT_M/\mu_0 dH$ and ΔS. ΔT_{ad} is not instantaneously proportional either to $dT_M/\mu_0 dH$ or ΔS. This is a immediate consequence of the fact that the minute relation $\Delta T_{ad} = (T_M/C_p^{Mart})\ \Delta S$ cannot be directly extrapolated to finite differences. This is because that the specific heat, $C_p(H, T)$ depends on magnetic field and the temperature.

In order to obtain an adiabatic temperature change ΔT_{ad} from the results of the thermal measurements, the model proposed by Procari et al. is used [19, 53]. In **Figure 11**, the gradient of the entropy curve between AC in zero fields is equal to $C_p/T = 1.5 \pm 0.1$ J/kg K^2 is considered, where C_p is the specific heat. According to Porcali's model, the ΔT_{ad} is obtained as −4.5 K, which was shown in an arrow. The error between the calculated value $\Delta T_{ad} = -3.2$ K from Eq. (1) and experimentally obtained value $\Delta T_{ad} = -4.5$ K from **Figure 10** is 30%. It is correct qualitatively. **Table 2** shows the adiabatic temperature change of the Heusler alloys. The absolute value of ΔT_{ad} of the alloys, which shows re-entrant magnetism and metamagnetism is larger than that of the alloys which shows the magnetostructural transition from martensite ferromagnet to austenite paramagnet. This result is due to the large $dT_M/\mu_0 dH$ value of $Ni_{41}Co_9Mn_{32}Ga_{16}In_2$ and $Ni_{41}Co_9Mn_{31.5}Ga_{18.5}$. Consequently, large MCE has been appeared in $Ni_{41}Co_9Mn_{31.5}Ga_{18.5}$.

Entel et al. studied about $Ni_{50-x}Co_xMn_{39}Sn_{11}$ for $0 \le x \le 10$ [16]. The experimental phase diagram of $Ni_{50-x}Co_xMn_{39}Sn_{11}$ resembles that of $Ni_{50-x}Co_xMn_{31.5}Ga_{18.5}$ [41]. The T_M of $Ni_{50-x}Co_xMn_{31.5}Ga_{18.5}$ gradually decreases with increasing content x and temperature above $x = 9$, T_M drastically decreases. The T_M of $Ni_{50-x}Co_xMn_{39}Sn_{11}$ also shows same x dependence. Around $x = 8.5$, T_M drastically decreases. Entel et al. also suggested that superparamagnetic behavior or superspin glass phase has been appeared in martensite phase. As observed for some nonmagnetic martensitic systems, disorder and local structural distortions can lead to strain glass in austenite. Wang et al. reported that both a strain glass transition and a ferromagnetic transition take place in a $Ni_{55-x}Co_xMn_{20}Ga_{25}$ Heusler alloys [22], which results in a ferromagnetic strain glass with coexisting short-range strain ordering and long-range magnetic moment ordering for $10 \le x \le 18$. As for $Ni_{50-x}Co_xMn_{31.5}Ga_{18.5}$, microscopic (X-ray diffraction, neutron diffraction, μSR, etc.), measurements should be needed to clarify these problems. The complex magnetic and structural properties of Co-doped Ni–Mn–Ga Heusler alloys have been investigated by using a combination of first-principles calculations and classical Monte Carlo simulations by Sokolovskiy et al. [54]. The Monte Carlo simulations with *ab initio* exchange coupling constants as input parameters allow one to discuss the behavior at finite temperatures and to determine magnetic transition temperatures. The simulated magnetic and magnetocaloric properties of Co- and in-doped Ni-Mn-Ga alloys were in good qualitative agreement with the available experimental data. A similar calculation is expected in $Ni_{50-x}Co_xMn_{31.5}Ga_{18.5}$.

http://dx.doi.org/10.5772/64375

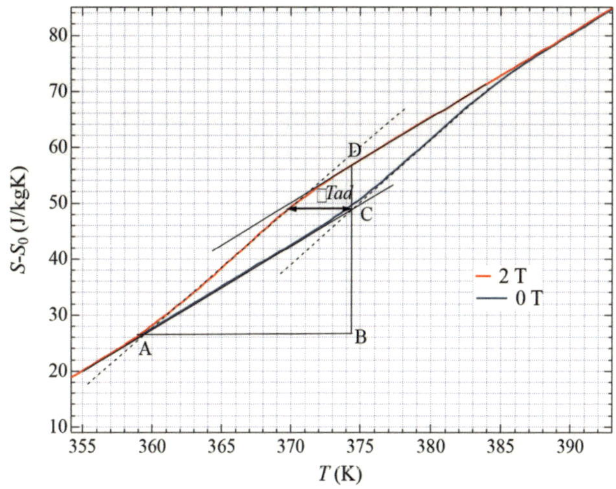

Figure 10. Real heating calorimetric $S(T, H)$ curves across the reverse martensite transition of $Ni_{41}Co_9Mn_{31.5}Ga_{18.5}$. The geometrical construction has superimposed on them.

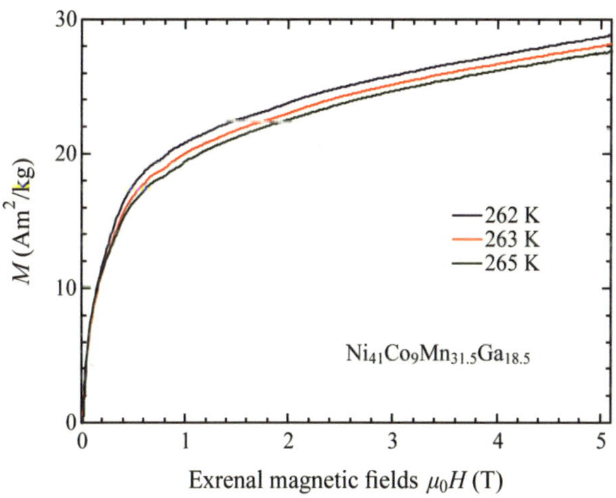

Figure 11. Magnetization process of $Ni_{41}Co_9Mn_{31.5}Ga_{18.5}$.

Sample	λ (kJ/kg)	ΔS (J/kg K)	ΔT_M (K)	T_M (K)	ΔT_{ad} (μ_0 $H[T]$) (K)	Reference
$Ni_{50}Mn_{30}Ga_{20}$	6.90	-3.7	+0.9	370	+0.8 (1.8 T)	[47]
$Ni_{52.5}Mn_{24.5}Ga_{23}$	6.78	-4.6	+1.5	348	+1.0 (1.5 T)	This work
$Ni_{41}Co_9Mn_{32}Ga_{16}In_2$	2.30	4.5	-11.3	320	-2.3 (1.8 T)	[42, 47]
$Ni_{41}Co_9Mn_{31.5}Ga_{18.5}$	2.34	7.2	-8.6	348	-4.5 (2.0 T)	This work

Table 2. The adiabatic temperature change of the Heusler alloys.

3.3. Itinerant electron magnetic properties of $Ni_{41}Co_9Mn_{31.5}Ga_{18.5}$

We performed the magnetization measurements by means of the pulsed magnetic fields in order to investigate the itinerant electron magnetic properties of $Ni_{41}Co_9Mn_{31.5}Ga_{18.5}$. Takahashi proposed a spin fluctuation theory of itinerant electron magnetism [44, 45]. The induced magnetization M is written as the formula of,

$$\left(\frac{M}{M_S}\right)^4 = 1.20 \times 10^6 \frac{T_C^2}{T_A^3 p_S^4} \cdot \frac{H}{M} \tag{2}$$

where, $M_S = N_0 \mu_B p_S$ is a spontaneous magnetization in a ground state. N_0 is a molecular number. $p_S = g_S$, where g is the Landé g-factor and S is a spin angular momentum. T_A is the spin fluctuation parameter in k-space (momentum space). Nishihara et al. measured the magnetization of Ni and Ni_2MnGa precisely [55]. Direct proportionality was observed in the M^4 vs H/M plot at the Curie temperature for Ni. The critical index δ (defined as $H \propto M^{\delta}$) for Ni and Ni_2MnGa is 4.73 and 4.77, respectively. The critical index δ for Fe, CoS_2 and ferromagnetic Heusler alloys, Co_2VGa is 4.6, 5.2 and 4.93, respectively ([45] and references there in).

In most cases, the critical temperature dependence was determined using the Arrott plot. The analysis is based on the implicit assumption that the linearity is always satisfied. Takahashi suggested that the Arrott plot is not applicable in much itinerant d-electron ferromagnets and the revision is necessary in the itinerant electron magnetism [45].

Figure 11 shows the magnetization process of $Ni_{41}Co_9Mn_{31.5}Ga_{18.5}$ around the T_C^M. The horizontal axis is the external magnetic fields. As for the ferromagnetic materials, the diamagnetic effect should be concerned. The effective field H_{eff} is written as the formula of,

$$\mu_0 H_{eff} = \mu_0 H - \mu_0 NM \tag{3}$$

where H is the external magnetic fields, M is the measured magnetization value, and N is a diamagnetic factor. As for this sample, $N = 0.11$.

Eq. (2) can be written as the formula of,

$$H_{eff} = DM^{\delta} \tag{4}$$

and $\delta = 5$, and D is the constant value. From Eq. (4), δ was demanded using such an expression as below [56].

http://dx.doi.org/10.5772/64375

$$\frac{H_{eff}}{H_{eff\,max}} = \frac{DM^{\delta}}{DM^{\delta}_{max}} = \left(\frac{M}{M_{max}}\right)^{\delta} \tag{5}$$

where $H_{eff\,max}$ is the maximum value of the effective magnetic fields, and M_{max} is the maximum value of the measured magnetization. δ can be demanded when a logarithm of the statement are taken, as the formula of, Eq. (5).

Figure 12 shows the logarithm plot of Eq. (5). The gradient of the X-Y plots indicate the critical index δ. **Table 3** shows the index δ and the standard deviation of δ around $T_C^M = 263$ K. Between 262 and 264 K, the error of δ is small. These results indicate that the critical index δ is 5.2(+-, plusminus sign) 0.2.

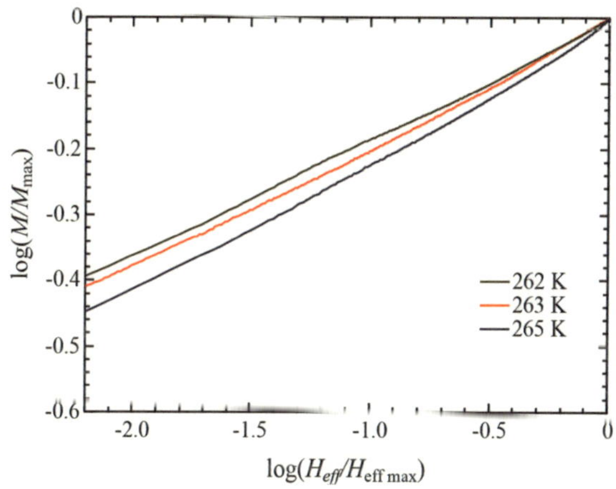

Figure 12. Logarithmic plot of $Ni_{41}Co_9Mn_{31.5}Ga_{18.5}$.

Figure 13 shows the M^4 vs H_{eff}/M plot of $Ni_{41}Co_9Mn_{31.5}Ga_{18.5}$ at $T_C^M = 263$ K. The M^4 vs H_{eff}/M plot crossed the coordinate axis at the Curie temperature in the martensite phase, T_C^M, and the plot indicates a good linear relation behavior around the T_C^M. The result was in agreement with the theory of Takahashi, concerning itinerant electron magnetism [21, 22]. From the M^4 vs H/M plot, the spin fluctuation temperature T_A can be obtained. The obtained T_A was 7.03×10^2 K and which was much smaller than Ni (1.76×10^4 K).

Table 4 indicates the values of p_s, T_C, and T_A estimated from magnetization measurements by means of Eq. (1). MnSi and URhGe are the compounds, which indicate small magnetic moment. The smallness of the moment p_s is due to the spin polarization. UGe_2 also indicate small magnetic moment compared to the full moment of U 5f electron, 3.6 μ_B/U. This is due to the large magnetic anisotropy, due to the spin polarization band [58–60]. The magnetic anisotropy energy is estimated as 6.17 meV = 107 T [58]. T_A is the spin fluctuation temperature, and it reflects width of the quasiparticle at the Fermi surface. Supposing that T_A is large, narrow quasi-fermion (electron) band is formed at the Fermi level and the correlations between quasi fermions are strong. The Zommerfeld coefficient γ also indicates the strength of the correlations of the fermions (electrons). The γ of URhGe and UGe_2 are 163 and 110 mJ mol^{-1} K^{-2}, respectively. The γ of normal metals, Cu, Ni is around 1 mJ mol^{-1} K^{-2}. Therefore, the γ is two orders larger than that of normal metals. Supposing that γ is large, the narrow quasi-fermion (electron) band is formed at the Fermi level. The density of states of the band is large, which indicates the correlations of the electrons are large in U compounds. As for $Ni_{41}Co_9Mn_{31.5}Ga_{18.5}$, the T_A is 7.03×10^2 K, 6.45×10^2 K for $Ni_{52.5}Mn_{24.5}Ga_{23}$, and 4.93×10^2 K for Ni_2MnGa. These values are comparable to that of UGe_2 (4.93×10^2 K), This result indicates that the correlations of the electrons are strong in $Ni_{41}Co_9Mn_{31.5}Ga_{18.5}$, $Ni_{52.5}Mn_{24.5}Ga_{23}$, and Ni_2MnGa.

T (K)	Critical index δ	Standard deviation (%)
258	5.80	0.950
260	5.77	0.208
262	5.42	0.169
263	5.25	0.119
264	4.95	0.187
265	4.60	0.210

Table 3. The critical index δ and the standard deviation of δ around $T_C^M = 263$ K.

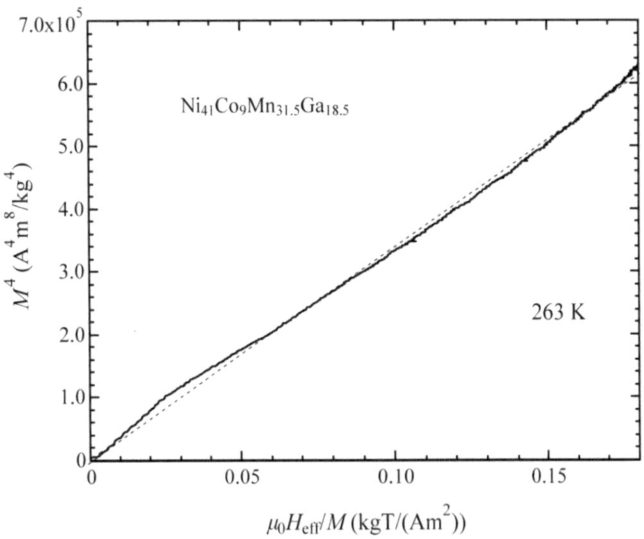

Figure 13. M^4 vs H_{eff}/M plot of $Ni_{41}Co_9Mn_{31.5}Ga_{18.5}$ at 263 K. Dotted line is the extrapolated line.

The value p_s of $Ni_{41}Co_9Mn_{31.5}Ga_{18.5}$, suggested in **Table 4**, is smaller than that of Ni_2MnGa. The small value of p_s has been observed at $Ni_{1.65}Co_{0.28}Mn_{1.31}Ga_{0.62}In_{0.67}$ [14]. The p_s of this alloy was 1.61 μ_B/f.u. at 5 T in ferromagnetic martensite phase. *M-H* curve shows metamagnetic transition at 42 and at 60 T, the magnetic moment reached to 5.0 μ_B/f.u. Karamad et al. point

http://dx.doi.org/10.5772/64375

to the Jahn-Teller effect as a source of the tetragonal distortion of the crystal structure of these alloys. However, they also suggested that the external magnetic field of 60 T seems to be high enough to suppress the Jahn-Teller distortion of crystal lattice of $Ni_{1.65}Co_{0.28}Mn_{1.31}Ga_{0.62}In_{0.67}$. Further experimental and theoretical investigations are needed to clarify this problem.

Compound	p_s (μ_B/f.u.)	T_C (K)	T_A (K)	Reference
Ni	0.6	623	1.76×10^4	[55]
MnSi	0.4	30	2.18×10^3	[45]
Co_2CrGa	3.01	488	1.0×10^4	[44]
Ni_2MnGa	4.5	363 (T_C^A)	4.63×10^2	[55]
$Ni_{52.5}Mn_{24.5}Ga_{23}$	3.75	350 (T_C^M)	6.45×10^2	[6, 57]
URhGe	0.32	9.6	8.56×10^2	[45]
UGe_2	1.44	53.5	4.93×10^2	[45]
$Ni_{41}Co_9Mn_{31.5}Ga_{18.5}$	1.74	263 (T_C^M)	7.03×10^2	This work

Table 4. Experimentally estimated values of p_s, T_C, and T_A from magnetization measurements.

4. Conclusions

We studied about the magnetocaloric properties of $Ni_{41}Co_9Mn_{31.5}Ga_{18.5}$ by means of differential scanning calorimetry (DSC) measurements. Magnetocalorimetric measurements and magnetization measurements of $Ni_{41}Co_9Mn_{31.5}Ga_{18.5}$ polycrystalline ferromagnetic shape memory alloy (FSMA) were performed across the T_R, at atmospheric pressure. When heating from the martensite phase, a steep increase in the thermal expansion due to the reverse martensite transition at T_R was observed by the thermal expansion measurements. These transition temperatures decreased gradually with increasing magnetic field. The field dependence of the reverse martensite transition temperature, $dT_R/d(\mu_0H)$, is -7.0 K/T around zero fields. From the DSC measurements in zero fields, the value of the latent heat λ was obtained as 2.6 kJ/kg, and in magnetic fields, the value was not changed. The entropy change ΔS was -7.0 J/(kgK) in zero fields and gradually increases with increasing magnetic fields. The relative cooling power (RCP) was 104 J/kg at 2.0 T, which was comparable with in-doped $Ni_{41}Co_9Mn_{32}Ga_{16}In_2$ alloy [47].

In order to investigate the itinerant electron magnetic properties of $Ni_{41}Co_9Mn_{31.5}Ga_{18.5}$, we performed the magnetization measurements by means of the pulsed magnetic fields. The M^4 vs H/M plot crossed the coordinate axis at the Curie temperature in the martensite phase, T_{CM}, and the plot indicates a good linear relation behavior around the T_{CM}. The result was in agreement with the theory of Takahashi, concerning itinerant electron magnetism [44, 45]. From the M^4 vs H/M plot, the spin fluctuation temperature T_A can be obtained. The obtained

T_A was 7.03×10^2 K and which was smaller than Ni (1.76×10^4 K). The value was comparable to that of UGe_2 (4.93×10^2 K), which is famous for the strongly correlated heavy fermion ferromagnet [58].

Acknowledgements

The authors thank to Dr. M. Mori for helping SEM microscope experiment. DSC measurements in steady magnetic fields were performed at High Field Laboratory for Superconducting Materials, Institute for Materials Research, Tohoku University, Japan.

Author details

Takuo Sakon[1*], Takuya Kitaoka[1], Kazuki Tanaka[1], Keisuke Nakagawa[1], Hiroyuki Nojiri[2], Yoshiya Adachi[3] and Takeshi Kanomata[4]

*Address all correspondence to: sakon@rins.ryukoku.ac.jp

1 Department of Mechanical and System Engineering, Faculty of Science and Technology, Ryukoku University, Otsu, Shiga, Japan

2 Institute for Materials Research, Tohoku University, Sendai, Miyagi, Japan

3 Graduated School of Science and Engineering, Yamagata University, Yonezawa, Yamagata, Japan

4 Research Institute for Engineering and Technology, Tohoku Gakuin University, Tagajo, Miyagi, Japan

References

[1] K. Ullakko, J. K. Huang, C. Kantner, R. C. O'Handley and V. V. Kokorin, Large magnetic-field-induced strains in Ni_2MnGa single crystals. Appl. Phys. Lett. 69 (1996) 1966.

[2] P. J. Webster, K. R. A. Ziebeck, S. L. Town and M. S. Peak, Magnetic order and phase transformation in Ni_2MnGa. Philos. Mag. B. 49 (1984) 295.

[3] P. J. Brown, J. Crangle, T. Kanomata, M. Matsumoto, K-. U. Neumann, B. Ouladdiaf and K. R. A. Ziebeck, The crystal structure and phase transitions of the magnetic shape memory compound Ni_2MnGa. J. Phys. Condens. Matter. 14 (2002) 10159.

http://dx.doi.org/10.5772/64375

[4] J. Pons, R. Santamarta, V. A. Chernenko and E. Cesari, Long-period martensitic structures of Ni-Mn-Ga alloys studied by high-resolution transmission electron microscopy. J. Appl. Phys. 97 (2005) 083516.

[5] R. Ranjan, S. Banik, S. R. Barman, U. Kumar, P. K. Mukhopadhyay and D. Pandey, Powder X-ray diffraction study of the thermoelastic martensitic transition in $Ni_2Mn_{1.05}Ga_{0.95}$. Phys. Rev. B. 74 (2006) 224443.

[6] T. Sakon, K. Otsuka, J. Matsubayashi, Y. Watanabe, H. Nishihara, K. Sasaki, S. Yamashita, R. Y. Umetsu, H. Nojiri and T. Kanomata, Magnetic properties of the ferromagnetic shape memory alloy $Ni_{50+x}Mn_{27-x}Ga_{23}$ in magnetic fields. Materials 7 (2014) 3715.

[7] Y. Sutou, Y. Imano, N. Koeda, T. Omori, R. Kainuma, K. Ishida and K. Oikawa, Magnetic and martensitic transformations of NiMnX (X=In,Sn,Sb) ferromagnetic shape memory alloys. Appl. Phys. Lett. 85 (2004) 4358.

[8] K. Oikawa, W. Ito, Y. Imano, Y. Sutou, R. Kainuma, K. Ishida, S. Okamoto, O. Kitakami and T Kanomata, Effect of magnetic field on martensitic transition of $Ni_{46}Mn_{41}In_{13}$ Heusler alloy. Appl. Phys. Lett. 88 (2006) 122507.

[9] R. Y. Umetsu, R. Kainuma, Y. Amako, Y. Taniguchi, T. Kanomata, K. Fukushima, A. Fujita, K. Oikawa and K. Ishida, Mössbauer study on martensite phase in $Ni_{50}Mn_{36.5}{}^{57}Fe_{0.5}Sn_{13}$ metamagnetic shape memory alloy. Appl. Phys. Lett. 93 (2008) 042509.

[10] V. V. Khovaylo, T. Kanomata, T. Tanaka, M. Nakashima, Y. Amako, R. Kainuma, R. Y. Umetsu, H. Morito and H. Miki, Magnetic properties of $Ni_{50}Mn_{34.8}In_{15.2}$ probed by Mössbauer spectroscopy. Phys. Rev. B 80 (2009) 144409.

[11] R. Kainuma, Y. Imano, W. Ito, Y. Sutou. H. Morino, S. Okamoto, O. Kitakami, K. Oikawa, A. Fujita, T. Kanomata and K. Ishida, Magnetic-field-induced shape recovery by reverse phase transformation. Nature 439 (2006) 957.

[12] F. Albertini, S. Fabbrici, A. Paoluzi, J. Kamarad, Z. Arnold, L. Righi, M. Solzi, G. Porcari, C. Pernechele, D. Serrate and P. Algarabel, Reverse magnetostructural transitions by Co and In doping NiMnGa alloys: Structural, magnetic, and magnetoelastic properties mater. Sci. Forum 684 (2011) 151.

[13] C. Seguí, E. Cesari and P. Lázpita, Magnetic properties of martensite in metamagnetic Ni–Co–Mn–Ga alloys. J. Phys. D: Appl. Phys. 49 (2016) 165007.

[14] J. Kamarád, J. Kaštil, Y. Skourski, F. Albertini, S. Fabbrici and Z. Arnold, Magneto-structural transitions induced at 1.2 K in Ni_2MnGa-based Heusler alloys by high magnetic field up to 60 T. Mater. Res. Express 1 (2014) 016109.

[15] C. Seguí, Effects of the interplay between atomic and magnetic order on the properties of metamagnetic Ni-Co-Mn-Ga shape memory alloys. J. Appl. Phys. 115 (2014) 113903.

[16] P. Entel, M. E. Gruner, D. Comtesse, V. V. Sokolovskiy and V. D. Buchelnikov, Interacting magnetic cluster-spin glasses and strain glasses in Ni–Mn based Heusler structured intermetallics, Phys. Stat. Sol. B 251 (2014) 2135.

[17] J. Kamarád, S. Fabbrici, J. Kasštil, F. Albertini, Z. Arnold and L. Righi, Pressure dependence of magneto-structural properties of Co-doped off stoichiometric Ni_2MnGa alloys. EPJ Web Conf. 40 (2013) 11002.

[18] J. Kasštil, J. Kamarád, K. Knížek, Z. Arnold and P. Javorský, Peculiar magnetic properties of Er conditioned $Ni_{43}Co_7Mn_{31}Ga_{19}$ at ambient and hydrostatic pressures. J. Alloys Compd. 565 (2013) 134.

[19] G. Porcari S. Fabbrici, C. Pernechele, F. Albertini, M. Buzzi, A. Paoluzi, J. Kamarad, Z. Arnold and M. Solzi, Reverse magnetostructural transformation and adiabatic temperature change in Co- and In-substituted Ni-Mn-Ga alloys. Phys. Rev. B. 85 (2012) 024414.

[20] C. Segui and E. Cesari, Composition and atomic order effects on the structural and magnetic transformations in ferromagnetic Ni–Co–Mn–Ga shape memory alloys. J. Appl. Phys. 111 (2012) 043914.

[21] T. Kanomata, S. Nunoki, K. Endo, M. Kataoka, H. Nishihara, V. V. Khovaylo, R. Y. Umetsu, T. Shishido, M. Nagasako, R. Kainuma and K. R. A. Ziebeck, Phase diagram of the ferromagnetic shape memory alloys $Ni_2MnGa_{1-x}Co_x$. Phys. Rev. B. 85 (2012) 134421.

[22] Y. Wang, C. Huang, J. Gao, S. Yang, X. Ding, X. Song and X. Ren, Evidence for ferromagnetic strain glass in Ni-Co-Mn-Ga Heusler alloy system. Appl. Phys. Lett. 101 (2012) 101913.

[23] S. K. Ayila, R. Machavarapu and S. Vummethala, Site preference of magnetic atoms in Ni–Mn–Ga–M (M = Co, Fe) ferromagnetic shape memory alloys. Phys. Stat. Sol. B 249 (2012) 620.

[24] S. Fabbrici, J. Kamarad, Z. Arnold, F. Casoli, A. Paoluzi, F. Bolzoni, R. Cabassi, M. Solzi, G. Porcari, C. Pernechele and F. Albertini, From direct to inverse giant magnetocaloric effect in Co-doped NiMnGa multifunctional alloys. Acta Mater. 59 (2011) 412.

[25] P. O. Castillo-Villa, D. E. Soto-Parra, J. A. Matutes-Aquino, R. A. Ochoa-Gamboa, A. Planes, L. Mañosa, D. González-Alonso, M. Stipcich, R. Romero, D. Ríos-Jara and H. Flores-Zúñiga, Caloric effects induced by magnetic and mechanical fields in a $Ni_{50}Mn_{25-x}Ga_{25}Co_x$ magnetic shape memory alloy. Phys. Rev. B. 83 (2011) 174109.

[26] J. Bai, J. M. Raulot, Y. Zhang, C. Esling, X. Zhao and L. Zuo, The effects of alloying element Co on Ni–Mn–Ga ferromagnetic shape memory alloys from first-principles calculations. Appl. Phys. Lett. 98 (2011) 164103.

[27] D. E. Soto-Para, X. Moya, L. Mañosa, A. Planes, H. Flores- Zúñiga, F. Alvarado-Hernández, R. A. Ochoa-Gamboa, J. A. Matutes-Aquino and D. Ríos-Jara, Fe and Co

selective substitution in Ni_2MnGa: effect of magnetism on relative phase stability. Philos. Mag. 90 (2010) 2771.

[28] X. Xu, W. Ito, R. Y. Umetsu, K. Koyama, R. Kainuma, K. Ishida, Kinetic Arrest of Martensitic Transformation in $Ni_{33.0}Co_{13.4}Mn_{39.7}Ga_{13.9}$ Metamagnetic Shape Memory Alloy. Mater. Trans. 51 (2010) 469.

[29] B. M. Wang, P. Ren, Y. Liu and L. Wang, Enhanced magnetoresistance through magnetic-field-induced phase transition in Ni_2MnGa co-doped with Co and Mn. J. Magn. Magn. Mater. 322 (2010) 715.

[30] S. Yan, J. Pu, B. Chi and L. Jian, Estimation of driving force for martensitic transformation in $(Ni_{52.5}Mn_{23.5}Ga_{24})_{100-x}Co_x$ alloys. J. Alloys Compd. 507 (2010) 331.

[31] S. Fabbrici, F. Albertini, A. Paoluzi, F. Bolzoni, R. Cabassi, M. Solzi, L. Righi and G. Calestani, Reverse magnetostructural transformation in Co-doped NiMnGa multifunctional alloys. Appl. Phys. Lett. 95 (2009) 022508.

[32] Y. Ma, S. Yang, Y. Liu and X. Liu, The ductility and shape-memory properties of Ni–Mn–Co–Ga high-temperature shape-memory alloys. Acta Mater. 57 (2009) 3232.

[33] L. Ma, H. W. Zhang, S. Y. Yu, Z. Y. Zhu, J. L. Chen, G. H. Wu, H. Y. Liu, J. P. Qu and Y. X. Li, Magnetic-field-induced martensitic transformation in MnNiGa:Co alloys. Appl. Phys. Lett. 92 (2008) 032509.

[34] V. Sánchez-Alarcos, J. I. Pérez-Landazábal, V. Recarte, C. Gómez-Polo and J. A. Rodríguez-Velamazán, Correlation between composition and phase transformation temperatures in Ni–Mn–Ga–Co ferromagnetic shape memory alloys. Acta Mater. 56 (2008) 5370.

[35] D. Y. Cong, S. Wang, Y. D. Wang, Y. Ren, L. Zuo and C. Esling, Martensitic and magnetic transformation in Ni–Mn–Ga–Co ferromagnetic shape memory alloys. Mater. Sci. Eng. A 473 (2008) 213–218.

[36] X. Q. Chen, X. Lu, D. Y. Wang and Z. X. Qin, The effect of Co-doping on martensitic transformation temperatures in Ni–Mn–Ga Heusler alloys. Smart Mater. Struct. 17 (2008) 065030.

[37] P. Entel, M. E. Gruner, W. A. Adeagbo and A. T. Zayak, Magnetic-field-induced changes in magnetic shape memory alloys. Mater. Sci. Eng. A 481–482 (2008) 258–261.

[38] Y. Ma, S. Yang, C. Wang and X. Liu, Tensile characteristics and shape memory effect of $Ni_{56}Mn_{21}Co_4Ga_{19}$ high-temperature shape memory alloy. Scr. Mater. 58 (2008) 918–921.

[39] S. Y. Yu, Z. X. Cao, L. Ma, G. D. Liu, J. L. Chen, G. H. Wu, B. Zhang and X. X. Zhang, Realization of magnetic field-induced reversible martensitic transformation in NiCoMnGa alloys. Appl. Phys. Lett. 91 (2007) 102507.

[40] I. Glavatskyy, N. Glavatska, O. Söderberg, S.-P. Hannula and J.-U. Hoffmann, Transformation temperatures and magnetoplasticity of Ni–Mn–Ga alloyed with Si, In, Co or Fe. Scr. Mater. 54 (2006) 1891–1895.

[41] T. Sakon, Y. Adachi, R. Y. Umetsu, H. Nojiri, H. Nishihara and T. Kanomata, Crystallography and Magnetic Field-Induced Strain by Co Doping NiCoMnGa Heusler Alloy, TMS2013 Supplemental Proceeding, pp. 967–974, 2013, Wiley, USA.

[42] C. Seguí and E. Cesari, Contributions to the transformation entropy change and influencing factors in metamagnetic Ni-Co-Mn-Ga shape memory alloys. Entropy 16 (2014) 5560.

[43] A, E. Clark, J. D. Verhoeven, O. D. McMasters and E. D. Gibson, Magnetostriction in twinned [112] crystals of $Tb_{0.27}Dy_{0.73}Fe_2$. IEEE Trans. Magn. Mag. 22 (1986) 973.

[44] Y. Takahashi, Quantum spin fluctuation theory of the magnetic equation of state of weak itinerant-electron ferromagnets. J. Phys.: Condens. Matter. 13 (2001) 6323–6358.

[45] Y. Takahashi, *Spin Fluctuation Theory of Itinerant Electron Magnetism;* Springer-Verlag: Berlin/Heidelberg, Germany, 2013.

[46] T. Sakon, K. Sasaki, D. Numakura, M. Abe, H. Nojiri, Y. Adachi and T. Kanomata, Magnetic field-induced transition in Co-doped $Ni_{41}Co_9Mn_{31.5}Ga_{18.5}$ Heusler Alloy. Mater. Trans. 54 (2013) 9–13.

[47] S. Fabbrici, G. Porcari, F. Cugini, M. Solzi, J. Kamarad, Z. Arnold, R. Cabassi and F. Albertini, Co and In doped Ni-Mn-Ga magnetic shape memory alloys: a thorough structural, magnetic and magnetocaloric study. Entropy 16 (2014) 2204–2222.

[48] F. -X. Hu, B-G. Shen, J. -O. Sun and G. -H. Wu, Large magnetic entropy change in a Heusler alloy $Ni_{52.6}Mn_{23.1}Ga_{24.3}$ single crystal. Phys. Rev. B 64 (2001) 132412.

[49] M. Pasquale, C. P. Sasso, L. H. Lewis, L. Giudici, T. Lograsso and D. Schlagel, Magnetostructural transition and magnetocaloric effect in $Ni_{55}Mn_{20}Ga_{25}$ single crystals. Phys. Rev. B. 72 (2005) 094435.

[50] A. K. Nayak, K. G. Suresh and A. K. Nigam, Magnetic, electrical, and magnetothermal properties in Ni–Co–Mn–Sb Heusler alloys. J. Appl. Phys. 107 (2010) 09A927.

[51] A. K. Pathak, I. Dubenko, J. C. Mabon, S. Stadler and N. Ali, The effect of partial substitution of In by X = Si, Ge and Al on the crystal structure, magnetic properties and resistivity of $Ni_{50}Mn_{35}In_{15}$ Heusler alloys. J. Phys. D: Appl. Phys. 42 (2009) 045004.

[52] D. Bourgault, J. Tillier, P. Courtois, D. Maillard and X. Chaud, Large inverse magnetocaloric effect in $Ni_{45}Co_5Mn_{37.5}In_{12.5}$ single crystal above 300 K. Appl. Phys. Lett. 96 (2010) 132501.

[53] G. Porcari, F. Cugini, S. Fabbrici, C. Pernechele, F. Albertini, M. Buzzi, M. Mangia and M. Solzi, Convergence of direct and indirect methods in the magnetocaloric study of

first order transformations: The case of Ni-Co-Mn-Ga Heusler alloys. Phys. Rev. B. 86 (2012) 104432.

[54] V. Sokolovskiy, A. Gruünebohm, V. Buchelnikov and P. Entel, *Ab Initio* and Monte Carlo approaches for the magnetocaloric effect in Co- and In-doped Ni-Mn-Ga Heusler alloys. Entropy 16 (2014) 4992.

[55] H. Nishihara, K. Komiyama, I. Oguro, T. Kanomata and V. Chernenko, Magnetization processes near the Curie temperatures of the itinerant ferromagnets. Ni_2MnGa and pure nickel. J. Alloys. Compd. 442 (2007) 191.

[56] M. Seeger, S. N. Kaul, H. Kronmüller and R. Reisser, Asymptotic critical behavior of Ni. Phys. Rev. B. 51 (1995) 12585.

[57] In Ref. [6] (Sakon et al., Materials 2014), The calculated result of T_A was incorrect. From the experimental results of the magnetic moment and the gradient of M^4 vs. H/M, which are shown in Ref. [6], the calculated spin fluctuation parameter T_A is 645 K.

[58] T. Sakon, S. Saito, K. Koyama, S. Awaji, I. Sato, T. Nojima, K. Watanabe and N. K. Sato, Experimental investigation of giant magnetocrystalline anisotropy of UGe_2. Phys. Scr. 75 (2007) 546–550.

[59] V. G. Storchak, J. H. Brewer, D. G. Eshchenko, P. W. Mengyan, O. E. Parfenov and D. Sokolov, Spin-polaron band in the ferromagnetic heavy-fermion superconductor UGe_2. J. Phys.: Conf. Ser. 551 (2014) 012016.

[60] F. Hardy, D. Aoki, C. Meingast, P. Schweiss, P. Burger, H. V. Löhneysen and J. Flouquet, Transverse and longitudinal magnetic-field responses in the Ising ferromagnets URhGe, UCoGe, and UGe_2. Phys. Rev. B. 83 (2011) 195107.